MONOGRAPHS ON GREENLAND
MEDDELELSER OM GRØNLAND
Vol. 105, no. 4

L.R. Wager &
W.A. Deer

Geological Investigations in East Greenland

Part III
The Petrology of the Skaergaard Intrusion, Kangerdlugssuaq, East Greenland

MUSEUM TUSCULANUM PRESS
UNIVERSITY OF COPENHAGEN
2010

L. R. Wager & W. A. Deer
Geological Investigations in East Greenland
Part III – The Petrology of the Skaergaard Intrusion,
Kangerdlusgssuaq, East Greenland

© Museum Tusculanum Press, 2010
Cover design: Erling Lynder
ISBN 978 87 635 1385 2 (Facsimile Edition)

Original print edition, Copenhagen, 1962

Monographs on Greenland | Meddelelser om Grønland
Vol. 105, no. 4
ISSN 0025 6676

www.mtp.dk/MoG

Published with financial support from
The Commission for Scientific Research in Greenland.

Museum Tusculanum Press
University of Copenhagen
126 Njalsgade, DK-2300 Copenhagen S
DENMARK
www.mtp.dk

MEDDELELSER OM GRØNLAND

UDGIVNE AF

KOMMISSIONEN FOR VIDENSKABELIGE UNDERSØGELSER I GRØNLAND

Bd. 105 · Nr. 4

THE SCORESBY SOUND COMMITTEE'S SECOND EAST GREEN-
LAND EXPEDITION IN 1932 TO KING CHRISTIAN IX.s LAND

Leader: Ejnar Mikkelsen

————

THE BRITISH EAST GREENLAND EXPEDITION 1935—36

Leader: L. R. Wager

————

GEOLOGICAL INVESTIGATIONS IN EAST GREENLAND

PART III.

THE PETROLOGY OF THE SKAERGAARD INTRUSION, KANGERDLUGSSUAQ, EAST GREENLAND

BY

L. R. WAGER AND W. A. DEER

————

WITH 68 FIGURES IN THE TEXT 27 PLATES
AND ONE MAP

KØBENHAVN

C. A. REITZELS FORLAG

BIANCO LUNOS BOGTRYKKERI A/S

1939
RE-ISSUE 1962

Færdig fra trykkeriet den 1. oktober 1962.

PREFACE TO THE RE-ISSUE

The account of the Skaergaard Intrusion, originally published in 1939, has long been out of print and the Committee for Scientific Research in Greenland, who publish the Meddelelser om Grønland, has decided to make it available again. The text of this re-issue has been reproduced photographically and the original blocks for the plates and figures have either been re-used or copied without change. The re-issue is thus a facsimile of the original, with the addition only of this preface, and a list of papers concerning the intrusion which have appeared since its original publication.

In 1953, a fourth period of field investigation of the intrusion was undertaken as part of the programme of the small British East Greenland Geological Expedition, (1953) which was led by the two authors of the original Skaergaard memoir and financed largely by the Royal Society of London. The new observations and collections then made have given fresh impetus to the study of the intrusion and have added considerably to our detailed knowledge but without making necessary any fundamental changes in interpretation from that originally given in 1939.

The further investigations, made by us and our many collaborators, deal either with the geochemistry of the rocks and minerals, or the detailed study of some of the mineral series, particularly the pyroxenes, iron-titanium oxide minerals and the accessory sulphides. Some further petrological work has also been done, mainly on the rocks of the border groups.

The hypothesis that the layered series is a bottom accumulate has become further substantiated by the subsequent investigations; so has the distinction between primary precipitate crystals (now called cumulus crystals) and the interprecipitate material (now intercumulus material); and so has the deduction that the upper and marginal border groups are the result of congelation of magma, with some suspended crystals, due to heat loss from the top and sides of the intrusion. The further work has made more secure the hypothesis that thermal convection currents existed during cooling, which produced the rhythmic layering and igneous lamination, and kept the magma stirred, thus permitting the development of the cryptic variation. Finally the general trend in the composition of the rocks and magma, first, towards strong iron enrichment and then, towards acid granophyre, has become more firmly established. With regard to details, however, it should be noted that there have been considerable additions to our knowledge of the pyroxenes and iron ores, so that these sections of the original memoir are seriously out of date. Re-investigation of the Basistoppen raft rocks, previously regarded as forming a large inclusion, indicates that they are part of a differentiated sheet formed from a body of magma which was injected into the solidified, but still hot, Skaergaard intrusion. There has also been some re-interpretation of the rocks previously called the unlaminated layered series; half are now seen to belong to the top of the layered series and half to the upper border group.

The discussion of the trend of differentiation, which forms the last chapter of the memoir, would undoubtedly be considerably modified if written at the present time, but it could not, even now, be written with any finality, despite the continuing interest in the problem of fractionation of basic magma. The evidence for an important stage of iron enrichment in the fractionation of the Skaergaard would still deserve emphasis. On the other hand the attempt to consider why the Skaergaard fractionation took this trend and the implications this has for igneous petrogenesis —matters which were treated in the last two sections of the memoir—are out of date. In particular the significance of the low oxidation state of the iron and the high aluminium content of the initial magma, both briefly mentioned in the memoir, require re-assessment.

During the last two decades layered intrusions in various parts of the world have continued to be a fruitful field of petrological investigation and several important papers have recently appeared, notably that by H. H. Hess on the Stillwater Complex, those by G. M. Brown and W. J. Wadsworth on the layered ultrabasic complex of Rhum, and that by B. G. Worst on the Great Dyke of Southern Rhodesia. These accounts, and the earlier ones of the Skaergaard, the Bushveld Complex of South Africa, the Bay of Islands complex in Newfoundland and the Ilímaussaq batholith of S. W. Greenland, show that layered intrusions vary considerably among themselves; but they also show that the Skaergaard is unique, in many respects.

A list of new papers relating to the Skaergaard intrusion is given as a postscript to this preface. A summary of the memoir is to be found on pages 325-35 and a list of contents on pages 344-52, immediately preceding the plates.

December 1961.

L. R. Wager, Oxford
W. A. Deer, Cambridge.

LIST OF PAPERS

RELATING TO THE SKAERGAARD INTRUSION PUBLISHED SINCE THE FIRST ISSUE (1939)

1943. WAGER, L. R. and MITCHELL, R. L. Preliminary observations on the distribution of trace elements in the rocks of the Skaergaard intrusion, Greenland. Miner. Mag., Vol. 26, pp. 283–296.

1947. WAGER, L. R. Geological investigations in East Greenland. IV. The stratigraphy and tectonics of Knud Rasmussens Land and the Kangerdlugssuaq region. Medd. om Grønland, Bd. 134, No. 5, pp. 1–64. [Geological setting of the intrusion]

1948. WAGER, L. R. and MITCHELL, R. L. The distribution of Cr, V, Ni, Co and Cu during the fractional crystallization of a basic magma. Internat. Geol. Congress, Report of 18th Session, Great Britain, Pt. 2, pp. 140–150. [Data on Skaergaard intrusion]

1951. MUIR, I. D. The clinopyroxenes of the Skaergaard intrusion, eastern Greenland. Miner. Mag., Vol. 29, pp. 690–714.
WAGER, L. R. and MITCHELL, R. L. The distribution of trace elements during strong fractionation of basic magma – a further study of the Skaergaard intrusion, East Greenland. Geochim. et Cosmochim. Acta, Vol. 1, pp.129–208.

1953. SCHWANDER, H. Bestimmung des relativen Sauerstoffisotopen-Verhältnisses in Silikatgesteinen und –Mineralien. Geochim. et Cosmochim. Acta, Vol. 4, pp. 261–291. [Data for Skaergaard rocks given]
VINCENT, E. A. Hornblende-lamprophyre dykes of basaltic parentage from the Skaergaard area, East Greenland. Quart. Journ. Geol. Soc. London, Vol. 109, pp. 21–49.
WAGER, L. R. Layered intrusions. Med. fra Dansk Geol. Forening, Bd. 12, pp. 335–349. [Brief summary of Skaergaard intrusion and comparisons]

1954. CARR, J. M. Zoned plagioclases in layered gabbros of the Skaergaard intrusion, East Greenland. Miner. Mag., Vol. 30, pp. 367–375.
HOLGATE, N. The role of liquid immiscibility in igneous petrogenesis. Journ. Geol., Vol. 62, pp. 439–480. [Liquid immiscibility hypotheses applied to the Skaergaard intrusion]
VINCENT, E. A. and PHILLIPS, R. Iron-titanium oxide minerals in layered gabbros of the Skaergaard intrusion, East Greenland. Part 1. Chemistry and ore-microscopy.
CHEVALLIER, R., MATHIEU, S. and VINCENT, E. A. Part 2. Magnetic properties. Geochim. et Cosmochim. Acta, Vol. 6, pp. 1–26 (Part 1) and pp. 27–34 (Part 2).

1956. GAY, P. The structures of the plagioclase felspars: VI. Natural intermediate plagioclases. Miner. Mag., Vol. 31, pp. 21–40. [Optical data for Skaergaard felspars]

GAY, P. and BOWN, M. G. The structures of the plagioclase felspars: VII. The heat treatment of intermediate plagioclases. Miner. Mag., Vol. 31, pp. 306–313. [Optical data for Skaergaard felspars]

HUGHES, C. J. Geological investigations in East Greenland. VI. A differentiated basic sill enclosed in the Skaergaard intrusion, East Greenland and related sills injecting the lavas. Medd. om Grønland, Bd. 137, No. 2, pp. 1–27.

REYNOLDS, D. L. Calderas and ring-complexes. Verhandl. Koninkl. Nederlandsch Geol. Mijnbouwkundig Genootscap, Deel 16 (Gedenkboek H. A. Brouver), pp. 355–379. [Metasomatism hypothesis]

WAGER, L. R. A chemical definition of fractionation stages as a basis for comparison of Hawaiian, Hebridean, and other basic lavas. Geochim. et Cosmochim. Acta, Vol. 9, pp. 217–248. [Albite and iron ratio plot for Skaergaard rocks and liquids]

1957. BROWN, G. M. Pyroxenes from the early and middle stages of fractionation of the Skaergaard intrusion, East Greenland. Miner. Mag., Vol. 31, pp. 511–543.

VINCENT, E. A., WRIGHT, J. B., CHEVALLIER, R. and MATHIEU, S. Heating experiments on some natural titaniferous magnetites. Miner. Mag., Vol. 31, pp. 624–655. [Mainly experiments on Skaergaard magnetites]

WAGER, L. R. and BROWN, G. M. Funnel-shaped layered intrusions. Bull. Geol. Soc. Amer., Vol. 68, pp. 1071–1074 [Form of Skaergaard intrusion discussed]

WAGER, L. R., VINCENT, E. A. and SMALES, A. A. Sulphides in the Skaergaard intrusion, East Greenland. Econ. Geol., Vol. 52, pp. 855–903.

YODER JR, H. S. and SAHAMA, T. G. Olivine X-ray determinative curve. Amer. Mineral., Vol. 42, pp. 475–491. [Data on Skaergaard olivines]

1958. CHEVALLIER, R. and MATHIEU, S. Susceptibilité magnétique spécifique de pyroxènes monocliniques. Bull. Soc. Chim. France, No. 5, pp. 726–729. [Pyroxenes from Skaergaard intrusion]

WAGER, L. R. Beneath the Earth's crust. Presidential Address to Section C, British Association, Glasgow Meeting, 1958. The advancement of Science, No. 58, 15 pp. [Hypothesis on origin of Earth's silicate shell based on analogies with Skaergaard layered intrusion]

WAGER, L. R., SMIT, J. van R. and IRVING, H. Indium content of rocks and minerals from the Skaergaard intrusion, East Greenland. Geochim. et Cosmochim. Acta, Vol. 13, pp. 81–86.

1959. CHEVALLIER, R. and MARTIN, R. Le moment magnétique de l'ion ferreux dans une série de pyroxènes monocliniques. Bull. Soc. Chim. France, No. 9, pp. 9–10, [Data on Skaergaard pyroxenes]

HAMILTON, E. The uranium content of the differentiated Skaergaard intrusion, together with the distribution of the alpha particle radioactivity in the various rocks and minerals as recorded by nuclear emulsion studies. Medd. om Grønland, Bd. 162, No. 7, pp. 1–35.

SHIMAZU, Y. A physical interpretation of crystallization differentiation of the Skaergaard intrusion. Journ. Earth Sci., Nagoya University, Vol. 7, pp. 35–48.

1960. BOWN, M. G. and GAY, P. An X-ray study of exsolution phenomena in the Skaergaard pyroxenes. Miner. Mag., Vol. 32, pp. 379–388.

BROWN, G. M. The effect of ion substitution on the unit cell dimensions of the common clinopyroxenes. Amer. Mineral., Vol. 45, pp. 15–38. [Data for Skaergaard pyroxenes]

HESS, H. H. Stillwater igneous complex, Montana: a quantitative mineralogical study. Geol. Soc. Amer. Memoir 80, 230 pp. [Comparisons with Skaergaard intrusion]

POLDERVAART, A. and TAUBENECK, W. H. Layered intrusions. Internat. Geol. Congress, Report of 21st Session, Norden, Pt. 13, pp. 239–246. [Cryptic and rhythmic layering of Skaergaard type discussed]

SMITH, J. R. Optical properties of low-temperature plagioclase. Appendix 3 of Stillwater igneous complex, Montana, by H. H. Hess, Geol. Soc. Amer. Memoir 80, pp. 191–219. [Optical data for Skaergaard felspars]

TAUBENECK, W. H. and POLDERVAART, A. Geology of the Elkhorn Mountains, Northeastern Oregon: Part 2. Willow Lake intrusion. Bull. Geol. Soc. Amer., Vol. 71, pp. 1295–1322. [Comparisons with Skaergaard layering]

VINCENT, E. A. Ulvöspinel in the Skaergaard intrusion, Greenland. N. Jb. Miner., Abh., Bd. 94 (Festband Ramdohr), pp. 993–1016.

— and BILEFIELD, L. I. Cadmium in rocks and minerals from the Skaergaard intrusion, East Greenland. Geochim. et Cocmochim. Acta, Vol. 19, pp. 63–69.

— and CROCKET, J. H. Studies in the geochemistry of gold. I. The distribution of gold in rocks and minerals of the Skaergaard intrusion, East Greenland. Geochim. et Cosmochim. Acta, Vol, 18, pp. 130–142.

WAGER, L. R. The major element variation of the layered series of the Skaergaard intrusion and a re-estimation of the average composition of the hidden layered series and of the successive residual magmas. Journ. Petrol., Vol. 1, pp. 364–398.

— BROWN, G. M. and WADSWORTH, W. J. Types of igneous cumulates. Journ. Petrol., Vol. 1, pp. 73–85. [Orthocumulates of Skaergaard intrusion discussed]

1961. GIRDLER, R. W. Some preliminary measurements of anisotropy of magnetic susceptibility of rocks. Geophys. Journ., Vol. 5, pp. 197–206. [Data for Skaergaard rocks given]

WAGER, L. R. A note on the origin of ophitic texture in the chilled olivine gabbro of the Skaergaard intrusion. Geol. Mag., Vol. 98, pp. 353–366.

WRIGHT, J. B. Solid solution relationships in some titaniferous iron oxide ores of basic igneous rocks. Miner. Mag., Vol. 32, pp. 778–789. [Data for Skaergaard ore minerals]

1962. ROBSON, G. R. and SPECTOR, J. Crystal fractionation of the Skaergaard type in modern Icelandic magmas. Nature, Vol. 193, pp. 1277–1278.

WAGER, L. R. and VINCENT, E. A. Ferrodiorite from the Isle of Skye. Miner. Mag., Vol. 33, pp. 26–36 [Comparison with Skaergaard ferrogabbro].

In the press:

TAYLOR, H. P. and EPSTEIN, S. O^{18}/O^{16} ratios in rocks and coexisting minerals of the Skaergaard intrusion, East Greenland.

In preparation:

DOUGLAS, J. A. V. The Basistoppen Sheet in the Skaergaard intrusion.

PREFACE

During the brief stay of the British Arctic Air-Route Expedition in Kangerdlugssuaq in 1930, a remarkable iron-olivine gabbro was discovered by the first named author on the Skaergaard Peninsula, and a preliminary investigation was made of the intrusion to which it belonged. In 1932, during the Scoresby Sound Committee's second East Greenland Expedition under Captain Ejnar Mikkelsen [1933 A and B][1]), the same author, with the assistance of Dr. H. G. Wager and Dr. Jens Jensen, made a preliminary map of the whole intrusion. Mention of some of the results obtained was made in the general review of the geological work of these two expeditions [Wager 1934].

A considerable amount of field work had been accomplished up to this time, but it was not sufficient to allow an adequate account of the intrusion to be written. It seemed unlikely that any general expedition would return to Kangerdlugssuaq in the near future, since the more interesting geographical discoveries in the region had already been made during Admiral Amdrup's first visit in 1900 and the two expeditions mentioned above. A small scientific expedition was therefore planned which would provide opportunities for detailed work on the geology and botany of the Kangerdlugssuaq region. This expedition, called the British East Greenland Expedition 1935—36, spent from July 1935 to August 1936 at Kangerdlugssuaq, and much of this time was devoted to the study of the Skaergaard intrusion. A general account of the expedition is to be found in the Geographical Journal [Wager 1937].

In high mountain country, where the valleys and corries are occupied by glaciers, it is necessary to travel in roped parties of at least two. About half of the detailed mapping here described was carried out by the authors in two separate parties with the help of Dr. E. C. Fountaine and Mr. P. B. Chambers, and we wish to thank most sincerely these two members of the expedition for their unstinted collaboration. The remainder of the mapping and the examination of all critical sections were carried out by both authors working together.

Study of the Skaergaard intrusion, both in the field and in the laboratory, was begun by the first named author (L. R. W.). Since then the two authors have collaborated fully, but certain of their respective

---

[1]) Dates in square brackets refer to books and papers cited in the list of references at the end of the paper.

fields of work can be delineated. The second named author (W. A. D.) has carried out all the chemical analyses of the rocks and minerals. The optical work on the minerals has been shared, except that the determinations of the high refractive indices were carried out by the second author at Cambridge and Manchester. The first named author has carried out most of the petrological examination of the thin sections. The same author is also responsible for the main outlines of the interpretation and for the presentation of the results, except for the detailed discussion of the pyroxenes in section XI. All aspects of the paper, however, have been fully considered by the two authors together, and they are in complete agreement over the conclusions which have been reached.

To the leaders of the two earlier expeditions, the late H. G. Watkins and Captain Ejnar Mikkelsen sincere thanks are here given for the facilities they provided, and also to Dr. Jensen and Dr. Wager who assisted with the work in the field.

During the summer of 1935 the British East Greenland Expedition was fortunate in being able to collaborate with Mr. Augustine Courtauld's party [see Courtauld 1935] and the many advantages which this conferred have already been acknowledged [Wager 1936].

The last expedition was only made possible by the goodwill and help of many individuals and Societies whose names have been given in the general account of the expedition. Here we wish to offer our sincere thanks to those who specifically helped with the geological side of the expedition, in particular to the late Prof. A. Hutchinson, Prof. Sir Albert C. Seward, Prof. P. G. H. Boswell, Prof. H. L. Hawkins, Dr. Lauge Koch, Prof. W. J. Pugh, Sir Franklin T. Sibly and Prof. C. E. Tilley. We wish particularly to thank Prof. Hawkins for his continued interest in the work and his share in furthering it, by arranging leave of absence for the first named author from his duties at Reading University during two separate years.

Generous grants were made towards the cost of the expedition by the Royal Society, the Royal Geographical Society, the Trustees of the Percy Sladen Fund and the Research Board of Reading University. The first named author (L. R. W.) has to thank the Trustees of the Leverhulme Fellowships for the award of a Leverhulme Fellowship for 1935—36, and the second author (W. A. D.), the Royal Commissioners for the Exhibition of 1851 for the award of a Senior Studentship.

Since returning to England we have greatly profited by discussion with Prof. C. E. Tilley and Dr. E. B. Bailey, and certain critical parts of the present account have been read by them when in manuscript. Dr. H. G. Wager has read the whole paper and given valuable help with the manner of its presentation.

L. R. WAGER. (Reading University).
W. A. DEER. (Manchester University).

I. INTRODUCTION

The rocks of the Skaergaard intrusion may be divided into two natural groups, a layered series and a border group. The stratification in the layered series is so marked that, when these rocks were first distantly seen from the deck of the ship in 1930, they were mistaken for well bedded red sandstones. The border group, considerably less in bulk than the layered series, occurs everywhere between the central layered series and the steep transgressive margin of the intrusion, and it also forms a partly eroded upper cover to the layered series. Fluxion structures parallel to the margin and roof are usual in the border group. In the marginal border group these are almost at right angles to the structures of the layered series, while in the upper border group they are parallel. The composition of the rocks of the border group and the layered series cover a wide range but all are the result of differentiation of a single body of magma. The present paper is largely a description of the differentiation and a discussion of the mechanism by which it was produced.

Once the difficulties of reaching this part of East Greenland have been overcome, the Skaergaard intrusion, among the known layered intrusions of the world, is probably the easiest to study because of its relatively small size and because the Tertiary rocks composing it are fresh, and magnificiently exposed in a region of high relief. The intrusion as a whole can be shown to have had originally the form of a tilted inverted cone, while the different units of the layered series had the form of a stack of saucers of steadily increasing size. A flexure affecting the coastal strip of this part of East Greenland took place after the intrusion had solidified, and has had the good effect of exposing to observation a much greater thickness of the layered series than would otherwise have been the case.

The undifferentiated magma is found as a chilled facies at all contacts with the surrounding rocks, and is close to average gabbro in composition. The lowest exposed rocks of the layered series are olivine gabbros of common types, but upwards there is a gradual passage into

more and more iron-rich rocks. The clinopyroxene, present throughout
the series, gradually approaches and finally becomes a member of the
hedenbergite-ferrosilite solid solution series; the olivine becomes increasingly rich in fayalite until a type with ninety-eight per cent.
Fe_2SiO_4 is reached; at the same time the plagioclase changes from basic
labradorite to acid andesine and quartz becomes a phase in equilibrium
with the iron-rich olivine and pyroxene. In the Skaergaard intrusion
an unusual series of iron-rich rocks have been developed as a result
of simple crystallisation differentiation of fairly normal gabbro magma.

To explain the structure and certain aspects of the differentiation
of the Skaergaard intrusion we have had to invoke a mechanism, involving the idea of convection, which has not been offered for any other
investigated plutonic masses so far as we know, though some aspects
of it have been considered in theoretical studies by Adams, Jeffreys,
Holmes and others. Since some of the rocks which have been developed
in the Skaergaard intrusion are unusual, the postulated processes of
differentiation may likewise be rare. On the other hand, there is evidence
that the process here so admirably preserved for examination, has
acted in some of the deep magma reservoirs from which the Plateau
basalts of the world have been derived. It also appears from this study
that straightforward crystallisation differentiation of basalt magmas
takes place in a direction which is not that of the normal calc-alkaline
series of igneous rocks.

II. SITUATION AND TOPOGRAPHY

The Skaergaard intrusion, Lat. 68°10′ north, Long. 31°40′ west, lies on the east side of the large fjord of Kangerdlugssuaq near its mouth. Part of the region was originally mapped on a small scale by Admiral Amdrup during his expedition along the coast from Scoresby Sound to Angmagssalik [Amdrup 1898—1900]. A coastal map was also made during the British Arctic Air-Route Expedition 1930—31 [Watkins 1932]; this and air photographs taken during the same expedition greatly assisted the early work on the intrusion. The final geological mapping has been done on a scale of 1:50,000, using for this purpose enlarged copies of a map drawn by the Danish Geodetic Institute from the air survey made by Knud Rasmussen's 7th Thule Expedition, and supplemented by the photogrammetric survey of M. S. Spender and Chr. Larsen which was carried out on the expedition led by Capt. E. Mikkelsen in 1932 [Spender 1933].

The intrusion, oval in shape and with an area of about fifty square kilometres, is deeply dissected (Pls. 1 and 2). Gabbrofjaeld in the northern part rises to a height of 1360 metres and two subsidiary peaks to 1277 and 1200 metres. The now shapely, spiked summits of these mountains have been overridden by ice, as erratics of gneiss from the metamorphic complex to the north were found within a few feet of the highest summit. The last inundation by ice must have been some time ago as the present small glaciers have been responsible for the production of the existing sharp ridges and peaks. In the southern half of the intrusion there are several conspicuous summits: Basistoppen, 897 metres, Brødretoppen, 1120 metres, and Tinden, 1200 metres. No gneiss erratics were found high on these mountains and we believe that they were not completely overridden by ice at the same time as Gabbrofjaeld. The western part of the intrusion forms relatively low ground but almost continuous, glaciated, rock surfaces provide excellent exposures, and two arms of the sea, Uttentals Sund and Skaergaardsbugt, cut into this area. Forbindelses glacier, rising only to a height of 450 metres, leads from Uttentals Sund to Mikis Fjord and separates the Gabbrofjaeld group from

Basistoppen and the other mountains to the south. The reason for this line of easy erosion which is continued in the eastwest part of Mikis Fjord and over another low pass to Jacobsens Fjord is not certain. It seems likely that it is connected with the earth flexure and accompanying dyke swarm which has already been described [Wager and Deer 1937], and which is briefly discussed below (p. 19). The zone of easy erosion occurs where the flexure first becomes definite; it does not continue further southwards, where the bending is still more severe, because the dense dyke swarm there increases the average resistance of the rocks to erosion.

During the course of the field work we were able to reach most of the places which seemed likely to yield critical evidence. Peaks 1360 metres and 1200 metres of Gabbrofjaeld, Basistoppen, Brødretoppen and Osttoppen were climbed, but the summits of Tinden and of Peak 1277 metres of Gabbrofjæld were not reached and their upper parts are not therefore known in detail.

III. GENERAL GEOLOGY OF THE REGION

The Skaergaard intrusion is markedly transgressive, cutting cleanly through the metamorphic complex and the overlying sediments, basalts and tuffs (see geological map at end of paper). Round the northern end of the intrusion the lavas are dipping at 7° to the south, and the dip increases steadily until, south of the intrusion the dip is 60°. In illustrating the present, and also the original, form of the intrusion it is necessary to consider the thickness and lie of the surrounding rocks in some detail, and it is necessary to describe briefly their petrography since many of them are found within the intrusion as inclusions whose distribution throws light on the mechanism of the differentiation.

(a) Gneisses of the Metamorphic Complex.

The metamorphic complex, which at the present level of denudation is in contact with the Skaergaard intrusion to the north and west, is a somewhat granulitised acid gneiss with subordinate bands of amphibolite. The convenient field name of grey gneiss will be used for the dominant acid part of the metamorphic complex. This rock type is dominant among the gneisses which form the coast of Greenland from Kangerdlugssuaq to Angmagssalik, and it has been already briefly described [Wager 1934, pp. 10, 11]. In the neighbourhood of the Skaergaard intrusion there are no typical pegmatites but sporadic occurrences of acid gneiss containing much pink orthoclase are probably pegmatites which were granulitised in a later phase of metamorphism. In the Kangerdlugssuaq region metamorphosed sediments in the metamorphic complex are seldom found. Porphyroblast schist occur on Uttentals Plateau as bands in the grey gneiss and these are probably derived from sediments. A definite paragneiss forming an extensive zone in the metamorphic complex occurs four miles to the northwest of the Skaergaard intrusion and consists of rocks containing garnet and sillimanite, thus being of high grade, as is to be expected. Ultrabasic rock of a type previously described [Wager 1934, pp. 13—14] are also found sparingly as inclusions in the grey gneiss of the present area. The gneisses

TABLE I.
ANALYSIS OF GREY GNEISS.

XXIV (1867)		Norm.		Mode (Vol. %)	
SiO_2	68.17	Qu	22.74	Qu	28
Al_2O_3	16.13	Or	18.90	Orth	15
Fe_2O_3	0.58	Ab	37.20	Plag	49
FeO	2.09	An	8.90	Biot	8[1]
MgO	1.82	Cor	2.14		
CaO	2.07	Hy $\left\{\begin{matrix}4.50\\2.38\end{matrix}\right\}$	6.88	Ratios.	
Na_2O	4.40			$\dfrac{FeO + Fe_2O_3 \times 100}{MgO + FeO + Fe_2O_3} = 60$	
K_2O	3.22	Ilm	1.22		
H_2O+	0.40	Mt	0.93		
H_2O-	0.16	Ap	1.01	$\dfrac{Fe_2O_3 \times 100}{FeO + Fe_2O_3} = 22$	
TiO_2	0.63	H_2O	0.56		
MnO	0.05			$\dfrac{K_2O \times 100}{Na_2O + K_2O} = 42$	
P_2O_5	0.31	Plag. $Ab_{81}An_{19}$			
	100.03	Hyp. $En_{66}Fs_{34}$			
S. G.	2.66				

[1] Includes some ore and chlorite.

COMPARISONS.

	XXIV	A	B
SiO_2	68.17	68.97	67.63
Al_2O_3	16.13	14.66	13.34
Fe_2O_3	0.58	1.36	1.91
FeO	2.09	1.92	3.11
MgO	1.82	1.88	1.71
CaO	2.07	4.28	1.93
Na_2O	4.40	4.18	3.97
K_2O	3.22	1.42	4.04
H_2O+	0.40	0.47	1.43
H_2O-	0.16	0.03	..
TiO_2	0.63	0.38	0.60
MnO	0.05	0.38	0.05
P_2O_5	0.31	0.13	0.29
CO_2	—	0.13	..
	100.03	100.19	100.01

XXIV. Grey gneiss, 1867. Average material, 100 m W. of contact, Mellemø, Kangerdlugssuaq.

    A. Hornblende-biotite-gneiss. Ben Hogh Series, Lewisian [Mem. Geol. Surv. Anals. of Igneous Rocks etc. 1931, no. 410, p. 106]. Anal. E. G. Radley.

    B. Adamellitic dyke, S. of Lake Tallsjøn, Halleförs, Sweden [Krokstrom, 1936]. Anal. N. Sahlbom.

on Kraemers Ø strike approximately E.N.E. and dip steeply N.N.W. The metamorphic complex is almost certainly Pre-Cambrian in age, but in a sketch map Koch [1935] has put forward another view.

A specimen of grey gneiss, 1867[1]), collected 150 m from the contact with the Skaergaard intrusion on Mellemø has been selected for analysis being considered to be as close to the average of the grey gneiss as can be obtained within the limits of a single hand specimen (Table I). It is a fairly fine-grained, granulitised gneiss mainly of quartz and plagioclase with vague layers of green ferromagnesian minerals and pink bands rich in orthoclase. A considerable amount was crushed down for analysis in order to obtain a reasonably average sample. A section of the rock shows approximately 70 per cent. of felspar, both orthoclase and plagioclase, the latter in excess, about 25 per cent. of quartz and 5 per cent. of melanocratic constituents. The plagioclase, which is andesine, shows fine scale lamellar twinning and irregularities of extinction due to strain. It occurs in crystals usually about 1 mm across, outside which is the quartz and orthoclase which are usually closely associated and occur in somewhat smaller crystals. The border of the plagioclase crystals is frequently broken down into small units due to granulitisation, and the larger quartz areas are also broken into variously orientated, interlocking grains from the same cause, although neither the quartz nor the orthoclase now show any strain effects. Both kinds of felspar are dusty with inclusions. The brown and strongly pleochroic biotite occurs in irregular streaks of many, small, and variously orientated grains, and it is extensively altered to chlorite, in which is a little sphene. Grains of iron ore accompany both the altered and unaltered biotite but more abundantly the latter. Hornblende is scarce and in the section examined, only one crystal was noticed. There seems to be no obvious effect due to the proximity of the Skaergaard intrusion unless it be the alteration of the biotite. Since, however, this rock is not far removed from the unconformable cover of sediment, the alteration of the biotite may be due to late Cretaceous weathering. Another example of the grey gneiss 1865 from near the analysed rock contains the same minerals in about the same proportions and with the same structures. Alteration of the biotite is also similar.

The analysis of 1867 confirms that plagioclase is dominant over orthoclase and gives a norm in which there is 9 per cent. of ferromagnesian minerals. The amounts of alkalies present are:— Na_2O, 4.40 per cent. and K_2O, 3.22 per cent. These values may be compared with the alkalies determined in 1865, which are Na_2O, 3.98 per cent. and

---

[1]) These numbers refer to specimens collected during the expeditions of 1930, 1932 and 1935—36. They are now housed in the Geological Museum of Reading University.

K_2O, 2.93 per cent. The composition of the grey gneiss is compared with an analysis of Lewisian Gneiss from Scotland and an adamellite from Sweden (Table I). The composition of the average grey gneiss will later be compared with that of the granophyre inclusions of the Skaergaard intrusion which were originally grey gneiss.

Within two metres of the contact with the Skaergaard intrusion the grey gneiss becomes obviously metamorphosed, the banding and all gneissose structures being obliterated. Two examples will be described:— 1870, collected 1 m from the contact on Mellemø and 2291, $1^1/_2$ m from the contact on Kraemers Ø near the southern end of the contact. The plagioclase of 1870 is andesine, and it occurs in rather larger crystals than in the unmetamorphosed rocks described. It has lamellar twinning on a fine scale, is slightly zoned and often has a margin of orthoclase. The quartz occurs, associated with orthoclase, outside the plagioclase crystals, and also as discontinuous cuspate pieces scattered through the plagioclase and having the same optic orientation over considerable areas. The ferromagnesian minerals, mainly segregated into patches and not streaks, are, in order of abundance hypersthene, hornblende, biotite and iron ore. The moderately pleochroic hypersthene forms aggregates of some hundreds of individuals, the largest of which is about 0.1 mm. Grains of iron ore are associated with the hypersthene in somewhat smaller masses and both are embedded in quartz. These hypersthene-rich patches are surrounded by a discontinuous rim of green hornblende which is an aggregate of small crystals, often with the same orientation over a considerable area. Hypersthene also occurs in sporadic crystals which are not closely associated with hornblende. Both brown and green mica occur, usually among the green hornblende, and some colourless hornblende appears to replace some of the hypersthene.

The other thermally metamorphosed grey gneiss, 2291, is similar to the above. The plagioclase has the same habit and encloses similar sieve-like quartz. The hypersthene forms rather larger grains and is a good deal replaced by a mineral probably related to bowlingite. Brown and green mica both occur and a little hornblende. Apatite is a rather more abundant accessory than in 1870.

It will be show later that the thermally metamorphosed grey gneiss from near the contact is different in structure and mineral composition from the granophyre patches produced from inclusions of grey gneiss which became immersed in the Skaergaard magma.

(b) The Kangerdlugssuaq Sedimentary Series.

Sediments are found to the east, north and west of the intrusion resting on a fairly even surface of gneiss. Within the area of the map

Fig. 1. Gabbrofjaeld Group from Mikis Fjord. The distant mountains are of layered gabbro in front of which are basalts and tuffs. The mass of white sediment high on the south-east ridge of Gabbrofjaeld is conspicuous

they are best exposed in Vandfaldsdalen which drains into Mikis Fjord. The sequence observed in 1932 was incomplete; the full sequence with revised estimates of thicknesses, is:—

Plateau Basalts.

Kangerdlugssuaq Series	Conglomerate	20—50 ft.
	Sandy shales	250 ft.
	False bedded sandstones often calcareous	100 ft.
	Ferruginous sandstone	200 ft.

(Base not seen).

Although the base is not seen in Vandfaldsdalen, the gneisses of the metamorphic complex cannot be far below the lowest sediments exposed. The sediments on the north-east side of Gabbrofjaeld have not been examined in detail, and to the west of the intrusion the only sediments preserved are a thin basal conglomerate and about four feet of highly metamorphosed calcareous shales which occur within a few yards of the contact on Mellemö. This occurrence is, however, of importance, as it is situated where it would be expected if the strike of the sediments in Mikis Fjord is extrapolated. This is one piece of evidence suggesting that the intrusion has not appreciably disturbed the lie of the surrounding rocks.

No recognisable fossils have been found within the area of the map but a few belemnites from near the base of the sediments some twenty kilometres to the north-west have been examined by Prof. H. H. Swinnerton, who assigns them to the upper Cretaceous, probably Senonian. Plant remains from near the top of the series, collected thirty kilometres to the north-west, have been examined by Prof. Sir Albert C. Seward, who regards them as of the same age as the Mull Leaf Bed, that is, late Cretaceous or early Eocene. As a close approximation the Kangerdlugssuaq Sediments may be taken as uppermost Cretaceous.

(c) Basalts and Tuffs.

The junction between the sediments and the immediately overlying lava in Vandfaldsdalen suggests that there was no significant time interval separating them. The basalts and tuffs are well exposed on both sides of the valley and dip at 10° in a southerly direction. (Fig. 1.) The following sequence was determined by aneroid on the north side of the steep glacier descending from the S.S.E. ridge of Gabbrofjaeld to the pass at the head of Vandfaldsdalen:—

Basalts...................... 300 ft.
Agglomerate................. 400 ft.
Basalts............... .... 175 ft.
Coarse Tuffs............. ... 100 ft.
Fine Tuffs.................. 200 ft.
Basalts..................... 300 ft.
Sediments .... 10 ft.

(base not seen)

The agglomerate forms much of the broad shoulder bounding Vandfaldsdalen on the west and from thence it descends towards Mikis Fjord Rejsehus; it was found to persist as far as the mountains at the head of Jacobsen Fjord twenty kilometres east-north-east.

In the walls of Mikis Fjord the sequence is difficult to determine because the dyke swarm, which is here dense, masks the lie of the lavas. It is impossible to see from a distance the individual lava flows, and detailed examination is impossible, except here and there, because of the danger of falling stones. Sampling of the cliffs showed that they consist dominantly of basalt with one thick tuff horizon and a few sills. The dip increases from 10° at Mikis Fjord Rejsehus to 45° at the mouth of the fjord.

No attempt was made to map the mountainous, inland country between Mikis Fjord and Hammerdalen, but the thick tuff horizon of Mikis Fjord was found on both sides of Hammerdalen Bugt dipping at 50° to the south, and again on the west side of Haengefjaeldet dipping at 25° south. Since this band, which may be called the Main Tuffs, was not mapped continuously, it is indicated on the map only where it

was observed. Above the main tuffs are more lavas with occasional thin
tuffs such as those observed on the point one kilometre east of Kap
Hammer. Traced southwards the lavas increase in dip, reaching 55°
on Kap Hammer and 60° on Haengefjaeldet.

The data collected are sufficient to allow the thickness of the lava
and tuff series to be calculated providing that faulting is absent. In
country of this kind faulting is difficult to prove or disprove but the
position of the main tuffs in Hammerdalen and Mikis Fjord is such as
would be expected if there is no faulting. On this assumption the thickness of the lavas and tuffs (with seven hundred metres of included sills
to be described below) is 4500 metres. Since the dyke swarm is approximately perpendicular to the lavas it does not appreciably affect the
estimate of the thickness. In Jacobsens Fjord, thirty kilometres to the
east, the thickness of lavas and tuffs still preserved is 6500 metres, and
it is likely that the thickness on the peninsula between Mikis Fjord and
Kangerdlugssuaq was at least as great as this before erosion removed
the upper part. The sequence and details of the thickness of the sediments,
lavas and tuffs, is given in the legend to the geological map at the end
of the paper.

About a dozen lavas within the area of the map have been sectioned,
but being a great deal decomposed they are not a pleasurable study
and they have so far only been examined superficially. Apparently some
never contained olivine, for example the twelve-metre flow which rests
directly on the sediments in Vandfaldsdalen. This lava is not strictly
a basalt but an augite andesite since the felspar is andesine. Other
lavas, collected just beyond the area represented on the map in the
easterly part of Mikis Fjord, formerly contained a considerable amount
of olivine but all is now replaced by chlorite. These lavas and others
collected from the southern point of Haengefjaeldet are decomposed;
they resemble spilites both in the field and under the microscope. The
flows are often only a few feet thick and in some cases the centres are
almost earthy. On Haengefjaeldet certain flows are intersected by a
three-dimensional network of cracks filled with calcite, epidote, chlorite
and carbonates, and the junction of the vein material and the rest of
the rock is indefinite. In other cases there is a rudimentary development of pillow structure. Under the microscope these lavas consist of
fresh augite, oligoclase or albite with inclusions of chlorite, and much
interstitial chlorite with a little epidote, carbonate and leucoxenised
iron ore. Some of the thin flows at the south point of Kap Hammer are
variolitic in texture and closely comparable with those described from
Mull [Bailey, Thomas etc. 1934, pp. 149—51]. Two of the lavas which
resemble spilites in appearance have been analysed by Dr. H. F. Harwood (Table II) but the analyses show that the rocks are not significantly richer in soda than normal basalts. Analysis A, allowing for the

decomposed state of the rock, is similar to two other basalts from the region further to the north-east which are about a mean between the average çomposition of Plateau Basalt and the Non-Porphyritic Central Basalt of the authors of the Mull Memoir. The felspar of the rock is, however, between oligoclase and albite, more exact determination from refractive indices being hindered by the presence of inclusions of chlorite. Analysis B corresponds with an olivine rich basalt but the minerals present are augite, oligoclase or albite, chlorite and zeolites. Some of the chlorite is replacing former olivine crystals but much has an interstitial arrangement. The percentage of soda is a little high for a normal olivine basalt but it is not sufficient to suggest that we are here dealing with an originally soda-rich rock.

The extensive decomposition of many of the basalts is not to be ascribed to weathering and is similar in many ways to the pneumatolitic

TABLE II.

ANALYSES OF BASALTS.

	A (383)	B (1519)	Norms A		Norms B	
SiO_2	47.89	47.61				
Al_2O_3	14.36	10.92	Or....	1.11		0.56
Fe_2O_3	1.63	3.80	Ab..	25.15		23.06
FeO	10.14	7.25	An..	25.30		17.24
MgO	7.28	13.49	10.44	20.53	9.16	17.44
			Di 5.60		6.70	
CaO	10.51	8.07	4.49		1.58	
Na_2O	3.01	2.74	Hy.... 4.40	7.96	13.80	17.10
			3.56		3.30	
K_2O	0.22	0.10	Ol..... 5.60	10.70	9.24	11.69
H_2O+	2.76	3.62	5.10		2.45	
H_2O-	0.10	0.58	Ilm........	3.34		3.04
CO_2	0.19	0.02	Mt.........	2.32		5.57
TiO_2	1.74	1.58	Ap.	0.34		0.34
P_2O_5	0.18	0.17	Calc.	0.40		.
MnO	0.19	0.19	H_2O........	2.86		4.20
Cl_2	tr.	0.02				
SrO	tr.	nil				
BaO	tr.	nil	Plag. ...$Ab_{50}An_{50}$		$Ab_{57}An_{43}$	
Li_2O	tr.	tr.	Diop...$Wo_{51}En_{27}Fs_{22}$		$Wo_{52}En_{39}Fs_9$	
NiO	tr.	0.06	Hyp. ..$En_{55}Fs_{45}$		$En_{81}Fs_{19}$	
S	tr.	tr.	Oliv....$Fo_{52}Fa_{48}$		$Fo_{80}Fa_{20}$	
	100.20	100.22				
S.G.	2.99	2.95				

COMPARISONS.

	A	B	C	D	E	F
SiO_2................	47.89	47.61	48.89	47.26	47.81	49.34
Al_2O_3...............	14.36	10.92	13.60	14.17	13.64	14.04
Fe_2O_3...............	1.63	3.80	4.72	4.68	4.50	3.41
FeO................	10.14	7.25	9.65	8.32	8.84	9.94
MgO...............	7.28	13.49	4.54	6.65	6.51	6.36
CaO...............	10.51	8.07	10.09	10.64	11.29	9.73
Na_2O...............	3.01	2.74	2.74	2.30	2.54	2.89
K_2O...............	0.22	0.10	0.30	0.81	0.35	1.00
H_2O+............	2.76	3.62	1.38	1.45	0.76	..
H_2O-............	0.10	0.58	0.14	0.90	1.26	..
TiO_2...............	1.74	1.58	2.64	2.69	2.51	2.59
MnO	0.19	0.19	0.24	0.19	0.22	0.21
P_2O_5	0.18	0.17	0.39	0.23	0.24	0.49
CO_2	0.19	0.02	0.02	0.04	nil	..
Cl_2................	tr.	0.02	tr.	0.01	tr.	..
SrO	tr.	nil	tr.	nil	nil	..
BaO...............	tr.	nil	tr.	nil	nil	..
Li_2O...............	tr.	tr.	tr.	tr.	tr.	..
NiO...............	tr.	0.06	..	..	..	..
S	tr.	tr.	0.02	0.01	0.01	..
	100.20	100.22	100.36	100.35	100.48	100.00

A. Altered variolitic basalt, 383. Middle of 18 ft. flow, south point of Haenge-fjaeldet Kangerdlugssuaq, East Greenland. Anal. H. F. Harwood.

B. Altered olivine basalt, 1519. Below the sills at the extreme end of Mikis Fjord on the south side, near Kangerdlugssuaq, East Greenland. Anal. H. F. Harwood.

C. Tholeiitic Basalt, 1057. Kap Dalton, East Greenland. Hill west of anchorage and 100 ft. below lowest sediments. [Wager, 1934]. Anal. H. F. Harwood.

D. Olivine Basalt, 1093. North side of Barclay Bugt, East Greenland. On Coast about one mile W.N.W. of Host's Havn. [Wager, 1934]. Anal. H. F. Harwood

E. Olivine Basalt, 1112. Kap Daussy, East Greenland. [Wager 1934]. Anal. H. F. Harwood.

F. Average of 50 plateau-basalts, calculated as water-free and to a total of 100.00 %. [Washington, 1922].

decomposition found in Mull [cf. Bailey, Thomas etc. 1934, pp. 141—43]; it may well be partly an effect of the Kangerdlugssuaq intrusions. On the other hand there is the following evidence that some of the lavas were poured out under water:—

1. "Red partings" indicating subaerial extrusion occur universally between the flows about Nansen Fjord, to the east-north-east, and in the inland mountains, to the north, while within the area of the map

they were only observed between some of the lowest lavas and again in some of the lavas just above the agglomerate horizon.

2. Between the spilites on Haengefjaeldet a thin glassy tuff is sometimes found which seems to be the quenched and broken upper surface of the lava underlying it.

3. The variolitic texture of some of the lavas suggests rapid chilling perhaps due to submarine extrusion.

4. The main tuffs are so evenly and well bedded that they can only have been deposited in water.

The extent to which the decomposition of the lavas is due to submarine extrusion and the extent to which it is due to a pneumatolitic effect of the Kangerdlugssuaq intrusion centre cannot at present be decided but it seems likely that both effects were significant.

The agglomerate near the base of the volcanic series consists of basalt fragments averaging 2—3 inches across but sometimes as much as 3 feet. The material is either wholly fine grained or fine grained except for conspicuous porphyritic augite crystals. Blocks 2—3 feet across, having a chilled crust so that they resemble huge bombs, make up the bulk of the agglomerate at the head of Jacobsens Fjord. In 1932 this agglomerate horizon was wrongly correlated with the fine-grained tuffs on Haengefjaeldet [Wager 1934, pp. 34—35], and the horizon was thought to vary in coarseness more quickly than is actually the case. The main tuff horizon consists, for the most part, of an extremely fine grained, flinty rock made of basaltic material where identification is possible. Towards the top of the zone on Haengefjaeldet a thin layer of coarse tuff is developed approaching the lower agglomerate in appearance.

(d) Sills.

Within the area of the map there are two thick, differentiated gabbro sills or thin laccolites and several thin sills, all of which are largely made up of a highly characteristic spotted gabbro. There is also a thin sill of columnar dolerite in Vandfaldsdalen and another at five hundred metres on the ridge to the west of Mikis Fjord Rejsehus. The uppermost, differentiated, gabbro sill on Haengefjaeldet is about one hundred metres thick on the west side and three hundred metres thick on the east. It is also found on the east side of Hammerdalen Bugt but there its thickness was not estimated. Below this sill on the west of Haengefjaeldet two other sills were noted, twenty and thirty metres thick. A similar gabbro sill averaging three hundred metres in thickness forms much of Tinden. An olivine-rich facies near the base of this sill was collected on the west and north-east ridges of Tinden. Material exactly similar to this sill forms the summit of Basistoppen and is a huge inclusion in the Skaergaard intrusion.

Abundant dolerite sills inject the sediments and tuffs which outcrop near the centre of the flat dome into which the rocks between Kangerdlugssuaq and the Watkins Mountains are bent. The sills in Vandfaldsdalen and west of Mikis Fjord Rejsehus are two examples where the suite is dying out. The thick sills of Haengefjaeldet and Tinden are different; they are plutonic in texture, and both the thick and thin sills are devoid of any columnar structure. The peculiar spotting of the latter rocks indicates a community of origin. The whole of the smaller of these spotted sills, and the dominant central material of the larger sills, is an olivine-free gabbro with poikilitic augite crystals 0.5—1 centimetre across which enclose well-shaped tabular labradorite crystals. The margins of the poikilitic augite crystals tend to be altered to uralite and outside the patches of poikilitic augite and the enclosed plagioclase there is a good deal of uralite and chlorite. The original water of the magma, present now in uralitic hornblende and chlorite, apparently became concentrated in patches lying outside the parts which consist of poikilitic augite and plagioclase. It is this feature which produces an indefinite but evenly distributed spotting of the rock which shows up both on weathered and fresh surfaces. In the upper part of the Haengefjaeldet sill there is a gradual decrease in number and increase in size of the light-coloured areas of uralite, chlorite, and felspar, and a zeolite makes its appearance so that the spots of high concentration of volatile constituents become still more conspicuous. Olivine appears towards the bottom of the sill, and at the base, the rock is a peridotite consisting of olivine embedded in poikilitic plates of augite and plagioclase. The tendency for hydrous decomposition products to be collected in spots is also found in the olivine gabbro stage. The peridotite developed at the base of these differentiated sills is quite distinct from that found as a differentiate of the Skaergaard intrusion.

(e) The Dyke Swarm.

Most of the dykes cutting the Skaergaard intrusion and neighbouring rocks are basic in composition. They form part of a definite dyke swarm already described [Wager & Deer 1934 pp. 41—44 and 1937], and which here runs east and west (Fig. 2). The dyke swarm[1] is younger than the Skaergaard intrusion since it cuts through it; it is in fact the latest manifestation of igneous activity within the area of the map.

---

[1] A remarkably similar dyke swarm also associated with a flexure of the crust has been described by du Toit [1930] from the Lebombo Range, South Africa.

Ten miles west-south-west on Kap Deichmann there is a quartz syenite complex which is not cut by the dyke swarm although lying in its path. Thus elsewhere in the igneous province plutonic intrusions occur which are later than the dyke swarm.

South of Tinden the dyke swarm is extremely dense, there being over a hundred full-sized dykes per mile. The dykes dip north, the average value being 60° on Haengefjaeldet, and this increases to about 80° on Tinden itself. Dykes are abundant for two kilometres to the north of Tinden and then become sporadic: at the same time the average dip increases until the dykes are vertical. The dykes therefore approach, although do not actually reach, perpendicularity to the lavas which they intersect. In the north-west part of the map, north-north-east-trending dykes occur, while east-west dykes are rare. Two of the north-running dykes on Uttentals Plateau were found to be earlier than the Skaergaard intrusion and to be metamorphosed by it.

Of the many hundreds of dykes which were observed, all were basic except for one keratophyre which occurs on the west side of Haenge-fjaeldet. Though the majority were non-porphyritic dolerites, there are many conspicuously porphyritic types, and some are augite and horn-blende lamprophyres. The petrology of the dykes has not yet been con-sidered in detail, but they are a regional phenomenon coming after the Skaergaard intrusion, and their petrology is not important for our present purpose.

(f) Structure of the Area.

The coastal region from Kangerdlugssuaq to Nansen Fjord, a distance of sixty kilometres, is structurally a flexure or monocline (Fig. 2). The dips of the inland lavas which are between two and three degrees become steadily greater as the coast is addached, and on some of the capes near Kangerdlugssuaq reach 50° and 60°. The flexure is more intense near Kangerdlugssuaq than further east, that is, there is a higher maximum value of the dips and at the same time the width of the zone of high dips is less. The dyke swarm is found only in associa-tion with the flexure, beginning where the dips reach 10° and becoming intense where the dips are greater than 20°. Because the dyke swarm and the strong coastal flexures are coextensive, and because the density of the swarm is proportional to the intensity of the flexure, we have argued that the two phenomena are related [1937, pp. 43—44]. We have suggested that the tension fn the upper crust which would be developed during the slow formation of the flexure, produced a suc-cession of tension fractures approximately perpendicular to the lavas,

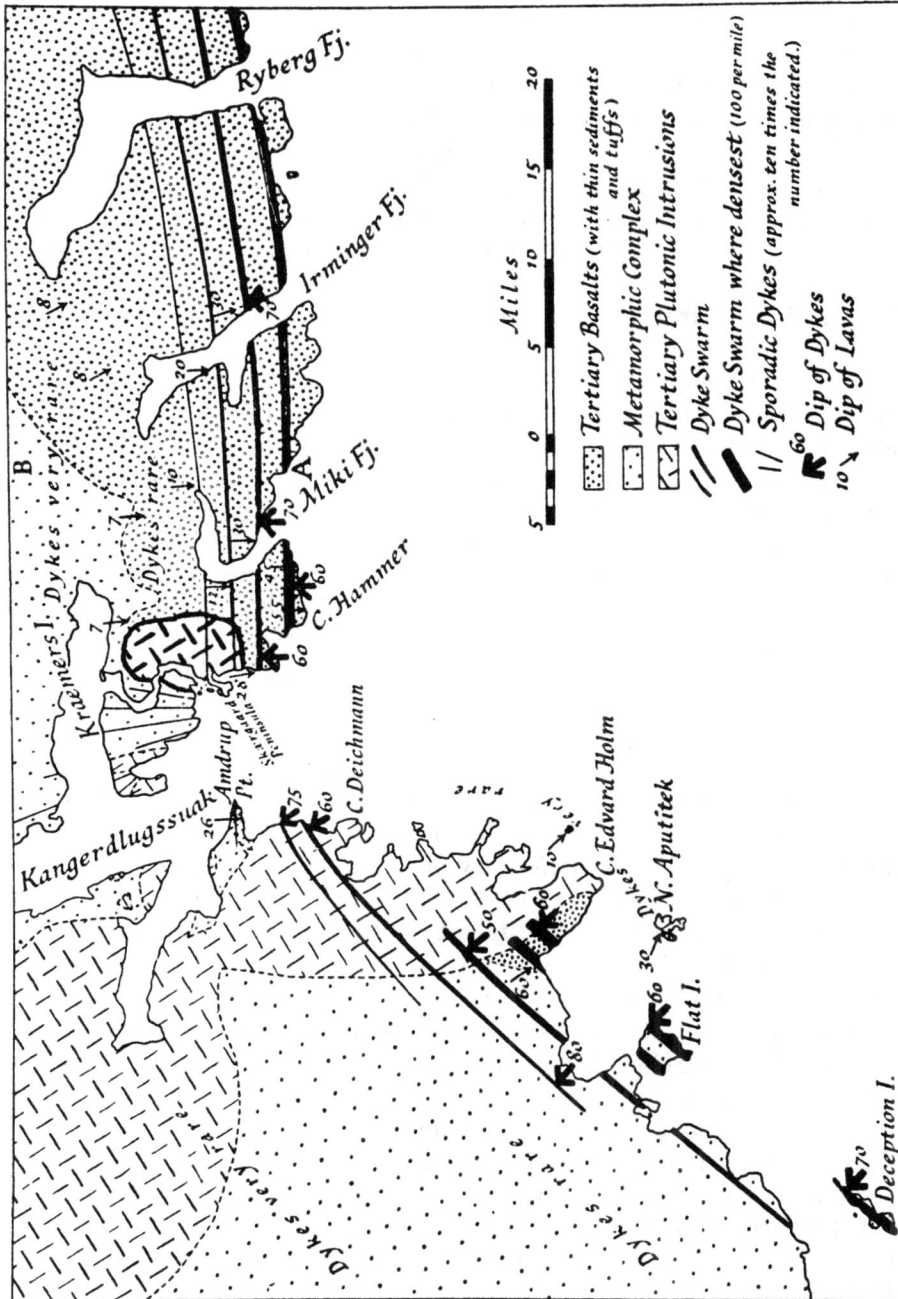

Fig. 2. The dyke swarm in relation to the Skaergaard intrusion which is shown by thickened boundary and pattern. (Reprinted with modifications from the Geological Magazine).

and that these have been filled by successive injections of basic magma, giving the observed dyke swarm.

The dip of the lavas appropriate to any particular zone of this flexure is not modified by proximity to the Skaergaard intrusion except along parts of the southern margin. Thus the 7° dip of the basalts immediately north-east of the intrusion is the average dip in a zone extending eastwards from this place. Likewise the dip of about 10° in Vandfaldsdalen is the unmodified regional dip in this part of the flexure. Close to the south and south-east margin, dips of 30°, 40° and 50° were recorded and these are higher than the dips of 28° and 35° about half a kilometre to the south. The latter are the values to be expected in this part of the flexure and the higher dips near the south and south-east margin are to be ascribed to a local effect of the intrusion considered later (p. 51). Likewise the dip of the small patch of meta-morphosed sediment on Mellemö and of the basalts on Ivnarmiut which lie within a few feet of the west margin of the intrusion, have been modified slightly from the regional dip of the zone of the flexure in which they are situated. The sediments nevertheless are in the position in which they would be expected were the observed strike in Vandfaldsdalen continued, and the slight modification of the regional dip probably extends only a few metres from the contact. A minimum estimate of the dip of the former lavas and sediments which lay west of the intrusion can be made from the height of the gneisses on Krae-mers Ö. Half a kilometre west of the boundary of the map the gneiss reaches a height of eleven hundred metres; the sediments must have been higher than this, while on Mellemö they are at sea level. From these facts a dip of 23° is obtained if the strike is considered to remain east-west. The unconformity between the metamorphic com-plex and lavas is found again on Amdrups Pynt, suggesting that the strike is deflected a little towards the south-west. Thus the probable minimum dip of the former lavas to the west of the Skaergaard intrusion is 20° in a direction 10° east of south.

The dips of the lavas and sediments near the Skaergaard intrusion—except locally along the southern margin—are those appropriate to their position in the coastal flexure. Therefore the formation of the intrusion did not greatly modify the lie of the surrounding rocks. There are two ways by which this disposition may have been brought about: the intrusion may have come after the flexure and without modifying the lie of the surrounding rocks, or the intrusion may have come into position and solidified before the flexuring movement tilted both the lavas and intrusion equally. The existence of the dyke swarm allows a decision to be made between these alternatives. Evidence has been given to show that the dyke swarm is the result of the flexuring and, since the

dykes cut through the intrusion, it must be assumed that the intrusion
has also been flexured. The intrusion was therefore formed before the
flexuring and with little disturbance of the surrounding rocks, and its
present disposition must be ascribed to the flexuring process. It might
have been expected that the huge mass of resistant gabbro which forms
the Skaergaard intrusion would have modified the flexure in extent
or direction, but apparently the flexuring was a major movement of
the Earth's crust which bent and fractured the Skaergaard gabbro with
the same ease as the lava series.

(g) Age of the Skaergaard Intrusion.

The late Cretaceous age of the Kangerdlugssuaq sediments, proved
by belemnites and plant remains, gives a lower limit of age for the
basalt series and for the whole igneous activity of the region. At Kap
Dalton sediments containing Middle Eocene fossils [Ravn 1904, 1933]
were deposited either at, or near the top of the basalt series, and there
was apparently no long time interval between their deposition and that
of the underlying lavas [Wager 1935]. The preservation of the sediments
at Kap Dalton is due to down faulting which is apparently an expression
of the same earth movement as gave rise to the coastal flexure; therefore the coastal flexure is probably post Middle Eocene. On this evidence
the dyke swarm, which was contemporaneous with the flexuring, is
also post Middle Eocene in age. The Skaergaard intrusion and others
of the district are cut by the dyke swarm while the Kap Deichmann
syenite intrusion is not. From these facts the following sequence of
events may be deduced for this part of East Greenland, two stages
being dated precisely by the fossil evidence:—

1. Submergence of a peneplained metamorphic complex by a
shallow sea and then deposition of thin sediments, a low horizon of
these being Middle Senonian and a high horizon, latest Cretaceous or
early Eocene.

2. Outpouring of basalts and deposition of tuffs without any
appreciable interval of time between the underlying sediments and the
volcanic rocks.

3. Deposition of the Kap Dalton sedimentary series of approximately Middle Eocene age at, or near, the top of the volcanic series.
Either before or contemporaneously there occurred at Kangerdlugssuaq:—

a) Injection of thick gabbro sills.
b) Intrusion of the Skaergaard and other basic plutonic masses.

4. Crustal flexuring and concomitant injection of the dyke swarm. On the Kap Dalton evidence these happenings were post Middle Eocene in age.

5. Intrusion of the Kap Deichmann and other syenite complexes. These are definitely later than the dyke swarm and therefore post Middle Eocene.

Vigorous igneous activity in East Greenland apparently lasted throughout most of the Eocene and perhaps a little longer. About half-a-dozen hot springs are known along the coast between Scoresby Sound and Kangerdlugssuaq, suggesting that the igneous activity may have persisted longer than in the British area. For the sake of brevity the whole of the igneous activity will be referred to as Tertiary, although the earliest lavas and tuffs may be very late Cretaceous.

IV. FIELD OBSERVATIONS ON THE INTRUSION

(a) The Outer Contact and the Rocks of the Border Group.

In this region of high relief and good exposure the outer contact of the intrusion is clearly seen to be a fairly even surface. Its varying dip and strike have been mapped with considerable certainty since it has been possible to rely on exposures usually a few tens of metres, and often several hundred metres, high. The rocks which we have classified as the border group are of variable composition but all may be classified broadly as gabbros. The distinction between border group and layered series is not based on the composition of the rocks but on structural and textural features. The marginal border group which is present everywhere between the outer contact and the layered series, has fluxion structures parallel to the outer contact of the intrusion—a direction approximately at right angles to the banding shown by the layered series. The upper border group lying above the layered series consists of rocks continuous with certain units of the marginal border group and having similar textures and fluxion structures. The direction of the fluxion structures is parallel to the presumed position of the original roof of the intrusion. The field observations on the various border rocks which are given in this section will be dealt with regionally.

1. Skaergaarden and Kraemers Ö.

The junction between the Skaergaard intrusion and the surrounding basalts is preserved on the extreme west point of Ivnarmiut in a cliff a few metres high. The dip of the junction is inwards at about 65° but the extent of the exposure is too small for any great weight to be given to this determination. The basalts at the junction are highly metamorphosed, and for about half a metre, lines of vesicles have been turned by slight flow so that they approach parallelism with the margin. Further away the lines of vesicles dip at approximately 40° to the S.E.; this is roughly the regional dip of the lavas to be expected at this point (see page 22). The marginal olivine gabbro is distinctly chilled and the grain size for about a metre from the margin is only a little coarser than, for example, the central parts of a 100 metres dolerite sill. For two to three metres from the contact on Ivnarmiut there are rounded blocks of olivine-rich material and about 4 metres from the margin rare inclusions of highly metamorphosed sediment were noted. Twenty to forty metres from the margin peculiar zones were found having very much elongated felspar crystals arranged with their length perpendicular to the margin of the intrusion (Pl. 3, fig. 1, and text-fig. 30). These zones may be only 10 cm wide and not extensive, or they may be as much as a metre wide and extend in a continuous sheet parallel

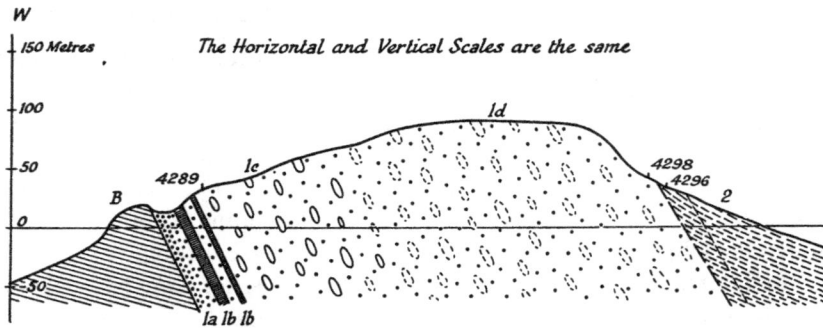

Fig. 3. Section across Ivnarmiut. 1. Border Group; 1a. Chilled olivine gabbro along contact; 1b. Zones of perpendicular felspar rock; 1c. Outer marginal border group where the granophyre inclusions (represented diagrammatically as lenticles) have fairly definite margins; 1d. Inner part where granophyre inclusions are much hybridised; 2. Layered series. A slight displacement along a fault parallel to the inner contact is shown.

to the margin for considerable distances. Between these zones the olivine gabbro is sometimes equigranular in texture and shows no obvious banding, but in other cases coarser wavy layers of pyroxene and felspar are developed extending roughly in a horizontal plane and like the felspars, at right angles to the margin (Pl. 3, fig. 2). The latter feature is also found where the perpendicular felspar bands are missing and it extends in places to about 50 metres from the margin. It seems to be a similar structure to the perpendicular felspar rock expressed in terms of pyroxene instead of felspar. Two bands of the perpendicular felspar rock are shown diagrammatically in the section across Ivnarmiut (Fig. 3).

On Mellemö the contact is with the gneisses of the metamorphic complex and with a few feet of sediments resting unconformably on the gneisses (see p. 13). The contact here is somewhat less steep than on Ivnarmiut; apparently the intrusion bulges out above the unconformity, an irregularity perhaps due to the sudden change in strength of the rocks at this level. The marginal rock is somewhat chilled, then follow

successive zones of perpendicular felspar rock as on Ivnarmiut. Where
the boundary is less steep the elongated felspars are still perpendicular
to it; apparently they formed at right angles to the cooling surface.
A rock having large pyroxenes at right angles to the margin occurs
here and grades into the rock with coarse wavy layers of pyroxene and
felspar (Pl. 3, fig. 2).

On Kraemers Ö the contact continues to be steep and an inward
dip of 70° was determined in the cliff at the north end of the contact.
Slight chilling at the margin is everywhere noticeable; coarse wavy
layers of pyroxene and felspar are common and zones with perpendicular
felspar occur sporadically. We found that these last two features are
significant marginal phenomena well developed along the eastern and
western contacts and found occasionally along the northern contact,
but not along the southern.

The zone with noticeable chilling and the special features just
described is about 50 metres wide, then begins the major unit of the
border series which extends until the inner contact with the layered
series is reach—on Ivnarmiut a distance of 300 metres. The material
forming this zone is extremely heterogeneous in detail, but in its field
relationships it behaves as a single unit and it has not been subdivided
on the map. In the Skaergaarden area and on Kraemers Ö the bulk
of the material of the border zone is olivine gabbro which is heterogeneous,
partly as a result of irregular banding, and partly as a result of inclusions
of acid material in all stages of digestion (Pl. 14, figs. 1 and 2). The banding
is due to variation in the relative amounts and relative sizes of the
constituents. It is peculiar, however, in that the trace of the banding
on a horizontal surface is tortuous, while on a vertical surface at right
angles to the margin it appears fairly regular (Pl. 4, fig. 1). The banding
due to variation in the amount of the dark and light minerals is sometimes broken into, or modified by patches of coarse material (Pl. 4,
fig. 2), but at other times the coarse material is streaked out and contributes to the banded appearance of the rocks. In the Skaergaarden
region the dip of the banding is inwards at about 70° and on Kraemers Ö
the dip averages 80° inwards; the banding is thus parallel to the margin
of the intrusion. The banding is to be regarded as a fluxion structure
and the fact that it is tortuous on horizontal surfaces and more even
on vertical surfaces indicates that the flow must have been upwards
or downwards and not horizontal.

The inclusions of acid material, which add greatly to the heter-
ogenity of the border group, are distributed fairly evenly throughout
this zone on Ivnarmiut and Mellemö. The one figured (Pl. 14, fig. 1)
is about 40 metres from the contact, and some of the surrounding gabbro
shows the wavy patches of coarse pyroxene and felspar arranged roughly

at right angles to the margin (this does not appear in the photograph). The boundaries of this inclusion are fairly definite but with increasing distance from the contact, the inclusions become streaked out parallel with the banding, and the original form is lost (an approach to this condition is shown in Pl. 14, fig. 2). The even-grained central portion of the inclusion shown in Pl. 4, fig. 1, is now granophyre. The border is a coarsely crystalline aggregate of augite and plagioclase, and long crystals of these minerals project into the granophyre area. The nature of this occurrence and of many others indicates that the inclusion had originally the bulk composition of the granophyre and that the coarse margins are the result of reaction with the surrounding olivine gabbro. It will be shown below that these inclusions are blocks of the acid part of the metamorphic complex; they will be referred to as granophyre when their present state is being considered and as gneiss when their origin is the aspect stressed. Further away from the outer margin of the Skaergaard intrusion the reaction between the inclusions and the magma has been greater and the boundaries of the inclusions become completely indefinite. There granophyre cores are preserved only in the larger inclusions, and they are no longer central but occur on the upper side or as a cumulose head to a lenticular patch of coarse hybrid material. An example of this advanced alteration of the xenoliths is figured in Pl. 4, fig. 2; the streak of coarse material is dipping away from the observer towards the left at 70° and the lighter material is thus on the upper side. In the field, because transitional stages can be found, it is quite clear that the coarse lenticular patches of pyroxene and felspar rock occurring in the inner part of the border group are the remains of completely digested acid inclusions. Granophyre inclusions, or patches of coarse material from their digestion, make up about 5 or 10 per cent. of the border group on Mellemö. The percentage is on the average less on Kraemers Ö and over much of the Skaergaard Halvö digestion of the acid inclusions has proceeded so far that it is not safe to venture on any estimate of their relative abundance. Here unmodified centres of granophyre are very rare except towards the western point of the peninsula which is near the margin.

Other inclusions in the western part of the border group are found only rarely. On Kraemers Ö 900 metres north of Mellemö several blocks of metamorphosed basalt, some with obvious vesicles, were found. On Mellemö near the margin, basic blocks occur which are probably metamorphosed amphibolite from the metamorphic complex, and here also a small xenolith of extremely tough asbestos was found. The small inclusions of sediment and the cognate inclusions of olivine-rich gabbro which are found here have already been mentioned. At the north end of the group on Kraemers Ö blocks of peridotite were found which are similar to those making up much of the northern border group.

2. Eastern Margin, north of Forbindelsegletschen.

The ridge running S.S.E. from Peak 1360 m. of Gabbrofjaeld which also forms the west side of Vandfaldsdalen (Fig. 1) is composed of the rocks of the Skaergaard intrusion on the west side, and of basalts and tuffs on the east (Fig. 4). Where the ridge reaches Forbindelsegletschen the agglomerate horizon forms the country rock and a small stream

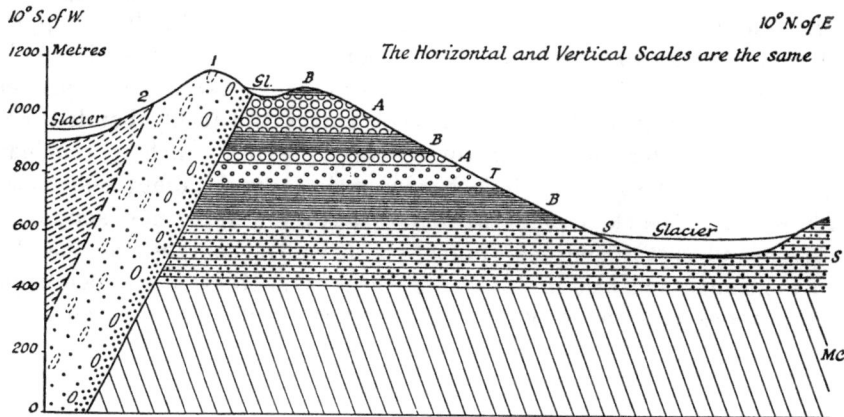

Fig. 4. Section across south-south-east ridge of Gabbrofjaeld. 1. Border group with chilled margin and sporadic granophyre inclusions; 2. Layered series; B. basalt; A. agglomerate; T. tuffs; S. sediments; M.C. metamorphic complex.

follows the contact which is exposed almost continuously for 300 metres. Standing on Forbindelsegletschen on the line joining the contact to the north and south of the glacier, the dip of the boundary over this height is seen to be constant and 65° inwards. The dip of the lavas and tuffs right up to the contact is the regional dip and it is clear that the intrusion has cut cleanly through the surrounding rocks without disturbing them. The gabbro at the contact is somewhat chilled, then follows coarser olivine gabbro with coarse, wavy pyroxene and felspar patches and zones of the perpendicular felspar rock. Further inwards the main unit of the border group is reached; here the banding is seen to dip inwards at about 70°; and acid patches in all stages of digestion are found but they are only about half as abundant as on Mellemö.

A kilometre to the north the boundary curves inwards and a gabbro mass not belonging to the Skaergaard intrusion is encountered. The mass is about 400 metres broad and cuts across the country to the N.E. as a huge dyke. The relations between this mass and the Skaergaard intrusion are partly hidden by moraine and partly exposed on a face so steep that it was not examined. It appears that the Skaergaard intrusion came after the dyke-like mass and that the boundary was somewhat deflected by it. A conspicuous patch of highly metamorphosed sediment (see fig. 1) was pushed up 400 metres by this dyke-like intrusion and not by the Skaergaard intrusion as was thought in 1932 [Wager 1934, p. 37].

To the north of the dyke-like gabbro mass the outer contact of the Skaergaard intrusion is for the most part obscured by glaciers but at two points it is seen cutting across the undisturbed and strongly metamorphosed basalts. The inward dip, although undoubtedly steep, was not clear enough to be measured. Immediately inwards from the chilled marginal gabbro the rock with horizontal wavy patches of pyroxene and felspar is well developed, and then comes the feebly banded main part of the border series with some coarse patches which seldom have granophyre centres. The whole border series here is about 300 metres wide which is the same as on Mellemö. It is remarkable how similar the western and eastern border groups are in width, composition, banding and inclusions and also in detailed structures, such as the perpendicular felspar zones and the wavy coarse pyroxene and felspar patches.

3. Northern Margin.

Except on Uttentals Plateau the northern margin is largely covered by glaciers. There, however, the contact is well seen (see Plate 2) and the inward dip of the boundary can be determined as approximately 45°. To the north of Gabbrofjaeld (Peak 1360 m.), the boundary dip is likewise inwards at about 45°, but no exact measurement was taken. The re-entrant angle of the intrusion boundary where it crosses Uttentals Sund is the result of the inward dip and not of an irregularity in the general shape of the boundary such as occurs in the Skaergaarden area. On Uttentals Plateau the width of the border series is about 150 metres, that is about half the width found on Mellemö; the value cannot be given precisely as the inner junction with the layered series is indefinite. The outcrop of the marginal zone narrows in descending to Uttentals Sund from both the top of Uttentals Plateau, and from the highest point of the boundary on Kraemers Ö, and this is due to an actual downward narrowing of the marginal zone. Gabbro-picrite[1]) blocks are abundant in the border zone on Uttentals Plateau and must be regarded as characteristic of the northern border group. They are also found at the lowest point of the northward projecting spur, east of Uttentals Plateau, and the marginal zone is therefore mapped as just touching this spur

---

[1]) Gabbro-picrite is defined on pp. 162—64.

but not the next to the east. North of the highest summit of Gabbro-fjaeld (Peak 1360 m) gabbro-picrite blocks were found in the marginal zone, although the region was only examined cursorily.

On Uttentals Plateau the border zone is excellently displayed for examination; this is fortunate as the zone is here a good deal different and more complex than further south. Half a kilometre from Uttentals Sund the actual margin is well exposed and is found to consist of a somewhat chilled olivine gabbro like that described from the other contacts. Inward and above the chilled gabbro there is a mass of gabbro-picrite about 20 metres thick which gives place to blocks of the same material embedded in olivine gabbro. Elsewhere there are no large masses of gabbro-picrite, but blocks of varying sizes and with indefinite margins, occur surrounded by the olivine gabbro. South of Uttentals Sund some gabbro-picrite masses were noted but as the border group is traced still further southwards they rapidly became less numerous and then disappear altogether. On Mellemö homologous inclusions consist of olivine-rich gabbro and not gabbro-picrite. Rare blocks of gabbro-picrite are also found in the border group on the S.S.E. ridge of Gabbrofjaeld.

At the top of Uttentals Plateau where the boundary turns round to the east, a reef of highly metamorphosed sediments about 30 metres long and several metres thick is found embedded in olivine gabbro and gabbro-picrite. A little east of this the olivine gabbro was found to develop locally the coarse wavy flecks of pyroxene and felspar characteristic of the outer border group elsewhere, and at one point a feeble development of perpendicular felspar rock was noticed. Occasional, digested, gneiss inclusions were found and also raft-like masses of a medium grained norite. Feeble banding, accentuated in places by the streaking out of the inclusions in the same direction, is just sufficient to allow dips to be taken at one or two points. The characteristic of the northern margin is the abundance of gabbro-picrite masses which makes up about half the border group. In the east and west marginal border rocks gabbro-picrite and olivine-rich gabbro masses are rare and they are not found in the southern marginal border group. The existence in the northern border group of chilled olivine gabbro at the contact and such features as the perpendicular felspar rock and horizontal, coarse pyroxene and felspar patches, are details which help to link the border rocks into one continuous unit although within the group there is considerable variation.

4. The Southern Margin and the Upper Border Rocks.

The direction of the contact between the Skaergaard intrusion and the basalts is vertical at the head of Udløberen, 80° outwards at the head

Fig. 5. Junction of basalts (right) and marginal group of the Skaergaard intrusion (left). West ridge of Hammersfjaeldet from the south-west.

of Hammerdalen (Fig. 5) and again vertical in the lower part of the west ridge of Tinden. Somewhat chilled olivine gabbro occurs at all these contacts, and small, chilled appophyses from the Skaergaard intrusion were noted at the coast on the east side of Skaergaardsbugt. In the chilled margin 100 metres above sea-level on Tinden, clusters of small inclusions of highly metamorphosed sediment were found, and also the usual granophyre patches with the beginnings of hybridization at the contacts. These inclusions of sediment and gneiss are over 1,000 metres above the gneiss-sediment junction at this point. Other features of the outer border zone found at the east, west and northern margins, such as the perpendicular felspar rocks, the horizontal wavy flecks of coarse pyroxene and felspar and the gabbro-picrite or olivine-rich gabbro inclusions, were not found in the southern border group.

Chilling of the gabbro can be detected for about 30 metres from the contact. Inwards the rocks of the southern part of the border group are, on the average, coarser in grain than elsewhere and banding is poor or absent. Streaks of coarse acid gabbro in which quartz can be detected in hand specimens, and which are the result of hybridization with acid gneiss inclusions, are abundant. The direction of the feeble banding and of the granophyre and coarse quartz-gabbro sheets is parallel to the contact. At the head of Hammersdalen the width of the zone with steeply dipping banding is about 200 metres. Elsewhere the width of the zone was not determined with certainty but it appears to be about 300 metres at the western foot of Tinden.

At Hammers Pas on the west side there is a heterogeneous and, for the most part, thoroughly coarse olivine-free gabbro. Certain layers contain platy felspars arranged so as to give a well-foliated rock, while others show no evidence of foliation. The dip of the foliation, where present, is outwards at 20° and streaks of granophyre and hybrid material lie in roughly the same direction. This direction is distinct from that of the outer contact and the fluxion structures of the marginal border group, and the rocks having these lower dips are distinguished as the upper border group, On Kilen, the name we have given to the spur running N.N.W. from Hammersfjaeldet, there are similar coarse gabbros with raft-like masses of hybrid rock dipping outwards at about 30°. In the east wall of Kilen the chilled gabbro is found at the contact. This changes gradually into coarser olivine gabbro with yet coarser streaks of material ascribed to hybridization with acid inclusions. About 200 metres from the estimated position of the contact which is here under the glacier, the feeble banding and coarse streaks are dipping at 70° outwards, and the rocks are classified as marginal border group. The general conditions on Kilen are shown in figure 6.

In ascending Brødretoppen from the west, the upper 300 metres are found to consist of coarse, olivine-free gabbros, sometimes showing slight foliation dipping south at 25°; there are also abundant reefs of more acid and leucocratic material parallel to the foliation. Four hundred

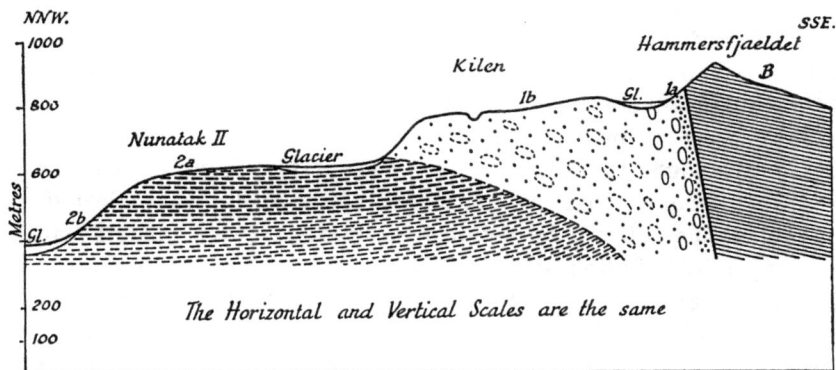

Fig 6. Section across south-east margin. 1a. Marginal border group with chilled margin, steeply dipping fluxion structures and lenticular inclusions of granophyre with fairly definite margins. 1b. Upper border group with indefinite reefs of acid material. 2a and b. Layered series. 2a. Purple band. B. Basalts.

metres of similar heterogeneous material were examined while ascending Osttoppen from the north. The lower 100 metres consist of coarse gabbro with no foliation and little streaking out of the abundant, partly digested, granophyre patches. This is followed by a middle section, probably the same horizon as at Hammers Pas, in which strongly foliated layers, dipping southwards at 30°, alternate with less foliated gabbro. The gabbro of the top 100 metres of Osttoppen is little foliated, but reefs and lenticles of hybridised granophyre are abundant, extending in a direction parallel with the fluxion structures of the lower rocks, that is at 30° to the south.

The variable gabbros along the southern border of the intrusion are all classified as belonging to the border group but they have flow structures in two different directions; there is first a zone of coarse gabbro, 200 to 300 metres wide, with steeply dipping banding and parallel streaks of ganophyre and hybrid material; secondly, there are somewhat coarser quartz gabbros which have fluxion structures dipping at 20° to 30° southwards. These low-dipping fluxion structures are believed to be approximately parallel to the upper border or roof of the intrusion, just as the steeply dipping flow structures are parallel to the marginal contacts. The rocks showing low dips are distinguished as the upper border group, while those with a foliation parallel to the marginal contact are distinguished as the marginal border group.

The nature of the marginal zone where it crosses Tinden is only slightly known because of the difficulties of climbing the mountain and because the dyke swarm, which is here dense, masks the differences in the rocks. Gabbro belonging to the early sills (pp. 18—9) which forms much of Tinden was encountered at the col three-quarters of a kilometre N.E. of the summit of Tinden, and there is presumably an inward bend of the boundary here. At lower levels to the N.E., coarse gabbro belonging to the Skaergaard intrusion occurs. On the N.W. face of Tinden a mass, 300 metres high, belonging to the metamorphic complex can be distinguished from a distance. A specimen of the gneiss was collected from the foot of the cliff but no detailed observations were made because of the danger from falling stones. This mass has been pushed up about 500 metres above the level of the undisturbed gneiss by the act of intrusion, and it will be discussed further when the mechanism of the intrusion is considered. Across the lower part of the west ridge of Tinden there is a conspicuous granophyre sill which is also found on a spur of Sydtoppen to the N.E. This sill, which is about 20 metres thick, has a southward dip of 30° and is therefore approximately parallel to the fluxion structures of the upper border group of Brødretoppen. Both above and below this sill, there is coarse gabbro which might belong either to the marginal or to the upper border group as fluxion structures are not clearly developed; when subsequently discussed it is assumed to belong to the upper border group.

(b) The Inner Contact between the Border Group and the Layered Series.

On Ivnarmiut, Mellemö and the southern half of the outcrop on Kraemers Ö, also east of the highest point of Gabbrofjaeld (Peak 1360 m.) there is a contact between the border group and the layered series, which can be defined within a metre. Elsewhere the contact is not so definite and is indicated on the map by a broken, wavy line. At the S.E. corner of Ivnarmiut, where the inner contact is well exposed (Pl. 5, fig. 1), there is no chilling of either group against the other. The character of the border group is maintained up to the inner contact which is parallel in direction to the banding of the border rocks. In the layered series, within a metre of the contact, the dip is about 50° inwards but this rapidly diminishes until twenty metres from the contact it is only about 35° (Fig. 3). The layered series thus behaves as though it were banked up against the border group. The actual contact can only be determined to about a metre because of the mingling of the two units and their structures. Outside this zone of mingling there is no trace of the injection of one unit by the other, or of blocks of one embedded in the other. At several places in the layered series close to the contact there are small contemporaneous faults. These faults (Pl. 5, fig. 2) are parallel to the inner contact and their downthrow, which is never more than half a metre, is always inwards. They were produced during a particular stage in the cooling of the layered series; under certain stresses the rock was just capable of fracture, and small faults were produced; under other stresses it was capable of being bent without fracture. The fault planes have been sealed by minerals of the magmatic phase, and in some cases have been occupied by coarse material resembling that of the border group. These small faults are probably an effect of the slumping of the layered series where it was banked steeply against the border rocks.

Traced northwards the inner contact becomes less definite until on Uttentals Plateau its position can only be defined roughly. The border group on Uttentals Plateau, as we have mapped it, consists of rocks which have the features of the outer half of the border group in the Skaergaarden area. Inwards for about a kilometre the rocks are transitional in structure between the border group and the more typical layered rocks, and we have classified them as the transitional layered series.

In the southern third of the intrusion the inner contact is also indefinite. South of Ivnarmiut for 500 metres a fairly definite junction is seen intermittently, but it becomes indefinite as it crosses the Skaergaard Halvö, the passage from the border group to the layered series taking about 10 metres. The contact is considered to be vertical from its direction as it crosses the hill, 300 metres high, which forms the end of the peninsula. On Kilen, the junction between the upper border group, having dips to the S.E., and the layered series, dipping gently S.W., is also not clear cut, and this is due to mingling of the two units over a distance of several metres. The field relations suggest that the dip of this indefinite junction is outwards at about the value of the dip of the upper border group and it is so indicated in the section (Fig. 6).

(c) The Layered Series.

1. Structural Features

The rocks for which the term layered series is used show two distinct structural features. The most conspicuous is a special type of banding simulating the stratification of a sedimentary series (Pls. 6 and 7). The banding is due to different proportions of the minerals, and not to changes in their composition; in this respect it is therefore normal. In two respects it is, however, unusual, firstly, single layers are perfectly even, uninterrupted surfaces extending over wide areas, and secondly, the banding is rhythmically repeated in a way which must be ascribed to the action of gravity. The other important structural feature of the layered series is a parallelism of the platy minerals which resembles the bedding of sediments rather than the foliation of gneisses and schists. The special type of banding and the parallelism of the platy minerals are not always found together and they must be considered as separate phenomena.

There is a gradual change in the composition of the minerals which make up the layered series, and successive average, layered, rocks differ slightly in bulk chemical composition. At certain horizons a particular mineral appears or disappears, and this phenomena is connected with the gradual change in the composition of the minerals and not with the different proportions of the minerals which give rise to the banding. The inconspicuous and usually gradual change in composition of successive layers is a much more fundamental feature of the layered series than either the banding or the parallelism of the platy minerals, and it takes place in a constant direction throughout the whole of the layered series.

The term layered has recently been widely used, especially by American geologists, in describing plutonic masses which resemble the Skaergaard layered series [see for instance Ingerson 1934, Cooper 1936, Peoples 1936, Buddington 1937]. In the case of the Bushveld Complex similar features have been described as rift, rifting or pseudostratification [cf. Hall 1932, pp. 264—5] and for the Ilimausak Batholith, Ussing [1912, p. 320] has made use of the term stratum or sheet. The latter is now definitely undesirable, since it has a well-established usage for separate injections of sheet form. We propose to use the term layering in the same general way as the American authors but we shall also distinguish three different features which are usually included under the one name. Thus the conspicuous banding, rhythmically repeated and constant over wide areas which gives layers forming almost plane surfaces (or which may be regarded as having been so originally), will be distinguished as rhythmic layering. Rhythmic layering of this kind is due solely to changes in the relative proportions of the minerals. The inconspicuous and for the most part gradual change in composition of successive layers which is due to the changing composition of the minerals, we shall call cryptic layering. This might have been called gradual or continuous layering (in contradistinction to the rhythmic layering) had it not to include a few abrupt changes due to the appearance or disappearance of particular minerals. When mapping in the field, we used the term bedding, for the parallel arrangement of the platy minerals as we wished to distinguish it from ordinary flow structures, and its similarity to the bedding of sediments was in our minds. In describing the remarkable kakortokite sheets of the Ilimausak Batholith, Ussing [1912, p. 43] has also been drawn to speak of them as bedded. Since our return Dr. L. Hawkes has pointed out the confusion that would arise if the term bedded were extended to igneous structures, and we now propose to use the term igneous lamination. Though the word lamina has so far been mainly used in petrography in describing sedimentary rocks it is not so common a word as bedded, and an extension of its connotation, especially with the qualifying word igneous, is not likely to cause confusion. It is much better that the term used to describe this structure of igneous rocks should be associated in the mind with sedimentation, by a process akin to which we shall show that it has been produced, than with metamorphism, where somewhat similar structures, produced by the flow of solid material under strong sheering stress, have received various special names. Of these three features of the layered series, cryptic layering is fundamental and found throughout. Rhythmic layering is imperfectly developed at low horizons in the series and is absent in the upper, while igneous lamination is present sporadically throughout the series except in the uppermost part where it is completely absent. In this paper where the term layered is used without qualification, reference is being made to what should properly be called cryptic layering.

The igenous lamination of the layered series is mainly due to the orientation of the plagioclase crystals which are tabular in form. In the lower half of the layered series the plates are equidimensional. In the upper part the plates become thicker and at the same time elongated in one direction. Here there is a slight tendency for the long direction

of the plates to have a parallel orientation within the plane of the
igneous lamination. It would have been of considerable interest to have
mapped this direction, but unfortunately it was not realised at the time
and must be left for future work. The small amount of evidence avai-
able on the orientation of the plagioclase in the plane of the igneous
lamination will be given when the possible causes of the phenomenon
are being considered.

2. Description.

Rhythmic layering is best seen in the steep faces of Peak 1277 m. of
Gabbrofjaeld. The top hundred metres of the south-west face (Pl. 6
and Pl. 7, fig. 1) is not obviously banded, then follow three conspicuous,
light layers, the two lower close together and the upper some way above.
Below these three conspicuous layers which we have called the triple
group, a very regular, fine-scale layering is to be seen. The same triple
group with the same spacing is seen in the east wall of the peak above
Gabbrofjaeld Glacier (Pl. 7, fig. 2) and there it can be followed continuously by eye for two and a half kilometres. The triple group also
appears about half way up Pukugaqryggen and shows faintly in Plate 7,
fig. 1. The outcrop of this band has been put down on the geological
map. The dip of 17° at the east end of Pukugaqryggen has been calculated from the height of the bands to the north of Peak 1277 m. and the
height where they reach Forbindelsegletschen; the southerly direction
of the dip was estimated in the south face of Pukugaqryggen. With
this the other dips determined during an ascent of Pukugaqryggen are
in close agreement. Out of a collection of typical specimens made at
about every fifty metres of altitude on this ridge, half show igneous
lamination while the others are essentially without directional structures.

In the Skaergaarden region rhythmic layering may be most easily
examined. Here weathering and the plucking action of ice has produced
extensive plane surfaces parallel to the layering or igneous lamination
(Fig. 7), and by using a clinometer on a board, one and a half metres
long, placed on these surfaces, the dips were read within one or two
degrees. The azimuth of the dips was put down by sighting some distant
feature marked on the map, this method being adopted as there was
erratic magnetic variation. Over the Skaergaarden area taken as a whole,
the direction and amount of the dip was found to vary regularly, but
there was no obvious meaning in the values as actually measured. The
amount of dip increases as the inner junction is approached, and this
seems to be the same effect as observed in a more emphatic way within
20 metres of the boundary (p. 35). However, this kind of variation is
not found near all contacts; thus on the Skaergaard Halvö, dips of
12° occur near the inner contact and dips of 15°—20° further away.

Fig. 7. Rhythmic layering, 50 metres north-east of northern house in the Skaergaarden area, showing the kind of surface parallel to the layering from which accurate dips were obtained. In the distance is Pukugaqryggen on which the triple group can be seen.

With regard to direction the dip is found to change as the inner contact is approached, so that it becames more nearly at right-angles to the contact. The significance of these changes in the dip and strike of the layering will become apparent when the directions which existed before the regional flexuring are calculated (p. 51).

The rhythmic layering at about the Skaergaarden horizon, which includes the Skaergaarden area itself and the upper part of Gabbro-fjaeld, is more definite than either above or below. A remarkable feature of the layering, conspicuous in the field and visible in Pl. 8, figs. 1 and 2 and Fig. 7, is that thin, highly differentiated layers are often separated by thicker layers of the average rock. The differentiated layers are characterised by having the heavy, melanocratic constituents concentrated at the bottom, while upwards they become gradually less abundant, eventually grading into a leucocratic rock, rich in light constituents which are mainly felspar. In passing upwards the change from leucocratic to melanocratic rock is always abrupt, while from melanocratic to leucocratic it is always gradual. In the field many thousands of cases of this were seen. Differentiated layers either directly succeed each other, or there is an intervening layer of average rock. Near the border of the layered series the upper, leucocratic part of a differentiated layer is often missing but the upward grading, if there is grading at all, is always from more melanocratic to less melanocratic material. The grading in any layer is connected with the specific gravity of the minerals, and it always has this relationship to the upward and downward directions; if a loose boulder of the layered series is examined, it is usually possible to decide which way up it had originally lain. This feature of the layered series may be ascribed without hesitation to an effect of gravity, and either of the recently proposed terms, gravity differentiated layer or gravity stratified layer (Peoples 1936, Buddington 1935), may be used in describing it.

Within the Skaergaarden area there is considerable variation in the perfection of the rhythmic layering and igneous lamination, and at some places both are lacking and the rock is massive. The zone of good banding and good igneous lamination (Fig. 7) which passes under the northern house is indicated on the large scale map (Fig. 11) by a zone with many dip arrows. South-eastwards there is a zone in which banding and igneous lamination is scarcely to be detected, and no dips are indicated on the map. The rock is here essentially massive, and forms higher ground as a result of greater resistance to erosion. Further south-eastwards there is a discontinuous zone of good banding with moderate lamination (indicated again by a row of dip arrows). This gives place upwards to a zone showing a series of peculiar features to which we have given the name trough banding; these are of sufficient importance to be dealt with in a separate section (pp. 45—50).

As the inner contact is approached on Ivnarmiut, Mellemö and Kraemers Ö there is a gradual change in the banding and indeed in the whole composition of the layered series. Well away from the contact, (e. g. about the houses) some of the rock is conspicuously layered and some is homogeneous. In the differentiated layers the proportion of the minerals is such that, if the constituents were evenly mixed, the resulting rock would have the composition of the adjacent homogeneous material. As the margin of the layered series is approached all the rocks show banding which, however, is not very even or distinct, and the rock which would be produced by evenly mixing this heterogeneous material would be decidedly richer in heavy minerals than the average rock remote from the margin (Pl. 5, fig. 1, and Pl. 9, figs. 1 and 2). The special feature of the banding near the margin is the absence of the more leucocratic layers. The base of a band is still the richest in melanocratic constituents and becomes less so upwards, but before a rock with any abundance of leucocratic minerals is developed, the sequence is begun again by a fresh melanocratic layer. The difference between the banding near the inner contact and well away from it, is shown diagrammatically in Fig. 8. This change in the type of banding and the mineral composition is quite gradual; the rock in usually appreciably more melanocratic for a distance of 200 metres from the inner contact.

The horizon of good rhythmic layering extends from the Skaer-gaard region through Pukugaqryggen and Peak 1277 m. of Gabbrofjaeld to Peaks 1200 m and 1360 m of Gabbrofjaeld. Below this horizon rhythmic layering is less perfectly developed, while igneous lamination may be either well developed or absent. On Uttentals Plateau, where the rocks are low in the sequence and also near the northern margin, the banding is irregular, and it does not warrant the use of the term layering (Pl. 10, fig. 1). Individual bands cannot usually be traced for more than

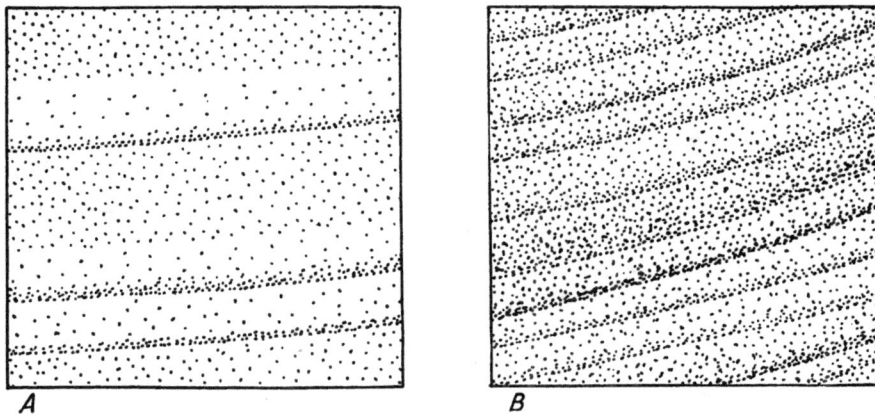

Fig. 8. The difference in type of rhythmic layering away from the inner contact (A) and near the inner contact (B).

forty or fifty metres, and the bands are not always gravity stratified. Cases were observed where the more melanocratic or leucocratic parts of the complex behaved as cognate xenoliths, and the banding was of the more usual kind, such as that in the Cuillin Gabbro of Skye or that of the large, basic intrusion of Kap Edward Holm on the west side of Kangerdlugssuaq. On Uttentals Plateau the dips are usually inwards at 25° to 30°, but there is a zone crossing the plateau where dips are less, usually about 15°. This zone of unexpectedly low dips continues to the east where it becomes intensified, and in the north-west ridge of Peak 1200 m. of Gabbrofjaeld the banding is dipping outward at 15°. There is perhaps a connection between the zone of unexpectedly low dips and the poorer layering with occasional cognate xenoliths. The rocks of Uttentals Plateau which show only poor rhythmic layering also have microscopic and chemical features which link them with the border group; they have been separated from the well-layered series and named the transitional layered rocks. Close to the inner contact on Uttentals Plateau, just as on Ivnarmiut and elsewhere along the east and west margins, there is a change in the bulk composition and the kind of banding;

the more leucocratic layers disappear, the rocks becoming more melanocratic, and at the same time the banding becomes less regular. Close to
the margin a strike section shows the banding as wispy layers, rich in
heavy minerals, alternating with others less rich (Pl. 9, figs. 1 and 2).
A sequence of such bands may be cut off by another set at a different
angle, thus crudely simulating the false bedding of sediments.

Above the Skaergaarden and Gabbrofjaeld horizon the rocks of
the layered series (best seen in the west face of Basistoppen, Pl. 10,
fig. 2), only occasionally show rhythmic layering and this is confined
to the first hundred metres above the trough-banding horizon. Most
of the rocks, however, show a moderate development of igneous lamination and there is a gradual change due to cryptic layering which can
be detected in the field and which becomes very clear with microscopic
and chemical examination. A layer about a hundred and fifty metres
thick which weathers a dark purple-brown colour, forms a resistant
band which can be traced three-quarters of the way round Basistoppen
and across the next two nunataks to the east. This band is a useful
horizon and was named the Purple Band. It grades downwards into
less purple-weathering rock but upwards it gives place rather abruptly
to quick weathering rocks which are unlaminated.

The unlaminated rocks above the purple band show no trace of
rhythmic layering but there are gradual changes in composition due to
cryptic layering. The lowest rock is an unlaminated gabbro with fayalite
crystals up to one centimetre across. Above is a light-brown-weathering
granophyric rock which gives place to a thin layer of hedenbergite-
andesinite. The latter is limited in distribution and of variable thickness
as it fills in the irregularities in the base of the huge inclusion of spotted
gabbro which forms the top hundred metres of Basistoppen. These
rocks, although grouped with the layered series, show various special
features and will be called the Unlaminated Layered Series (U.L.S.).

Irregularity in the direction of banding and igneous lamination in
a zone to the north of Gabbrofjaeld has been mentioned. There is also
some irregularity on that part of Kraemers Ö facing Forbindelsegret-
schen. Here a huge mass of slightly earlier consolidation occurs, and
the layering tends to be banked up against it, or to follow the top surface.
This accounts for the southerly dip of 15° marked on the map where
the average dip for the region is 23° to the south-west. Masses, apparently similar, were seen from a distance to the north-east of Gabbrofjaeld (Peak 1277 m.) and on the tongue dividing Gabbrofjaeld Glacier,
but these examples were not visited.

Observations in the field were sufficient to show that the gravity-
differentiated layers are to be ascribed to variation in the abundance of
the various minerals and not to the actual nature of them. Also the general

increase in melanocratic constituents as the inner contact is approached is due to a decrease in the leucocratic constituents and not to any change in the nature of the minerals present. The more significant, cryptic layering, which takes place throughout the whole thickness of the layered series is, on the other hand, the result of gradual changes in the composition of the minerals, and it has proved difficult to map. The lower, olivine-gabbro of Uttentals Plateau is a fairly normal type which continues eastwards to the foot of the north-west ridge of Peak 1200 m. of Gabbrofjaeld. Above this comes a zone, which we were able to map roughly, in which the rocks are transitional to the abnormal gabbro of the Skaergaarden region with its iron-rich olivine. Above this horizon the purple band and the unlaminated layered rocks are more definite, and were mapped with some precision. The fundamental changes due to the cryptic layering are much obscured by the more obvious rhythmic layering, and in the field only rough boundaries could be determined. From the thin sections and refractive index determinations made during the winter in Greenland we knew that the felspar of the layered series changes from basic labradorite at the base to andesine near the top; similarly we knew that there is a gradual upward change from normal olivine to an iron-rich olivine, and an equally striking but gradual change in the nature of the pyroxene. This knowledge did not, however, provide a field method of subdividing the series, because these changes are gradual and in the field the exact composition of the minerals could not be determined. From our winter laboratory work we also found that hypersthene occurred in the lower gabbros and quartz in the upper, but these minerals could not be detected in hand specimens and they also failed to provide a field method of subdivision. The boundaries within the layered series which appear on the map, except for the triple group and the purple band, have been determined since our return by examination of a large number of thin sections. The data are sufficient to show that the limits of the different types are parallel to the layering, and this fact has been used in putting the boundary lines on the map.

The nomenclature adopted for the different parts of the Skaergaard intrusion is shown diagramatically in Fig. 9 while the chief structural features are summarised in the north-south and approximately east-west sections of plate 26. In these sections the lie of the layering is expressed by the linear pattern used for the layered series, and in the border group the direction of flow structures is suggested by the orientation of the lenticular shapes which indicate diagrammatically the gneiss inclusions and their approximate amounts relative to the gabbro. Where shown by a broken line, the gneiss inclusions have indefinite boundaries and are partly hybridised. The subsequent flexure is responsible for much of the tilt of the layered series seen in the north-south section and for the dip of the lavas, except to the south of the intrusion where the sharp upward bend is greater than that due to the flexure alone; here some upturning due to the act of intrusion also took place (see p. 22). Since the flexure trends east and west, the approximately east-west section shows dips which only differ slightly from those which would have been seen in this direction before the flexure. In this section the tilting of the layering which is due to the flexure is expressed by the difference in height of the unconformity on the

HIGH UPPER BORDER GROUP
LOW UPPER BORDER GROUP
UNLAMINATED LAYERED SERIES

HIGH LAYERED SERIES

LOW LAYERED SERIES

HIDDEN LAYERED SERIES

INNER MARGINAL BORDER GROUP
OUTER MARGINAL BORDER GROUP
HIGH M.B.G.
LOW M.B.G.

Fig. 9. East-west section showing diagrammatically the nomenclature adopted for the different parts of the Skaergaard intrusion.

two sides of the intrusion. The greater dip of the layering near the inner contact is an original feature.

A discussion of the origin of the layering will be left for a later section but one possibility will be here considered, namely the hypothesis of a succession of separate injections, since this is eliminated mainly on field evidence. Without seeing the rocks in the field some of the layering, especially for instance that shown in Pl. 8, figs. 1 and 2, might be put down as the result of horizontal fractures which developed in the average rock, and which were subsequently invaded by magma, either from an extraneous source, or, as Bowen has suggested for other cases of banding, from the interstices of the partially solidified surrounding rocks [1919 pp. 417—22]. Within such an injected sheet, strong gravitative differentiation might be postulated of the sort observed in a crucible by Bowen [1915], and by many observers from Darwin onwards in lava flows and sills. In this way the rock complex, as now observed, might have been produced. Such a hypothesis becomes so unlikely as to be safely rejected when time and time again a single layer is followed for a hundred or so yards over magnificently exposed, glaciated surfaces. There is never the slightest indication of two distinct times of solidification of adjacent layers; for instance, no transgression of one layer across another is ever found, and no inclusions of the material of one layer in another. Again, the layers on Peak 1277 m. of Gabbrofjaeld which have been called the triple group can be traced over four square kilometres, and there is no appreciable change of thickness or of the spacing of the layers. It seems unlikely that any method of successive fracturing of a massive rock, whether hot or cold, could give rise to such accurately parallel fractures and such uniform separation of the two walls of the fracture. Besides field evidence of this kind there is the cryptic layering which gives rise to gradual variation in the composition of the minerals; this would not be expected on the simple hypothesis of successive injection, nor would it be expected on Bowen's especial form of the hypothesis. In the latter case, the injecting residual material should consist of low temperature minerals as is found to be the case in segregation veins which are formed in this way. However the minerals of adjacent layers of the Skaergaard intrusion, as will be shown below, are the same whether they are gravity differentiated or not. The field evidence, on which we base our rejection of the hypothesis of separate injections is the negative evidence that certain structures have not been found; however great weight can be attached to this evidence because of the perfection of the exposures on steep rock faces and recently glaciated, lichen-free rock surfaces.

(d) Trough Banding.

Near the houses and on the neck of the Skaergaard Halvö the even layering is interrupted by banding in the form of long narrow troughs (Fig. 10). In cross section the trough banding appears as a series of gravity differentiated bands, with arcuate form, resting one above the other (Plate 11, figs. 1 and 2). The best development of the banding occurs near, but not at, the bottom of the trough, while towards the margins the banding becomes less definite, and then fades out into unbanded, massive rock which separates the zones of trough banding. In longitudinal section (Plate 12, fig. 1) the bands are seen to be of even thickness and of considerable continuity. The zones of trough banding, as at present exposed, are seen to have a considerable length, and a height up to twenty times their average width (Plate 13, fig. 1).

On the basis of a plane-table map on a scale of twelve inches to a mile made by Dr. E. C. Fountaine, a detailed survey of the features of the layered series at the trough banding horizon was made (Fig. 11). The axes of the various troughs (lettered to facilitate reference) are indicated as continuous lines. On the whole the outcrop of the axes of the troughs is fairly straight but in the case of trough banding G there is significant curvature where it cuts across a small valley N.W. of the main house. The curvature indicates that the axial plane has a

Fig. 10. Trough banding G near the main house.

hade of 10—20° to the north, and a similar hade is the cause of the
slight curve of trough banding H. The pitch of the trough banding was
measured at intervals, and values from 14—23° obtained. Some of the
apparent variations may be errors of observation due to the difficulty
of deciding on the true pitch because of the confusing effect of the
asymmetry of the structures, but there is probably a slight, but real,
increase in the pitch in passing from N.E. to S.W. The maximum dips
of the flanks of the troughs were also recorded. No precise meaning can
be attached to these values as they depend on the rapidity with which
the banding fades out, and this is very variable at different heights in
the same example of trough banding (see for example Plate 13, figs. 1
and 2). However, in an overwhelming majority of cases the dips of the
north flank are higher than those of the south, and this asymmetry
of the flank dips is to be correlated with the fact that the best development of the gravity differentiated bands is not at the present lowest
point of the troughs but slightly on the northern flank. In the case of
the G trough banding, where the hade of the axial plane is 10°—20°
north, the banding and the flank dips are symmetrical about the axial
plane. In other cases the asymmetry of the banding and the flank dips
also suggests that the axial planes of the trough bandings have a hade
between 10° and 20° north. The lie of the structures before flexuring,

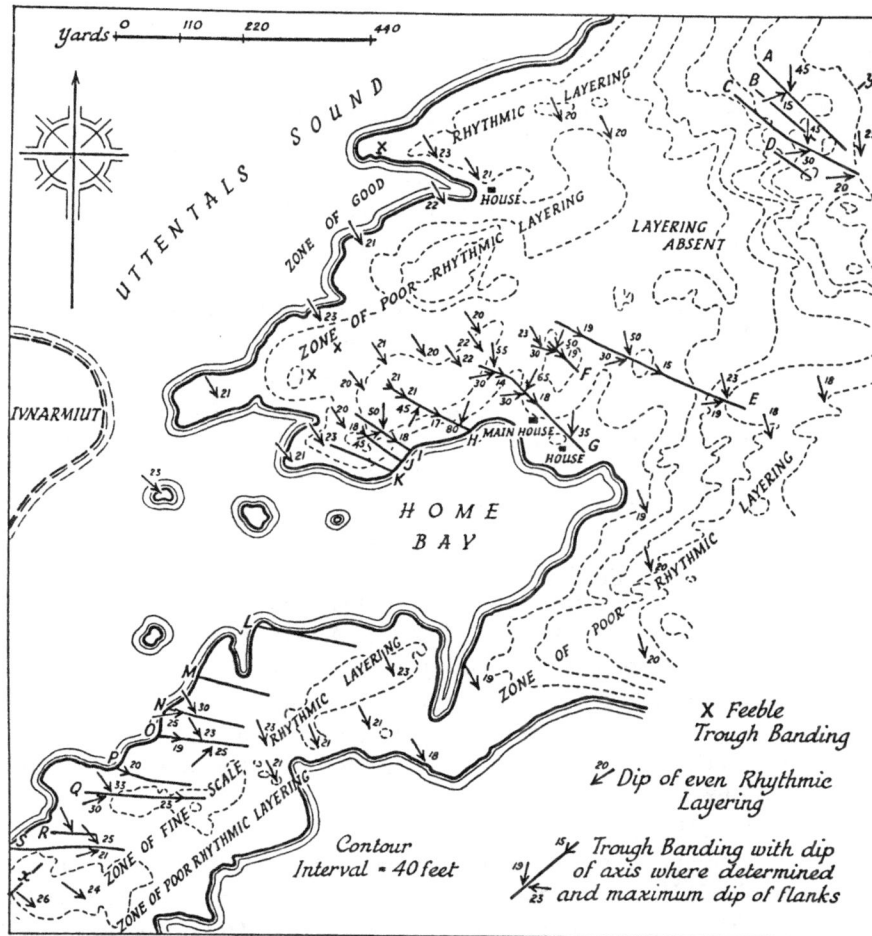

Fig. 11. Area near the Base Houses, showing trough banding structures (A, B, C, etc.), and the direction and quality of the layering. (xy marks the position of the section given in figure 12).

which involved the whole intrusion after its solidification, is discussed below (pp. 51—3). However, the chief result reached may be stated here namely, that the trough banding structures, before flexuring, were symmetrical about a vertical axial plane.

Between the zones of trough banding, which may be taken as extending on the average about 20 feet on both sides of the axes indicated on the map, the rock is usually free from noticeable banding or igneous lamination. Also between the northern group A—D and the middle group E—K the rocks are well exposed and trough banding is clearly not developed. To the N.E. of the group A—D there is no trough banding for 300 yards, then comes another group of examples. It is possible that the material eroded in the making of Home Bay was also free from trough banding. The known development of the trough banding is at a constant horizon in the layered series. It has also a limited areal distribution at this horizon; thus it is only found in the Skaergaarden area, while on Pukugaqryggen rocks of the same horizon occur which are free from trough banding.

Below the group E—K there is good rhythmic layering which is indicated on the map by the frequent dip arrows. The layering is even, and has the normal direction of dip to be expected at this place. The even layering begins directly below the trough banding, without any line of discontinuity, and this shows conclusively that the trough banding is not due to folding of even layers but is a primary structure formed during the solidification of the intrusion (Even layering can be seen below the small trough banding F on Plate 12, fig. 2). The zone of good rhythmic layering found immediately below trough bands E—K does not persist to the N.E.; instead this horizon is composed of massive rock. Massive rocks, where no dips could be taken, occur below the zone of good banding, then follows a zone of good layering, which is indicated on the map by frequent dip arrows, and which passes through the northern house (p. 38 and Fig. 7.) Above the trough banding A—K is a zone of poor rhythmic layering, but here and there accurate dips could be taken. Above the trough banding L—S there is a horizon of good, fine-scale, rhythmic layering which gives place upwards to poor layering; to the S.W. of the map at the horizon of the trough banding there is continuous, fine-scale, rhythmic layering without any trough banding.

It is evident that the various zones of trough banding once continued to the N.W. and that they have now been removed by erosion; to the S.E. their limit on the ground is not due to the dying out of the structures but to the pitch which carries them below higher horizons of the layered series. On the reasonable assumption that the various structures persist for several hundred yards in the direction of the regional dip, it is possible to draw a cross section of the trough banding and adjacent horizons (Fig. 12). In this section the hade of the axial planes of all the trough banding zones is taken as 15°, the approximate value proved for G and H.

There is much variation in the nature and extent of the trough banding. Trough banding E has a height of nearly 200 metres (Plate 13, fig. 1). The adjacent example F starts at a slightly higher horizon and dies out upwards in about 30 metres. Trough banding G has considerable height; near its base a subsidiary trough has developed on the south flank but upwards this merges into the main trough. In the group A—D the trough B, which is independent at lower horizons, becomes

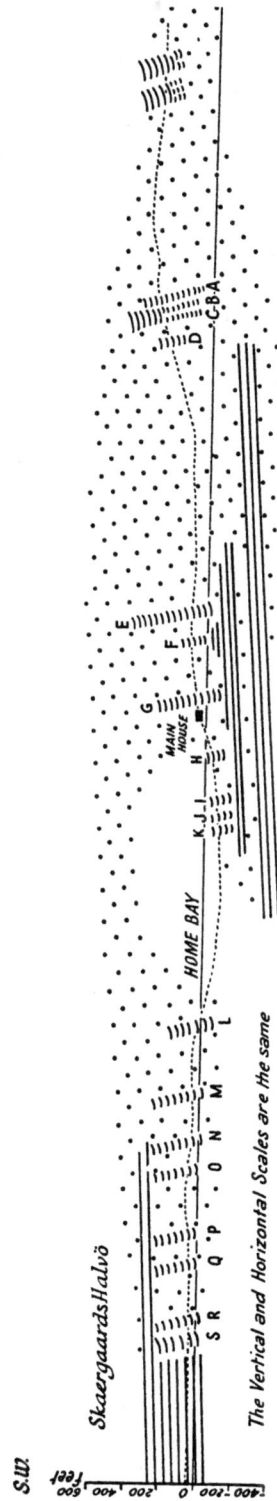

Fig. 12 Section across the Skaergaard region (along *xy* of fig. 11) showing diagrammatically the trough banding (small arcs), good layering (continuous lines) and rock without layering and with usually little igneous lamination (dots).

merged higher up into the more extended trough banding C. Trough banding D, like F, is of small vertical extent. The group not shown on the map but included in the section (Fig. 12) consists of several trough bandings near the base which upwards merge into two large trough bandings. The trough banding on the neck of the Skaergaard Halvö differs from the other examples in having narrower bands (Plate 13, fig. 2). Incipient trough banding (see Fig. 11) was noted between F and G, arcuate curves being picked out by slight concentration of the heavy constituents. The complementary light layers are not developed and the upward extent of the incipient banding is small. Similar incipient trough banding occurs sporadically in the zone below the main trough banding horizon and again on the point N.W. of the northern house. Irregularity in the layering which is due to the beginning of trough banding is shown in Plate 8, fig. 1.

V. FORM AND STRUCTURE OF THE INTRUSION

(a) Elimination of the effects of post-consolidation flexuring.

After consolidation, the intrusion has been subjected to the same crustal flexuring as the surrounding rocks (pp. 21—3). The amount of tilting undergone by the intrusion can be estimated from the dip of the surrounding basalts, excluding, however, those within half a kilometre of the S. and S.E. margin. In the latter place dips are higher than are usual in the particular zone of the flexure to which they belong, this being regarded as due to bending of the rocks by the act of intrusion and not by the later flexuring (see p. 22). There is only one general value of the dip of the sediments and lavas from the west of the intrusion and this is an estimated minimum dip of 20° (p. 22). On the basis of these values for the dip of the surrounding rocks, the Skaergaard intrusion has been divided into a series of zones showing to 5° the dip imposed by the flexuring. These zones do not run E.—W. which is in agreement with the observations from a wider area that the intensity of the flexure increases from east to west [see Fig. 1 in Wager and Deer 1937, p. 40].

The original dips of the boundaries, fluxion structures and layering may be estimated from the observed dips and the amount of tilt due to the post consolidation flexuring. A graphical method for finding the new values of the dip has proved quicker than the trigonometrical and sufficiently accurate. The direction of the surface represented by any dip arrow is plotted on a stereogram and then, by the usual method, the new position of this plane is found by a rotation of the appropriate amount about an E.—W. horizontal axis in the reverse direction to the tilt produced by the flexuring. In this way the boundary dips and selected dips of the fluxion structures and layering, which are typical of the various situations, have been modified from the observed values (Fig. 13) to the values which they had before the flexuring (Fig. 14).

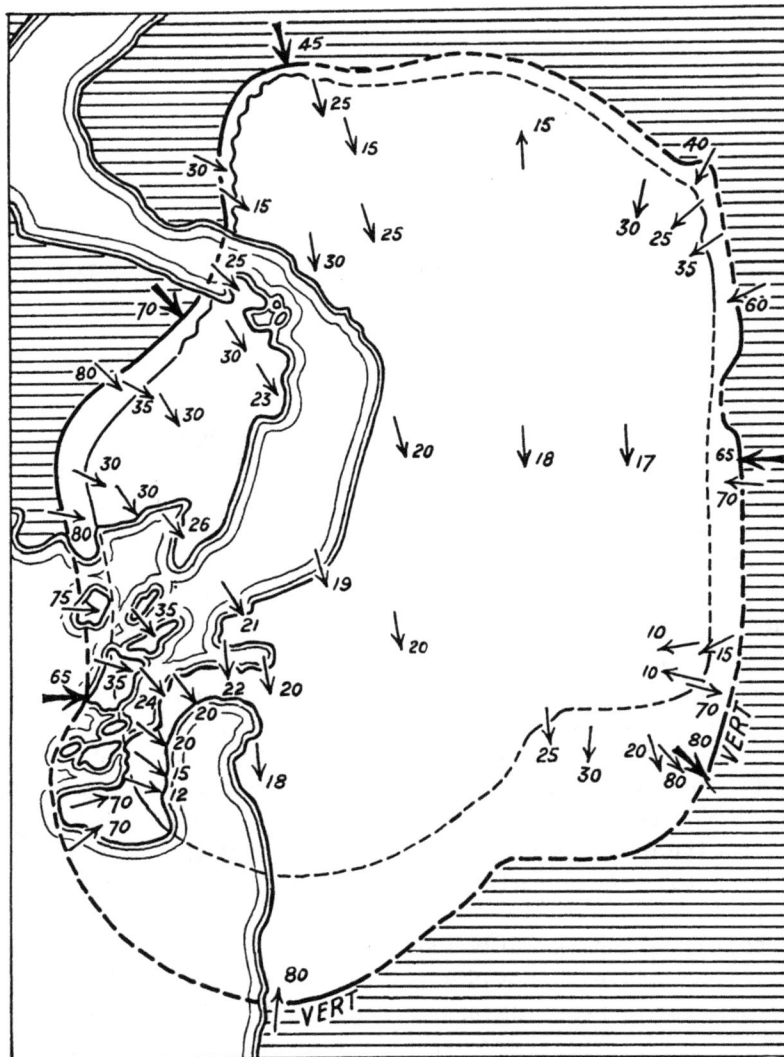

Fig. 13. Dips of the margin (thick arrows), the fluxion structures in the border
group and the layering (thin arrows).

(b) The Outer Form of the Intrusion.

The boundary dips shown on the map and in figure 13 were mainly
taken in high rock faces, and they represent values reliable to 5° or 10°.
At many other points the boundary is well enough exposed to corroborate
these values and in no case gave evidence opposed to them. At those
points where the dip of the boundary and the dip of the fluxion structures near the boundary were both taken, they correspond within the
limits of accuracy of the observations. For the Skaergaard intrusion

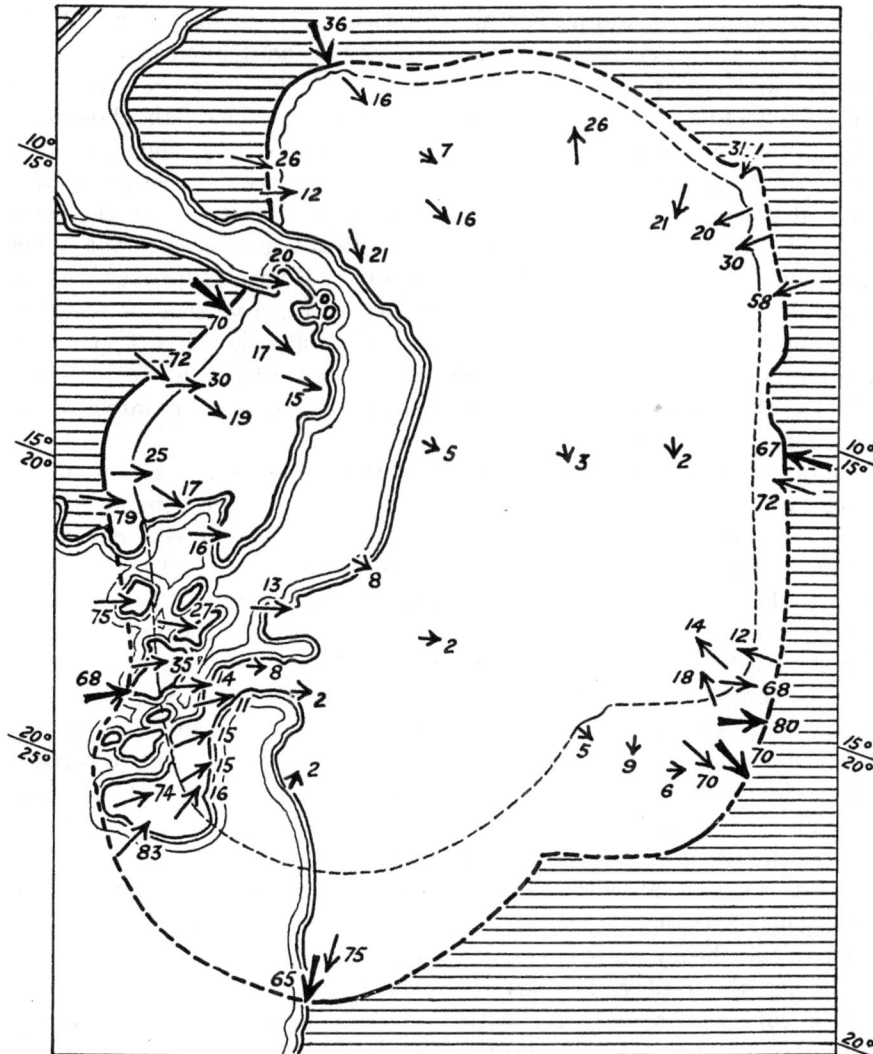

Fig. 14. Dips shown in figure 13, modified to the values which existed before flexuring. The assumed tilt due to flexuring is shown outside the margin. Dips of less than 10° are indicated by short arrows since the direction of such low dips may not be significant.

it seems that the dip of the fluxion structures near the margin gives a reliable indication of the dip of the margin itself, a generalization already made by Grout [1918 A, pp. 445—51] from his observations on the Duluth gabbro.

In considering the outer boundary, the first point to be noted is that it is an even, transgressive surface. Such a surface must have originated as a sudden, clean-cut fracture. The second point is that the present outcrop is approximately elliptical. The inward bend of the margin in the Skærgaarden region and again on Tinden is a real, but slight variation from elliptical shape, but the other irregularities are only a reflection of the relief. The third point is that the reliable boundary dips, which are well spaced round the whole periphery, show, without any doubt, that the boundaries of the intrusion converge downwards. This at once marks the Skærgaard intrusion as different from the usual type of transgressive plutonic intrusion which is produced in the upper crust in regions free from mountain building. Where sufficient evidence is available these have usually been shown to have vertical or outward dipping boundaries, and the space for the intrusion has apparently been produced by piecemeal stoping or downward subsidence of large blocks. Such processes cannot be invoked to explain an intrusion of the Skærgaard type.

In considering the original form of the intrusion the values of the boundary dips after elimination of the effects of the crustal flexure may be used (Fig. 14). By extrapolation of these a roughly cone-shaped space is delineated (Fig. 15)[1]) which has an apex about 12 kilometres below the unconformity between the Kangerdlugssuaq Sedimentary Series and the metamorphic complex. The cone is tilted which accounts for the oval form of the outcrop. A cone-shaped fracture of this sort may be compared with the fractures which have allowed the formation of the cone sheets of the British Tertiary igneous province. However, the usual cone sheets are approximately symmetrical about a vertical axis, whereas the central axis of the cone-shaped Skærgaard intrusion dips at approximately 40° to the south. Furthermore, the cone sheets of the British area converge at a depth of about 5 kilometres if the dips at the present surface are extrapolated, while the walls of the Skærgaard intrusion converge at a depth of about 12 kilometres. The mechanical theory of cone sheet formation developed by Anderson [1936, see also Bailey, Thomas etc. 1924, pp. 11—12] suggests that the dip of cone sheets decreases with increasing depth; in a similar way it is possible that the boundary of the Skærgaard intrusion is curved so as to converge to a point somewhat higher in the crust than 12 kilo-metres[2]). In accounting for the cone sheets of Mull, Anderson postulates

---

[1]) In figure 15 we have reconstructed the form of the intrusion before the flexure, hence the basalts, etc. lie horizontally, and the present sea level is curved in the opposite direction to the later flexure. The thickness of the basalts shown is probably close to that actually present at the time of the intrusion (see p. 15). The thickness of the upper border group is little more than a guess based on the position of the gneiss block in Tinden. The caldera form which is indicated for the original surface is simply a possible hypothesis.

[2]) From indirect evidence given later (pp. 218—19) it is suggested that a curvature of the boundary walls in the reverse direction is more likely.

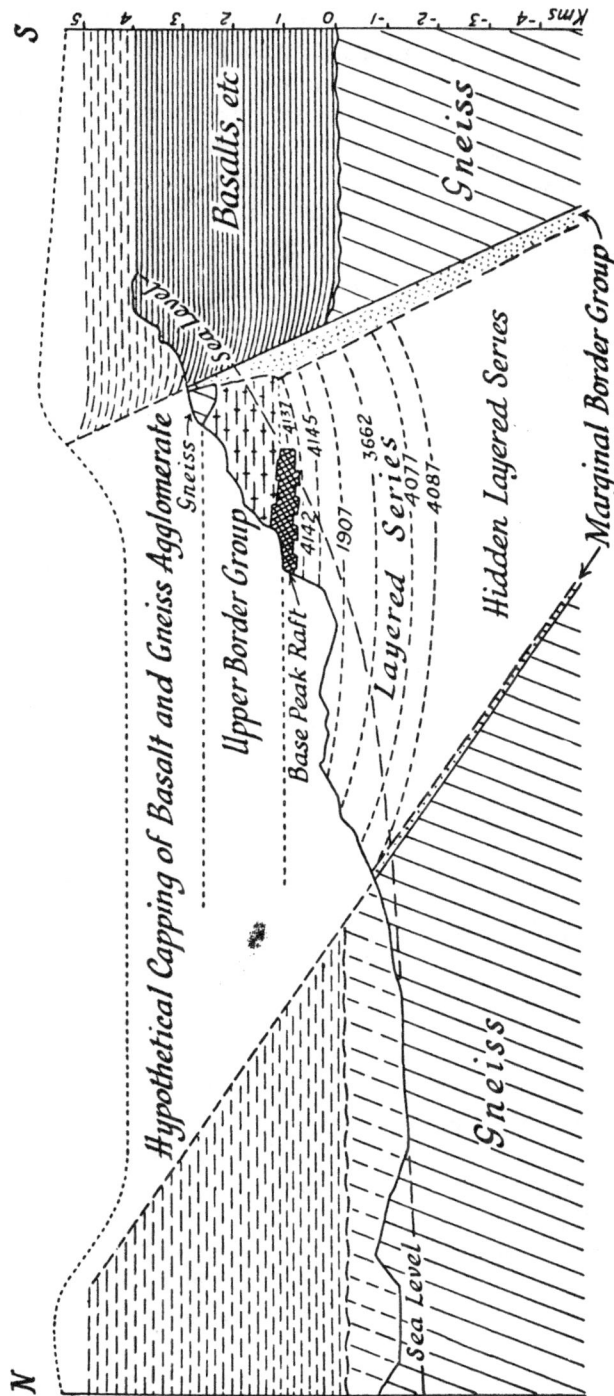

Fig. 15. Hypothetical, north—south section of the original Skaergaard intrusion before the flexuring and subsequent erosion.

a paraboloid magma reservoir at a depth of several miles. He shows that, if there be an increase in the magma pressure in the reservoir, the stresses in the roof would be relieved by cone-shaped fractures. Similarly the formation of the cone-shaped fracture giving rise to the Skaergaard intrusion must be ascribed to powerful magmatic pressures. It is difficult to understand why the Skaergaard cone fracture is strongly inclined. It may be due to regional stresses in the crust on which the effects of the magmatic pressures were superimposed, and it was thought at first that strains set up by the beginning of the flexuring movement might have provided the mechanical conditions required. Later consideration has suggested that such strains would not produce an inclination of the cone fracture in the direction observed. On the other

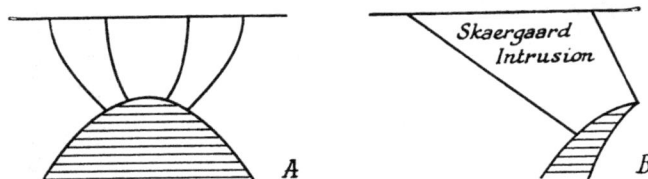

Fig. 16. Cone sheet fractures and magma reservoir as postulated by Anderson (A) and north—south section of Skaergaard intrusion with a possible form of the magma reservoir (B).

hand, the asymmetry might be connected with the form of the magma reservoir. If this were wedge shaped and not greatly extended in the E.-W. direction, and if it dipped northwards at 45°, the effective magmatic pressure would be exerted over the upper wall of the wedge and would be in a northerly and not vertical direction. Such pressure might perhaps be relieved by an inclined cone-shaped fracture, like that of the Skaergaard intrusion (Fig. 16).

Since the boundaries of the Skaergaard intrusion converge downwards, the crustal material replaced by the intrusion must have been expelled upwards, or incorporated in the invading magma. A certain amount of acid gneiss and other rocks occur as inclusions in the intrusion, and a small amount may have been so thoroughly incorporated as not to have been detected. However, the composition of the Skaergaard intrusion precludes the possibility of sufficient material having been incorporated to effect significantly the problem, of what has become of the original crustal rocks occupying the place of the intrusion. It seems to us that the most reasonable hypothesis to account for the removal of this mass of rock is that the initial explosion which produced the cone fracture, shattered the crustal rocks, and expelled the greater part so that they were distributed over the surrounding area. Some would fail to be expelled or would fall back and form a cover to the intrusion which would probably float in hydrostatic equilibrium in the basic Skaergaard magma. This view is summarised diagrammatically in figure 15, where it is also assumed that the original surface form of the intrusion was a caldera. It is pictured in this way because there are certain caldera-like structures on the surface of the earth at present, such as the Rieskessel, which may possibly represent the superficial phenomena associated with intrusions of the Skaergaard type.

In the first attempts to imagine the mechanism of the Skaergaard intrusion only partial shattering of the replaced rocks was assumed. The upper part of the replaced rocks, roughly the basalts, were considered to have been pushed up, without shattering, to form a mound on the surface, and only the rocks of the metamorphic complex were regarded as having been shattered or melted down to form an under-capping to the magma. Some breaking down of the replaced rocks would be necessary otherwise the conditions would be those giving rise to cone sheets. In the case of cone sheet formation the central block has remained as a single unit despite repeated cone-shaped fractures; after each fracturing and injection of basic magma it has settled down into the magma and in this way only a thin sheet intrusion has formed. It is necessary to postulate some break up of the central block in order to obtain a massive intrusion like that of the Skaergaard. If the upper part of the replaced block is pictured as remaining whole and simply pushed upwards, it is also necessary to assume that the elevated position of the cover is maintained by magmatic pressures, the cover in some way sealing the intrusion so that magmatic pressures could be effective. These difficulties make us prefer the hypothesis of complete shattering of the replaced rocks, and the distribution of some of the material over the surrounding area.

On Tinden there is a huge block of gneiss which seems to be part of the original cover of the intrusion (see map and fig. 15). This mass is over 500 m. long, at least 600 m. high and of considerable width. It is now situated two and a half kilometres above the estimated position of the gneiss-basalt unconformity at this place, and it must have been pushed into its present position by the act of intrusion. This mass gives strong support to the view that the replaced crustal rocks have been expelled upwards.

Another feature, namely the huge inclusion of sill gabbro forming the top of Basistoppen, supports the supplementary hypothesis that the replaced crustal rocks were shattered by the intrusion. At 700 m. on Basistoppen the layered series is capped by rocks which are so similar to the spotted gabbro sill of Tinden and Haengefjaeldet (pp. 18—19) that they may unhesitatingly be assigned to these sills or closely similar ones. The uppermost rock of the layered series, the pyroxene andesinite,

injects the base of the spotted gabbro series, and blocks of the latter
are found detached and embedded in the layered series. Although the
general dip of the contact is roughly parallel to that of the layered series,
it is in detail highly irregular and in places can be seen to step up
vertically for 50 m. (In the sections (Pl. 26) these irregularities are
shown diagrammatically). The field evidence leaves no doubt that the
spotted gabbro of Tinden was solid rock at the time of the intrusion
of the Skaergaard magma. The same spotted gabbro material which
forms the summit of Basistoppen, also forms the lower part of the
broken ridge lying to the south of Brödretoppen-gletschen. Here overlying and injecting the spotted gabbros, are rocks which from their
peculiar mineralogical composition, must belong to the Skaergaard
intrusion. On the first nunatak east of Basistoppen the lower parts
of two spotted gabbro masses are found and they also have suffered
injection by the rocks of the layered series. The field evidence is therefore clear, that the huge mass forming the summit of Basistoppen and
the other masses enumerated, must be inclusions in the Skaergaard
intrusion. On Basistoppen the spotted gabbro forms the local top of
the layered series, and this seems to be the case elsewhere, except
for two small blocks, about 30 and 50 metres across, of metamorphosed
spotted gabbro which occur as inclusions in the layered series on the
neck of the Skaergaard Halvö.

The lower part of the spotted gabbro series on Basistoppen is a
peridotite with large poikilitic plates of plagioclase surrounding the
olivine. This grades upwards into a pyroxene-rich, olivine-gabbro in
which spotting begins to appear, the spots being due to uralitization
of the augite. Typical spotted gabbro, identical with that described
from the sills, occurs above. This sequence of peculiar rock types is the
same as that described from the sills on Tinden and Haengefjaeldet
(p. 19), and it is concluded that the inclusion forming the summit of Basistoppen came either from the Tinden or Haengefjaeldet sill or from some
similar sill at another level. The spotted gabbro inclusions now have no
basalt or gneiss attached to them but, in our view they are huge blocks
derived from sills which cut the basalt or gneiss now replaced by the
intrusion. It is considered that they remained as large masses during
the shattering because of the strong and resistant character of the
sill gabbro. Being dense they gradually found their way through the
surrounding brecciated fragments into the Skaergaard magma, and in
some cases they sank a considerable way into it.

Only one intrusion has come to our notice, namely, the Peekskill
norite [Balk 1927], whose form can be closely compared with that of
the Skaergaard intrusion, although the transgressive basic intrusions
of Tabankulu mountain and Insizwa described by du Toit may possibly

be similar. The Peekskill norite has apparently the form of an inverted cone though the relief of the region is not great enough to make this as certain as it is in the case of the Skaergaard mass. The intrusion is oval, the larger axis being 12 kilometres. The presumed apex of the cone has a depth between 6 and 11 kilometres below the present surface. In its form, general dimensions and in the basicity of the magma, the Peekskill mass seems closely comparable with that of the Skaergaard.

With regard to the method of formation of the intrusion, a comparison may be made with the well-known Rieskessel in Germany. This depression which is 12 kilometres across seems to have been the result of a violent explosion, blocks of the deep-seated continental rocks having been widely distributed over the surrounding area. Magnetic anomalies suggest that a basic mass underlies the region, and recent gravity determinations suggest the existence of a funnel of explosion [see Daly 1933, pp. 382—4]. Other calderas also have points of resemblance, but since their origin is still more a matter of speculation, no further comparisons will be made.

The Skaergaard intrusion has a very definite form for which it is desirable to have a precise name. The inverted cone-shaped fracture which bounds the intrusion is similar to that producing cone-sheets, and it seems likely that these two types of intrusion are related in their manner of origin. It is desirable that this connection should be apparent in any name for the Skaergaard type of intrusion and we had therefore considered using the term "cone-intrusion". Such a term, meaning a solid plutonic intrusion of inverted cone shape, would have been sufficiently distinct to prevent any danger of confusion with the terms "cone-sheet" or "cone-sheet intrusion". However, Balk has described the Peekskill intrusion as a "Trichter pluton" or funnel intrusion, and since this term fits the Skaergaard case there is not sufficient justification for using a new one.

(c) The Border Group.

For about 30 metres from the margin the rocks of the border group are somewhat fine-grained due to chilling; then the main part of the group is reached consisting of variable, coarse gabbros with fluxion banding. The fluxion structures of the marginal border group are parallel to the contact and their character indicates upward or downward flow (p. 27). The upper border group, preserved only as remnants on Brödretoppen and neighbouring summits, is continuous with parts of the lateral border group and consists of coarse gabbros and hybrid rocks. If the effects of the late flexuring are eliminated the fluxion structures of the upper border group are approximately horizontal and parallel to the presumed roof of the intrusion. The border group is thus an envelope lining the intrusion cavity and having flow structures parallel to the walls of the intrusion.

Where the border group crosses Uttentals Sund, the mapping indicates that it decreases in width downwards, and this seems to be a general feature in the northern part of the intrusion. In the southern part there is some evidence that the reverse happens, and that the border group decreases in width upwards. Because of the flexure, higher levels of the intrusion are encountered in passing from north to south, and the outcrop of the border group, where it is narrowing downwards, should increase in width in that direction. If allowances are made for the effects of relief, the map shows that along the western border from Uttentals Plateau to the Skaergaard Peninsula the outcrop of the border group is widening; therefore the low marginal border group is apparently narrowing downwards. On the other hand, where the outcrop of the border group decreases in width in passing from north to south, there an upward decrease in width of the border group must be postulated. The outcrop of the border group is wider on the end of the Skaergaard Halvö than on the west ridge of Tinden or on Hammersfjaeldet; this suggests that here the high marginal border group is narrowing upwards. The border group is apparently widest opposite the upper part of the layered series, and from this horizon narrows both downwards and upwards.

The character of the inner contact shows that the border group is not an earlier intrusion pierced by the layered series but a peculiar marginal facies produced during the cooling of a single body of magma. In our view, for which more evidence will be given below, the marginal border group is partly material solidified by loss of heat from the sides of the intrusion, while the layered series and upper border group are the result of heat loss from the roof. The rate of heat loss from the underlying northern border of the intrusion, which is a foot wall, would be less than the rate of heat loss from the southern margin, which is a hanging wall, and it would therefore be expected that the southern part of the border group would be wider than the northern. Since the inclination of the east and west margins of the intrusion is similar we should expect a similar width of the border group. The hypothesis of different rates of heat loss from the differently inclined margins may partly explain the width of the border group, but it is likely that the main variation is due to differences in the time available for the inward growth of the border group before the layered series or upper border group was laid down against it. The sequence of solidification of the different parts of the intrusion can be established partly from the structures and partly from the mineral composition of the various rock types, and when this has been done we shall again consider the reason for the different widths of the border group at the different parts of its outcrop.

Another feature of the border group which is apparently related to the form of the intrusion may be noted here. Along the northern margin, gabbro-picrite is developed in the border group, and olivine-rich patches occur along the east and west margins although not along the southern margin. Because of the low inclination of the northern border an opportunity might well occur there for the accumulation, under the influence of gravity, either of heavy mineral constituents, such as olivine, or of blocks of early formed rocks. No such accumulation should occur along the southern margin because it is a hanging wall, and only slight accumulation should occur along the east and west borders. The gabbro-picrite blocks in the border series have therefore a distribution which is related to the original inclination of the walls in such a way as to suggest that the rock was formed by some such process.

(d) The Layered Series.

The estimated dips of the layering before the flexuring (Fig. 14) must be regarded as only approximate, since, in making the estimates, errors of observation of the dip have been added to the errors in evaluating the extent of the late flexure[1]. Nevertheless they are sufficient to show the general form of the original layering which is not apparent when the observed dips are considered (Fig. 13). Apparently the layering in the central area of the intrusion was originally approximately horizontal. As the inner contact between the layered series and the border group was approached, the dips increased to about 30°, and the direction of dip was always approximately at right angles to the margin. The data also suggest that the lower horizons were more strongly upturned at the margins than the higher. The layering was thus originally saucer-shaped, the only big irregularity being found in the northern part of the intrusion where the layering is imperfect and the rocks have some of the features of the border group.

When corrected for the post-consolidation flexuring, the axial planes of the trough banding become approximately vertical, and the zones of trough banding prove to be structures which are symmetrical about a vertical plane. The late flexuring has had little effect on the direction,

---

[1] When the actual dips were taken in the field no particular symmetry in their direction or amount could be seen, and thus there was no tendency for the observer to develop any particular prejudices about them. The symmetry apparent after eliminating the effects of flexure, suggests the essential correctness of the mapping and of the assumptions made in calculating the original from the observed dips.

as mapped, of the axial planes because these structures were vertical, and we may regard their present average direction as being sufficiently close to their original direction for practical purposes. The general directions of trough banding summarised in figure 17 are approximately at right angles to the margin and converge to a point some way south of the centre of the present outcrop of the intrusion. The data suggest

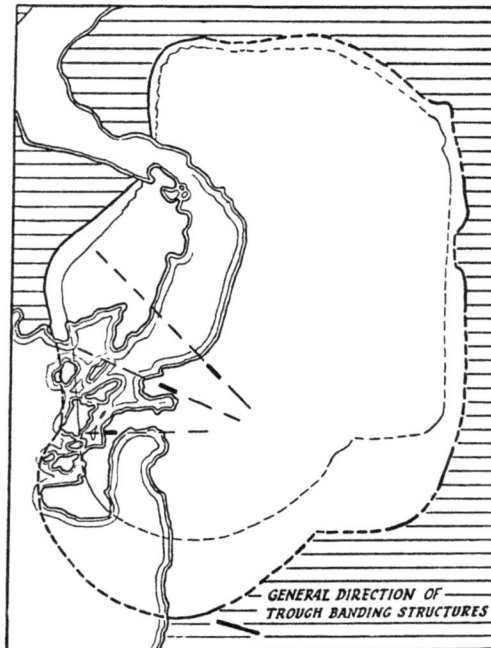

Fig. 17. The general direction and position of the trough banding in relation to the margins of the intrusion.

that the saucer form of the layering is also asymmetric, the lowest point of the saucer being the same as that to which the trough banding structures are directed (Fig. 14). The triple group of the layered series and the trough banding structures are at the same horizon; the trough banding structures are not found north of Forbindelsegletschen, nor the triple group to the south. It would appear that even at one horizon the structures of the layered series differ slightly in different parts, the differences being arranged symmetrically with respect to the point a little south of the centre of the intrusion, towards which the trough banding structures trend.

The layering of the Skaergaard intrusion must either have developed originally in a saucer shape, or it must have been bent into that form as a result of some special process connected with the intrusion; at present there seems to be little direct evidence to decide between these two alternatives. The absence of any regular system of fractures is perhaps against the view that the saucer form was produced by modification of a horizontal layering, unless of course it is imagined that the bending took place while the rock was at a high enough temperature to bend without fracture. On the other hand, the small faults at, or near the inner contact on Ivnarmiut and Mellemö (Pl. 5, fig. 2) is due to down sinking of the layered series after its formation, and it is precisely such down sinking, if wide-spread, which would give the saucer form. In attempting to decide on the reason of the saucer shape we get little help from other described examples of layered intrusions. The Bay of Islands Complex [Ingerson 1934, and Cooper 1936] and the Stillwater Complex [Peoples 1936] are much disturbed and not yet mapped in detail. The Bushveld Complex [Hall 1932] and the Duluth Gabbro [Grout 1918], which are on an altogether different scale, have a different form without extensive transgressive margins. Harker's description of Rum [1908] suggests that there the sheets have slight inward dips, but it is not at present known whether these rocks may be fairly compared with layered intrusions elsewhere. In the Cuillin gabbro of Skye the dip of the banding, which, however, cannot be closely compared with true layering, behaves in a different way from the layering of the Skaergaard intrusion, being steeper near the presumed centre of the mass [Harker 1904, pp. 90—92]. The closest similarity to the saucer form of the layering of the Skaergaard intrusion is found in the agpaitic rocks of the Ilimausak batholith of S.W. Greenland [Ussing 1912, figs. 26 and 27], and a somewhat similar form is surmised for the sheets of the Borolon Complex [Shand 1909 and 1910]. Along part of the margin of the Ilimausak batholith an augite-syenite is developed which, from Ussing's descriptions, resembles the border group of the Skaergaard intrusion. Within this there are layered rocks, and Ussing has shown that the form of the layering is that of a shallow saucer. Ussing believed that the layers were originally horizontal, and that reduction in pressure, or the drawing away of a still-liquid, underlying magma, caused the layers to sag into the saucer form. The same mechanism is also used to explain the lujavrite and its remarkable relations to the naujaite. Despite similarity in the form of the layering in the Ilimausak and the Skaergaard intrusions, we do not hold a similar view of its origin. In our attempted synthesis of the mechanism of the Skaergaard intrusion we shall show that an original saucer form for the layering is more likely than the hypothesis of bending of once horizontal sheets.

VI. PETROLOGY OF THE LAYERED SERIES

(a) General Features and Nomenclature.

The rocks of the layered series vary in two independent ways. The variation most conspicuous in the field is in the relative abundance of the light and heavy minerals. This gives rise to the rhythmic layering and the melanocratic facies near the inner contact. The less conspicuous, but more significant variation, which gives rise to what has been called the cryptic layering, is due to gradual changes in the composition of certain minerals which are solid solutions; these are high temperature varieties in the lowest exposed horizons and they change to lower and lower temperature varieties as the layered series is ascended. At certain horizons of the layered series a new mineral suddenly begins to be present, and at others, a mineral which has been present, suddenly ceases to occur. In the evolution of the rock series such abrupt changes in the minerals present were apparently related to the continuous changes in composition of the minerals, which are solid solutions, and the two sets of changes are grouped together and called the major variation. The term minor variation will be used for that which gives rise to the rhythmic layering and the increase in melanocratic constituents near the inner contact. The minor variation is an effect superimposed on the major and therefore the latter will be dealt with first. This involves treatment of average rocks which were carefully selected in the field as free from abnormal concentration of either the light or heavy constituents.

The composition and other features of the layered rocks may be conveniently plotted against their height in the intrusion as calculated from their position and the dips given on the map. The height is measured conveniently from the lowest layered rock at present exposed, 4087. Ten rocks representative of the whole sequence have been particularly fully investigated and chemically analysed. Results for these and for others lying near or at the limit of certain phases, are summarised diagrammatically in figure 18.

The average rock of the lower horizons is a hypersthene-bearing, olivine-gabbro of a fairly normal type. In ascending the layered series

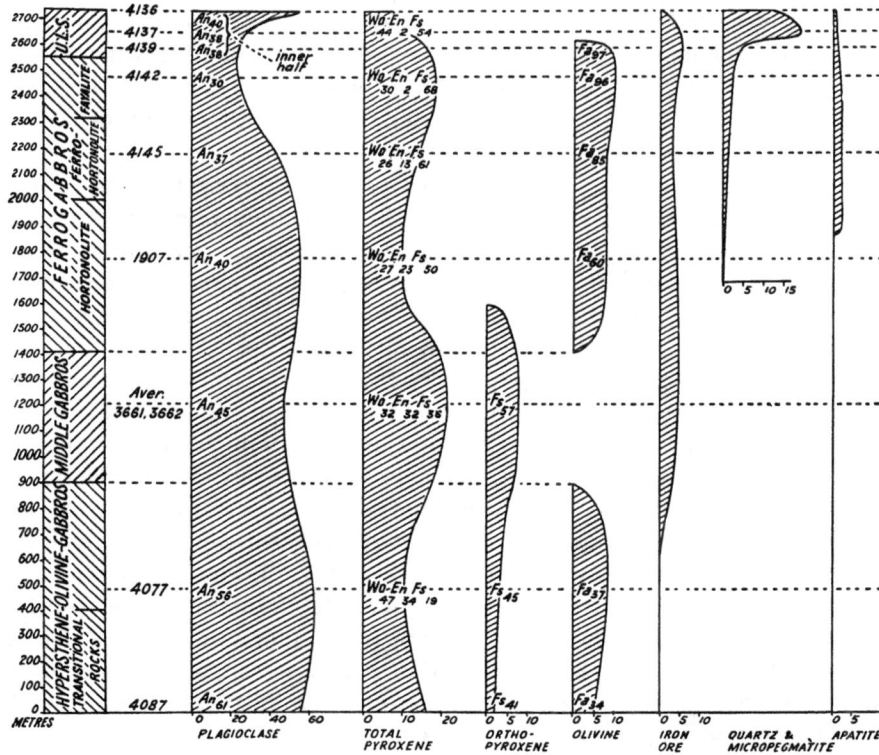

Fig. 18. Modal composition of average rocks of the layered series and the composition of certain of the minerals.

the plagioclase shows a slow decrease in anorthite content and the pyroxenes and olivine a slow increase in the amount of the iron component. At about 900 metres in the layered series the rocks suddenly become olivine-free and remain so for·500 metres, above which olivine again becomes an important constituent. Over this range the plagioclase and pyroxene continue to become richer in their more fusible component. The 500 metres which are described as olivine-free, are not strictly so, as a little olivine is developed in reaction rims between the pyroxene and iron ore. This, however, we regard as a secondary feature, developed after the accumulation of the primary phases and at a slightly lower temperature. At 1400 metres olivine reappears as a primary phase; a little higher hypersthene disappears, and a little higher still quartz comes in for the first time. The rocks here are of thoroughly abnormal composition, being unusually rich in ferrous iron which is mainly present in the pyroxene and olivine. Above this horizon, changes in the composition of the various minerals become rapid and by the level of the purple band, 2500 metres, the plagioclase contains 30 per cent. anorthite, the pyroxene only 2 per cent. $MgSiO_3$, and the olivine only 2 per cent. Mg_2SiO_4. The highest rocks of the layered series on Basistoppen are unlaminated, and seem to have been produced by rather different physical processes from those responsible for the differentiation of the rest of the layered series.

This summary of the characteristics of the layered series is sufficient to show that it includes some hitherto unrecorded rock types, and the difficulty of finding satisfactory names at once arises. The rocks of the lowest 900 metres may legitimately be called hypersthene-olivine-gabbro. By the time the olivine disappears the rocks are beginning to be abnormal,—the pyroxene is already rich in iron, and the absence olivine is not due to increasing acidity but to some other cause connected with crystal fractionation. It is not easy to find an adequate name for this part of the series, and for the present the non-committal term "middle gabbro" will be used. Above this, the rocks fall definitely outside the range of any normal gabbros. They keep the low silica percentage of olivine gabbro but the plagioclase is andesine and the ferromagnesian minerals are abnormally rich in iron. The analyses, when plotted on a triangular diagram (Fig. 19), showing normative amounts of the important constituents arranged in a way discussed below (p. 231), lie in a region which is well outside the normal range of the gabbros, and indeed, outside the range of any known igneous rocks. Since these rocks are related to gabbros and are characterised by richness in iron, we have ventured to use for them the new name ferrogabbro, a term more precisely defined below (pp. 98—91).

The further subdivision of the ferrogabbros may reasonably follow two directions. The classification may be based on significant special phases present in the rock, giving for example, quartz-free ferrogabbro or quartz-ferrogabbro. Such names will sometimes be used and the limit of these two particular types has been put down on the map. Since quartz first appears in very small amounts, this boundary could not be drawn without recourse to microscopic examination. On the other hand, the nomenclature might take into consideration the composition of the ferromagnesian minerals as determined by analysis or by optical data. It has proved convenient to use the composition of the olivine for subdivisions of this kind and such names as hortonolite-ferrogabbro, ferrohortonolite-ferrogabbro[1]) and fayalite-ferrogabbro will be used.

In this nomenclature, the relative abundance of the different minerals is not considered, the subdivisions being based on variation in the minerals present or their composition, that is, on the major variation. Variation in relative abundance of the minerals, the rhythmic variation, is not of such fundamental genetic significance and should

---

[1]) The new name ferrohortonolite here adopted for a certain range of iron-rich olivines is defined in the succeeding section.

only play a secondary role in nomenclature, a general principle which is recognised in those classifications of igneous rocks which make the composition of the plagioclase the prime factor of subdivision. From the point of view of general appearance, mineralogical composition and bulk

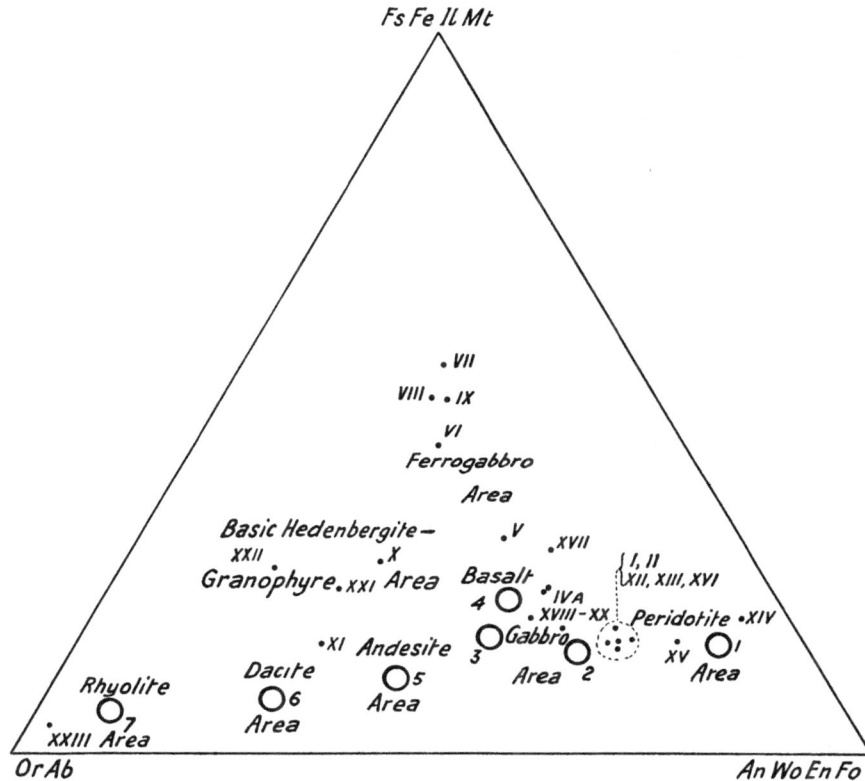

Fig. 19. The field of the ferrogabbros, basic hedenbergite-granophyres, and other Skaergaard rocks compared with Daly's averages of calc-alkaline rocks.

chemical composition, the extreme rock types developed in some of the layers are very different from the average rocks adjacent to them, and they might legitimately be called peridotite, anorthosite and andesinite etc. For the most part, however, such names will not be used and instead the rocks will be described as melanocratic or leucocratic varieties of the average rock.

(b) The Component Minerals.

1. Plagioclase.

The plagioclase of the main layered series is only zoned slightly and in a rather special way; the crystals are homogeneous except

68   L. R. WAGER & W. A. DEER.   IV

TABLE

OPTICAL PROPERTIES AND CHEMICAL COMPOSITION OF MINERALS

	Specimen	Height in metres	Plagioclase Mol. % An.	Olivine γ	Olivine % Fa.
U. L. S.	4136	2750	40	none	
	4138	2700	40	none	
	4137		38	none	
	4139	2600	38 (inner half)	(2 V = 48°) 1.879	97
Ferro-gabbros	1881	2500	30	(2 V = 48°) 1.875	96
	4142	2500	30	..	..
	4143	2400	31	..	..
	1974	2400	..	..	..
	4145	2175	37	(2 V = 53°)	85
	4146	2100	..	1.840	81
	1907	1800	40	1.795	60
	3651	1800	..	1.808	66
	2580	1700	..	1.800	62
	4296	1700	41	1.797	61
Middle gabbros	3655	1500	..	none	
	3658	1300	..	none	
	3661	1200	..	reaction rim 1.770	48
	3662	1150	45	none	
Hyp. ol. gabbros	4063	900	..	1.768	47
	1690	850	53	1.760	44
	4077	500	56	1.748	37
	4087	0	61	1.739	34

for a narrow, sporadically developed fringe of strong zoning, amounting to not more than five per cent. of the crystal. In the unlaminated layered series and the transitional layered series, neither of which are typical layered rocks, the plagioclase may show zoning in the outer half or even throughout. The most satisfactory method of estimating the composition of the almost homogeneous plagioclase of the main layered series is by means of refractive index determinations on cleavage fragments using Tsuboi's curves, and this method has usually been adopted. The composition of the plagioclase in selected rocks of the layered series is shown in Table III and figure. 25. It will be seen that the plagioclase in the lower rocks is medium labradorite, and that the content of albite increases gradually up to the purple band, where it is on the border between andesine and oligoclase. The zoned plagioclase crystals of the unlaminated layered

III.

OF LAYERED SERIES IN ORDER OF HEIGHT IN THE INTRUSION.

Clinopyroxene		Orthopyroxene		Normative Wt.% An	$\dfrac{(FeO + Fe_2O_3) \times 100}{MgO + FeO + Fe_2O_3}$
γ or γ'	Comp.	γ	% Fs.		
$\gamma = 1.753$	$Wo_{44}En_2Fs_{54}$	..	..	29	98
..	..	..	..	..	..
..	..	..	..	42	94
..	..	..	..	40	99
$\gamma = 1.772$	$Wo_{30}En_2Fs_{68}$	..	..	32	96
$\gamma' = 1.770$	..	..	..	36	99
..	..	..	..	..	..
$\gamma' = 1.769$ approx.	..	..	..	..	..
$\gamma' = 1.755$	..	..	..	41	94
..	..	..	..	..	..
$\gamma = 1.749$	$Wo_{27}En_{23}Fs_{50}$	..	..	50	79
..	..	..	..	..	..
..	..	..	..	..	..
$\gamma' = 1.739$	..	..	..	..	..
$\gamma' = 1.738$	..	..	..	..	..
..	..	1.735	60	..	..
..	..	..	..	60	74
$\gamma' = 1.735$	$Wo_{32}En_{32}Fs_{36}$	1.732	57	53	70
..	..	..	..	..	..
$\gamma' = 1.733$	..	..	..	..	..
$\gamma = 1.727$	$Wo_{47}En_{34}Fs_{19}$	1.715	45	66	55
$\gamma = 1.701$	$Wo_{24}En_{57}Fs_{19}$	1.705	41	67	49

rocks have a fairly homogeneous inner half, which is approximately An_{40}. In these rocks and in the border group estimates of the plagioclase composition have been made either by refractive index tests or by extinction angles; owing to zoning, the results are much less precise than for the plagioclase of the main layered series.

The rock analyses show that the main layered series is low in potash (Table XXXI). Some of the potash must be present in biotite which is common as small flakes in the lowest layered rock, and is found throughout the series in stray pieces In the ferrogabbros some of the potash is probably present in the interstitial micropegmatite. These considerations suggest that the proportion of potash felspar in solid solution in the normal plagioclase of the layered series is very small.

The plagioclase is usually in the form of tabular crystals, and to this habit of the felspars the igneous lamination is largely due. In the lower rocks the plagioclase crystals are roughly square tablets parallel to (010) and the thickness is about $^1/_{10}$ of the breadth. As the series is ascended, the tablets remain flattened parallel to (010) but they become elongated along the X crystallographic axis until this

TABLE IV.

LENGTH OF THE LARGER FELSPARS IN ROCKS OF THE LAYERED SERIES WELL REMOVED FROM THE MARGIN.

	Specimen	Height in metres	
Hedenbergite-andesinite..................	4136	2750	1.0 cm
Fayalite-ferrogabbro..................	4139[1])	2600	0.6 -
	4142	2500	0.6 -
Ferrohortonolite-ferrogabbro.............	4145	2150	0.6 -
	4146	2100	0.8 -
	4147	2050	0.8 -
Hortonolite-ferrogabbro	3649	1900	0.8 -
	3651	1800	0.8 -
Middle gabbro	3660	1300	0.9 -
	3662	1150	0.9 -
Hypersthene-olivine-gabbro.............	1690	850	1.0 -
	4077	500	1.0 -
	4076	450	1.2 -

[1]) In this rock one quite exceptional felspar measured 1.8 cm.

direction is two or three times that along the Y axis. The remarkable plagioclase laths developed in the perpendicular felspar rock of the border group show this same tendency strongly exaggerated (p. 144).

Variation in the size of the plagioclase crystals has been determined roughly with a pair of dividers, and it shows certain regular trends (see Tables IV, V). The figures given are the greatest length of the half-dozen largest crystals in the hand specimen. In the hypersthene-olivine-gabbro this length is from 1.0—1.2 cm and it decreases fairly steadily until, in the fayalite-ferrogabbro, it is 0.6 cm (Table IV). In some of the upper unlaminated rocks there are sporadic, large felspars 1 cm long, and such crystals form the bulk of the hedenbergite-andesinite. The average dimensions of the larger plagioclase crystals in the hypersthene-olivine-gabbro, 1690, are $1.0 \times 1.0 \times 0.1$ cm, while in the ferrogabbro, 4145, they are $0.6 \times 0.3 \times 0.1$ cm.

When comparisons are made at a constant horizon in the layered series, there is found to be a decrease in size of the felspars as the margin is approached (Table V). Although a small point, this affords confirmatory evidence for certain of our conclusions on the origin of the layered series (see p. 263). An exception to the usual decrease in size is found in the case of the highly melanocratic rock, 4296, which

TABLE V.

LENGTH OF THE LARGER FELSPARS IN THE MIDDLE GAB-BROS AND IN THE HYPERSTHENE-OLIVINE-GABBROS.

Middle gabbros (order from east to west across intrusion)	1815	0.5 cm
	1814	0.7 -
	1697	0.8 -
	3662	0.9 -
	1919	0.7 -
	1920	0.5 -
Hypersthene-olivine-gabbros (order from the northern margin inwards)............	4087	0.5 -
	4085	0.5 -
	4084	0.8 -
	4076	1.2 -

occurs within a few feet of the inner contact on Ivnarmiut. Here plagioclase is rare, but of the crystals present, several reach 1.1 cm which is greater than at the same horizon remote from the margin.

2. Olivine.

An extended range in composition of the olivines in the layered rocks was proved by rough optical methods. It was therefore decided to separate four representative examples for analysis and detailed optical study; some of the results of this work have already been published [Deer & Wager 1939]. The olivine in the analysed rock, 4077, being in fairly large and definite grains, free from inclusions, was ideal as a representative of the olivine from the lower layered rocks. The hortonolite-ferrogabbro, 1907, was found to be suitable for separation of both the olivine and pyroxene and a first, hand-picking of the crushed rock was made during the winter in Greenland. The olivine of the higher rocks usually contains very abundant, orientated, iron ore inclusions. As this is particularly the case with the olivine in the analysed ferrogabbro, 4145, olivine, decidedly freer from ore, was separated from the slightly lower rock, 4146[1]),

---

[1]) The positions and relative height, of all described specimens from the layered series is given in Plate 25.

although the rock itself has not been analysed. The fourth olivine was separated from 4139, the highest olivine-bearing rock of the layered series which contains conveniently large crystals. Except for the carefully hand-picked material from 1907, a first separation was made with methylene iodide and clerici solutions. Further purification of all four was made with an electro-magnetic separator, and this followed by a final hand picking. A chemical method was used for separating the abundant, minute iron ore inclusions from the olivine of 4139; the mineral was dissolved in dilute hydrochloric acid which left the iron ore inclusions and gelatinous silica. The silica and iron ore were ignited and weighed, and the silica then estimated by volatilisation with hydrofluoric acid. The results of the analyses and the calculated formulae are given in Table VI.

The refractive indices of the olivines were determined by the immersion method using a series of liquids with high dispersion. For those olivines with refractive indices between 1.735 and 1.770 a solution of sulphur in methylene iodide was used; for those between 1.770 and 1.852 two mixtures of phenyldi-iodoarsine ($C_6H_5AsI_2$) and methylene iodide were used; while for those above 1.852 pure phenyldi-iodoarsine was used [Anderson and Payne 1934]. The refractive indices of these liquids were obtained for varying wave lengths on a single circle Fuess goniometer by the minimum deviation method using a small, hollow, 30°

TABLE VI.
ANALYSES, OPTICAL PROPERTIES ETC. OF OLIVINES FROM THE LAYERED SERIES.

	I (from 4077)			II (from 1907)		
	Wt.%	Mol.%	Atomic ratio to 4 oxygens	Wt.%	Mol.%	Atomic ratio to 4 oxygen
SiO	38.11	0.6345	1.028	33.72	0.5614	1.006
Al_2O_3....	nil	..	..	nil	..	..
TiO_2.....	trace	..	..	trace	..	..
Fe_2O_3....	0.15	0.0009	0.003	0.05	0.0003	0.002
FeO.....	31.48	0.4382	0.710 ⟩ 1.94	47.91	0.6669	1.190 ⟩ 2.00
MnO.....	0.22	0.0031	0.004	0.41	0.0058	0.010
MgO	30.50	0.7564	1.225	18.07	0.4482	0.799
CaO.....	0.02	..	..	nil	..	..
	100.48			100.16		

(Mg, Fe, Mn)$_2$[SiO$_4$]   (Mg, Fe, Mn)$_2$[SiO$_4$]
Fo$_{64}$Fa$_{36}$ Fo$_{41}$Fa$_{59}$

	III (from 4145)			IV (from 4139)		
	Wt%	Mol.%	Atomic ratio to 4 oxygens	Wt.%	Mol.%	Atomic ratio to 4 oxygens
SiO_2.....	31.85	0.5303	1.008	30.15	0.5019	1.002
Al_2O_3....	trace	..	..	0.07	0.0007	0.003
TiO_2.....	0.01	..	..	0.20	0.0025	0.006
Fe_2O_3....	0.11	0.0007	0.003	0.43	0.0027	0.010
FeO.....	58.64	0.8162	1.552 } 1.98	65.02	0.9049	1.807 } 1.99
MnO	0.85	0.0120	0.023	1.01	0.0149	0.030
MgO	8.49	0.2105	0.400	1.05	0.0260	0.052
CaO.....	0.18	0.0032	0.006	2.18	0.0389	0.078
	100.13			100.11		

$(Mg, Fe, Mn, Ca)_2[SiO_4]$
$Fo_{20}Fa_{80}$

$(Mg, Fe, Mn, Ca)_2[SiO_4]$
Fo_3Fa_{97}

Olivine	I	II	III	IV
2 V (negative)[1].............	79°	65°	58°	48°
Refractive indices[2]	α 1.710	1.752	1.788	1.827
	β 1.733	1.781	1.828	1.869
	γ 1.748	1.795	1.840	1.879
	$\gamma - \alpha$ 0.038	0.043	0.052	0.052
Colour in thin section.	Faint yellow.	Pale yellowish amber		Yellowish amber
Pleochroism...............	As for IV but less corresponding to colour			$\alpha = \gamma$ pale yellow $\beta =$ orange yellow
Colour as small grains	Lemon yellow	Deep amber		Deep reddish brown[3]
Specific gravity.............	3.69	3.88	4.15	n.d.

[1] Determinations ± 2°.   [2] Determination ± .002.   [3] Partly the result of ore inclusions.

prism. To obtain monochromatic light of varying wave lengths Tutton's monochromatic illuminator [Tutton 1922] was used, the source of the light being a lilliput arc. The monochromator was standardised with the yellow sodium lines and the yellow, green and violet mercury lines. From a large number of readings, graphs giving the dispersion of the liquids were drawn. Measurements of the optic axial angle were made on a universal stage. The specific gravities were determined

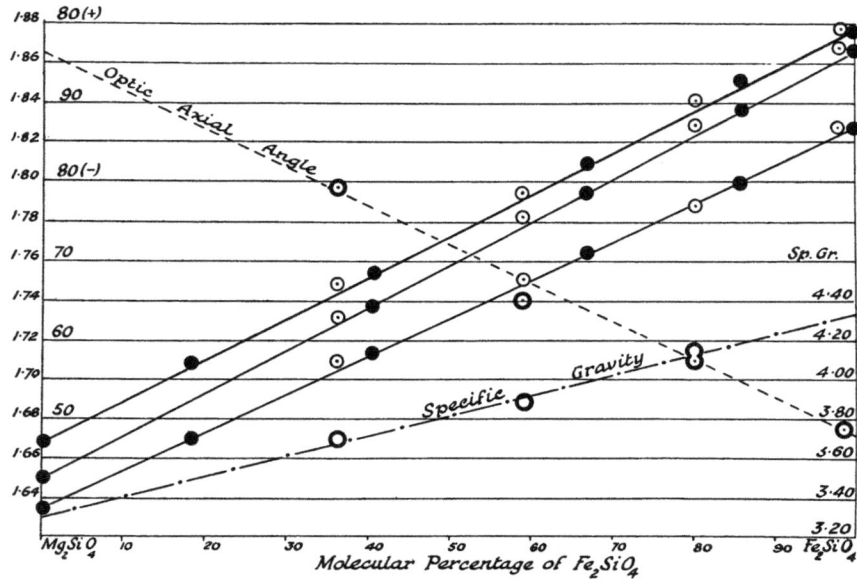

Fig. 20. Graph of the optical properties of the olivines (reprinted from the American Mineralogist 1939).

by suspension in clerici solution diluted with water, the specific gravity of the solutions being obtained with a hydrostatic balance.

The optical data are given in Table VI and they have also been plotted (Fig. 20) against chemical composition and superimposed on the refractive index diagram for the synthetic Mg—Fe olivines of Bowen and Schairer [1935, p. 198], the small amounts of impurities in the natural olivines having been disregarded and the analyses recalculated in terms of Mg_2SiO_4 and Fe_2SiO_4. The refractive index of the natural minerals is, in every case, a little higher than that of the artificial but the agreement is close. There is also close agreement, except at the fayalite end of the series, with the refractive index curves drawn by Winchell [1933, p. 191] from the previously known but less complete series of natural minerals.

The graphs show that the composition of the olivines occurring in the Skaergaard intrusion may be estimated with considerable accuracy from optical determinations. For this purpose it has been found more satisfactory to determine γ than 2V. The values of γ for various olivines of the layered series and the composition deduced by means of figure 20, are given in Table III and figures 18 and 25. The olivines range from Fa_{34} to Fa_{97} with a break in the middle gabbros from about Fa_{50} to Fa_{60}. A break in the solid solution between forsterite and fayalite had previously been suggested by Hawkes [1924, p. 564] but the work of Bowen and

Schairer [1933] proved that this does not occur under laboratory conditions. The gap in composition shown by the olivines of the layered series, is partly bridged by the olivine from a reaction rim in the middle gabbro, 3661, which is Fa_{48} ($\gamma = 1.770$), and partly by an olivine from the border rock, 2275, which is Fa_{51} ($\gamma = 1.775$). The olivine-free gabbros are not due to lack of solid solution between forsterite and fayalite, but to peculiarities in the reaction relations with pyroxene. The conditions are elucidated to some extent by the work of Bowen and his collaborators, as we shall show below.

The subdivision and nomenclature of the olivine group, as suggested by various authors and as finally adopted by us, is given in our previous paper; it is summarised again here in figure 21. The names adopted will recur repeatedly in the present paper, since we have partly subdivided the layered series on the basis of the olivine composition.

The olivine of the hypersthene-olivine-gabbros occurs in idiomorphic grains, sometimes showing a little of their crystal shape; it is a fairly common type, as will be seen from the analyses and optical properties of the olivine from 4077 (Table VI). The more iron-rich olivines have a distinct cleavage parallel to (010), and a somewhat stronger colour which is considerably enhanced in small grains because of the abundant ore inclusions. The inclusions of iron ore which are tabular in shape, are usually very abundant parallel to (100), and somewhat less

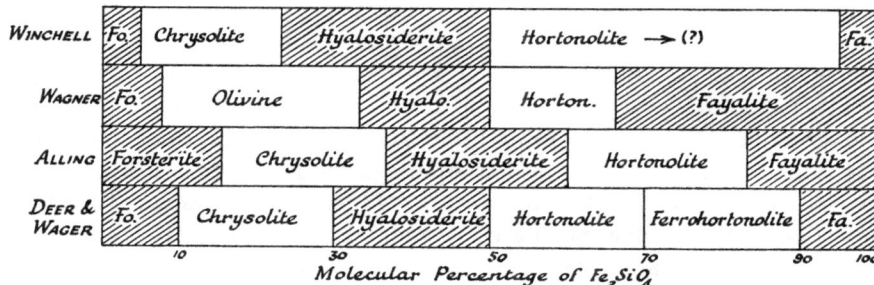

Fig. 21. Proposed classification of the Mg_2SiO_4—Fe_2SiO_4 olivines compared with the subdivisions of Winchell, Wagner and Alling (reprinted from the American Mineralogist 1939).

abundant parallel to (001). The iron-rich olivines show no rounding due to resolution; indeed they have an interstitial relationship to the plagioclase and the other minerals showing that they continued to crystallize until a late stage.

The minerals in equilibrium with the olivines will be discussed in the section dealing with the petrography of the layered rocks. However, it may be mentioned here that quartz is in equilibrium with the more iron-rich olivines. Experimental work on the system $MgO—FeO—SiO_2$ has shown that quartz is in equilibrium with olivines between Fa_{68} and fayalite. We find that quartz is present in the rocks of the layered series in which the olivine is dominantly Fa_{60}, and in the border group a very small amount is present in a rock containing olivine which is dominantly Fa_{42}. Zoning in the olivines is difficult to detect, and it may be that there is a thin outer zone of more iron rich olivine with which the quartz is really in equilibrium.

3. Clinopyroxene.

The same plan has been adopted for studying the clinopyroxenes as for the olivines. Pyroxenes from 4077, 3662, 1907, 1881, and 4136, which are regarded as reasonably representative of the whole range in the layered series, have been separated, analysed, and studied in detail optically. Results for two of these pyroxenes (1907 and 1881) and a discussion of their mineralogical significance have been already published [Deer and Wager 1938]. Separation and purification of the pyroxenes for analysis were carried out as for the olivines, and the same methods used for determination of their physical constants. There is present in the two lower rocks, 4077 and 3662 (and also in 4087 from the transitional layered series the analysis of which, is given in Table XIX), a small amount of orthopyroxene which could not be separated with certainty from the clinopyroxene; the mixed samples consisting of clino- and orthopyroxene had therefore to be analysed. The composition of the orthopyroxene present in each sample was estimated from a determination of the γ refractive index (see the following section) and the relative amounts of the ortho- and clinopyroxene were found from the mode, determined on the integrating stage. From this data the composition of the clinopyroxene has been estimated by subtraction of the modal orthopyroxene of the composition deduced from its refractive index.

The analyses, the calculated atomic ratios to 6(0,0H) atoms and the estimated formulae are presented in Tables VII and VIII. It will be seen that in all cases there is some replacement of Si'''' by Al''' and that in the titaniferous pyroxene III from 1907 there must also be replacement of Si'''' by Ti''''. The latter replacement has seldom been observed in pyroxenes and the matter has been discussed more fully in our earlier paper [1938, pp. 15—18]. Examination of the analyses shows that they are all fairly low in sesquioxides, and as a first approximation, the composition may be estimated in terms of the wollastonite (Wo), enstatite (En) and ferrosilite (Fs) molecules. By so doing their compositions may be expressed on a **triangular** diagram and their optical properties may be graphically presented (Figs. 22 and 23).

TABLE VII.

ANALYSES, OPTICAL PROPERTIES ETC. OF CLINO-PYROXENES FROM THE LAYERED SERIES.

Analysis	I and Ia[1] (from 4077)		II and IIa[2] (from 3662)		III (from 1907)	IV (from 1881)	V (from 4136)
SiO_2	49.89	48.98	47.92	47.55	42.32	42.62	46.54
Al_2O_3	3.74	4.67	4.60	5.01	2.25	5.24	5.77
TiO_2	0.93	1.16	1.71	1.86	4.42	1.69	1.22
Fe_2O_3	0.21	0.25	0.71	0.78	4.72	3.74	1.88
FeO	12.81	9.89	18.85	17.56	25.13	31.54	24.65
MnO	0.27	0.33	0.35	0.38	0.22	0.78	0.74
MgO	14.59	12.72	12.03	11.63	8.33	0.47	0.79
CaO	16.83	21.02	12.95	14.12	12.07	12.27	17.70
Na_2O	0.43	0.53	0.51	0.55	0.61	1.02	0.62
K_2O	0.13	0.16	0.13	0.14	trace	0.23	0.15
H_2O+	0.16	0.20	0.20	0.22	0.12	0.48	0.39
H_2O-	0.07	0.09	0.18	0.20	0.25	0.22	0.18
	100.06	100.00	100.14	100.00	100.44	100.30	100.53
2 V (positive)[3]	43°		40°		44°	58°	49°
$\gamma \wedge C$	41°		40°		39°	47°	44°
Refractive indices[4] α	1.695		1.707		1.721	1.743	1.726
β	1.702		1.715		1.728	1.751	1.732
γ	1.722		1.735		1.749	1.772	1.753
Pleochrism.	Weak		Weak (Colour: light brown)		α Yellow-brown β Violet-brown γ Violet-brown	Pale-green Green Green	Weak (Colour: light greenish brown)
Dispersion	r > v		r > v		r > v	r > v	r > v
Specific gravity	3.38		3.42		3.50	3.65	3.48
Weight %[5]) Wo	46.6		32.3		27.2	30.0	43.7
En	34.0		32.1		22.5	1.5	2.4
Fs	19.4		35.6		50.3	68.5	53.9

[1]) Anal. 1 after subtracting 20.0 % hypersthene (Fs_{45}) and recalculating to 100.00 %.
[2]) Anal. II after subtracting 8.4 % hypersthene (Fs_{55}) and recalculating to 100.00 %.
[3]) Determinations ± 2°.
[4]) Determinations ± .002.
[5]) Neglecting other constituents and recalculating to 100 %.

Tomita [1934] has drawn up a valuable series of diagrams showing the variation in optical properties of pyroxenes in which the sesquioxides can reasonably be disregarded and the composition be expressed in terms of Wo, En and Fs. With the new data from artificial melts supplied

TABLE VIII.

FORMULAE OF CLINO-PYROXENES FROM THE LAYERED SERIES.

	Ia (from 4077)		IIa (from 3662)		III (from 1907)	
	Mol. Ratios	Atomic ratio to 6(0.0H)	Mol. Ratios	Atomic ratio to 6(0.0H)	Mol. Ratios	Atomic ratio to 6(0.0H)
SiO_2.....	0.8155	1.822 ⎫	0.7917	1.833 ⎫	0.7046	1.720 ⎫
Al_2O_3....	0.0458	0.205 {0.178 2.00 / 0.027}	0.0491	0.227 {0.167 2.00 / 0.060}	0.0220	0.107 ⎬ 1.96
TiO_2.....	0.0145	0.032	0.0233	0.054	0.0552	0.135 ⎭
Fe_2O_3....	0.0156	0.007	0.0049	0.023	0.0296	0.144 ⎫
FeO....	0.1377	0.308	0.2444	0.566	0.3497	0.853
MnO....	0.0046	0.010 ⎬ 1.97	0.0053	0.012 ⎬ 2.01	0.0031	0.008
MgO....	0.3157	0.705	0.2884	0.668	0.2066	0.503 ⎬ 2.08
CaO.....	0.3747	0.837	0.2518	0.583	0.2152	0.525
Na_2O....	0.0085	0.038	0.0089	0.033	0.0098	0.048
K_2O.....	0.0017	0.007 ⎭	0.0015	0.007 ⎭	..	.. ⎭

	IV (from 1881)		V (from 4136)	
	Mol. Ratios	Atomic ratio to 6(0.0H)	Mol. Ratios	Atomic ratio to 6(0.0H)
SiO_2..........	0.7096	1.785 ⎫	0.7749	1.867 ⎫
Al_2O_3..........	0.0513	0.258 {0.215 2.00 / 0.043}	0.0575	0.278 {0.133 2.00 / 0.145}
TiO_2..........	0.0211	0.053	0.0153	0.037
Fe_2O_3..........	0.0234	0.118	0.0118	0.057
FeO..........	0.4390	1.105	0.3431	0.827
MnO..........	0.0110	0.028 ⎬ 2.02	0.0104	0.025 ⎬ 1.96
MgO..........	0.0117	0.029	0.0197	0.048
CaO..........	0.2188	0.550	0.3156	0.761
Na_2O..........	0.0164	0.083	0.0100	0.048
K_2O..........	0.0024	0.012	0.0016	0.008

I, II, IV and V. (Mg, Fe″, Fe‴, Ca, Mn, Ti, Al)₂ [(SiAl)₂O₆].
III. (Mg, Fe″, Fe‴, Ca, Mn)₂ [(Si, Al, Ti)₂O₆].

by Bowen, Schairer and Posnjak [1933] and that from the pyroxenes of 1907 and 1881, we have modified Tomita's diagram showing variations of α and γ [1938, p. 20]. On the revised diagram for γ the remaining analysed pyroxenes have now been plotted (Fig. 22). The determination of the higher refractive index of cleavage fragments resting on a 110 face (γ') can be more readily determined than γ. As Tomita has shown

[1934, p. 47] this value can be calculated if sufficient of the optical constants of a monoclinic crystal are known. For the pyroxene of 4077 the value of γ', calculated from our previous data, is 1.717 while the value determined directly is 1.722. Similar values for the pyroxene of 1907 are: 1.746 and 1.749. If 0.004 is added to the observed values

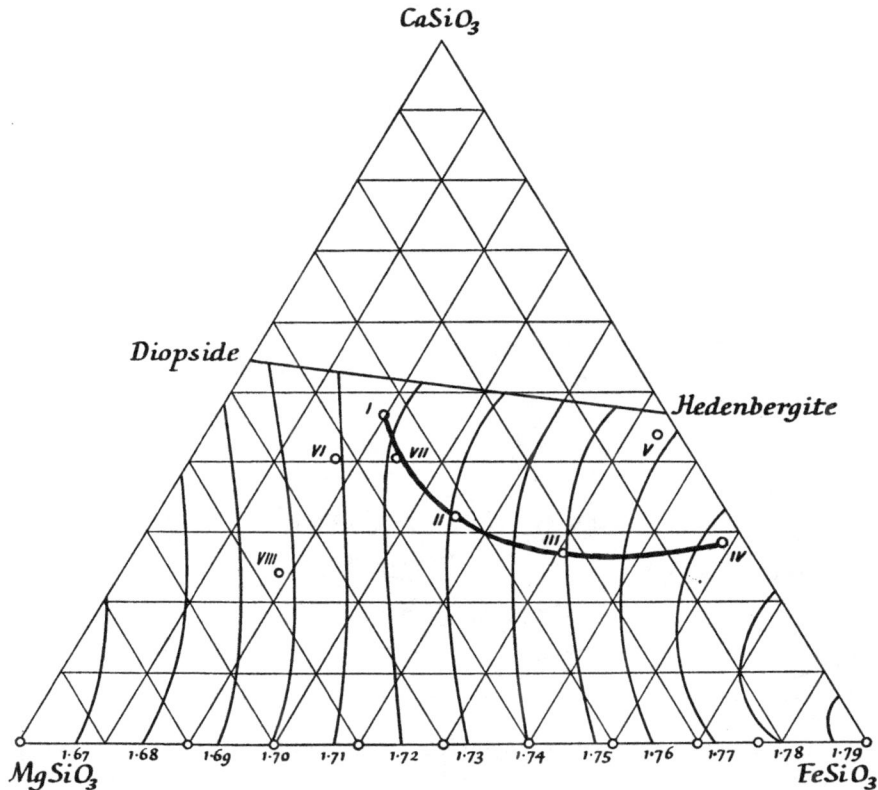

Fig. 22. The analysed pyroxenes of the Skaergaard intrusion superimposed on curves showing variation of refractive index γ for clinopyroxenes of the system diopside, hedenbergite, clinoenstatite and clinoferrosilite. Trend of pyroxenes from layered series shown by thickened line. [Refractive index curves from Deer & Wager 1938.]

of γ' an approximation to the value of γ is obtained which is sufficiently accurate for the determination of the composition from figure 22, of other pyroxenes of the layered series. The method adopted has been to assume that the general trend of variation in the pyroxenes of the layered series is defined by the analysed pyroxenes from 4077, 3662, 1907 and 1881 and is as shown in figure 22. The composition of other pyroxenes of the layered series (Table III), in terms of Wo, En and Fs, is then estimated from the point at which the γ value crosses the trend line. Without the assumption that there is a steady change in composition in these pyroxenes, it would be necessary to determine some other property such as 2V or $\gamma \wedge c$ which can be expressed as curves crossing those for γ (Fig. 23). We find, however, [1938, pp. 21 and 22] that there appears to be an erratic variation of these optical constants, probably because these

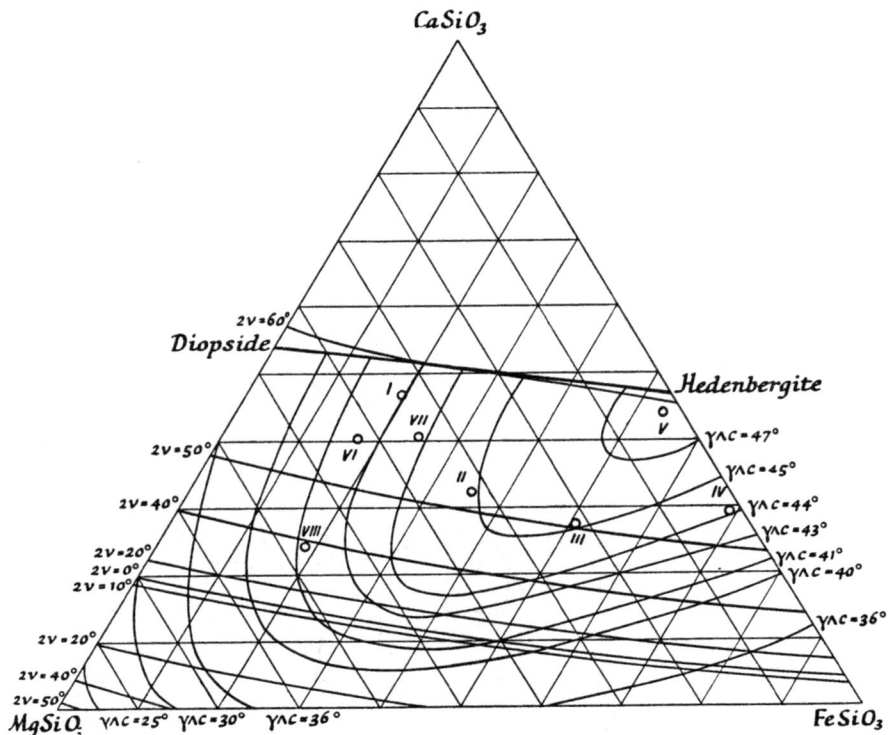

Fig. 23. Position of the analysed pyroxenes from the Skaergaard intrusion super-imposed on curves showing 2V and $\gamma \wedge c$ for clinopyroxenes of the system diopside, hedenbergite, clinoenstatite and clinoferrosilite [curves from Deer & Wager 1938].

values are sensitive to the presence of relatively small quantities of other constituents, and that a better estimate of composition can be obtained for the clinopyroxenes of the layered series on the assumption, which seems abundantly justified, that the trend of mineral composition is a steady one.

From the compositions obtained by analysis and estimated from optical properties, it is apparent that, in the layered series proper, there is a change from a pyroxene approximating to diopside in composition (I from 4077) to one very close to a hedenbergite-clinoferrosilite solid solution (IV from 1881). The general range in optical properties of the pyroxenes can be seen from Table VII, but a few other characteristics are worth comment. Besides the (110) cleavages there is often a fairly good cleavage parallel to the optic plane (010) and a less good cleavage parallel to (100). Twinning on (100) is fairly common. The pyroxenes of the layered series, from the lowest rocks exposed up to examples a little higher than 1907, always contain very numerous regularly spaced sheets of inclusions. These inclusions which are parallel to the basal plane are only about 0.003 mm thick, and they consist of a colourless mineral sometimes with iron ore as inclusions within them. We believe that the sheet inclusions are orthopyroxene produced by exsolution. Somewhat similar associations of hypersthene and clinopyroxene have often been noted; for example as early as 1897 Henderson [1897, pp. 17—18] described from rocks of the Bushveld complex what he believed to be a submicroscopic intergrowth of hypersthene sheets parallel to (001) of the clinopyroxene host. In the case of 3662 there are thicker sheets of definite orthopyroxene which may perhaps be homologous with the thin sheets. In this particular example the γ direction of both the clino- and orthopyroxenes is coincident, and the a direction of the orthopyroxene coincides with the β direction of the clinopyroxene. We think that this is probably the orientation of the material of the thin sheets and hope that it will be possible later to test this view by X-ray work. The thickness of the sheet inclusions is greatest in the lower layered rocks and becomes steadily less up to the 2,000 metres level, above which no inclusions can be seen. In the middle gabbros and ferrogabbros the thin sheets produce pale interference colours in ordinary light. This would be expected as they have a different refractive index from that of the host and their thickness is usually less than .003 mm. Besides the sheet inclusions which are believed to be orthopyroxene, there are orientated, iron ore inclusions parallel to (001), and iron ore inclusions may also be feebly developed parallel to (100).

The usual habit of the clinopyroxenes of which I, II and III from 4077, 3662 and 1907 are examples, is that of independent grains of the same period of crystallisation as the olivine and felspar. At the purple band the clinopyroxene develops a special habit which, we have suggested (p. 111), indicates that it has been formed by inversion from iron-rich β-wollastonite; pyroxene IV from 1881 is an example of this. Although believed to have formed from iron-rich β-wollastonite the composition of clinopyroxene IV is such that it continues the general trend of composition shown by the typical layered pyroxenes. Pyroxene V from the unlaminated rock, 4136, does not lie on the main trend of composition being richer in the wollastonite molecule. Evidence from the texture of the rock suggests that this pyroxene was not formed by inversion but directly precipitated from the magma (see p. 116). In the field the unlaminated rocks show features which distinguish them from the typical layered rocks, so that we are not surprised that the composition of the pyroxene from these rocks lies off the main trend.

4. Orthorhombic pyroxene.

Our knowledge of the relations between optical properties and composition in the orthorhombic pyroxenes has recently been restated and extended by Henry [1935] and we have considered it sufficient to rely on optical determinations for estimating the composition of the orthopyroxene in the Skaergaard rocks. Determination of γ by immersion methods has proved more satisfactory than determination of 2V on the universal stage. Both measurements, however, are interfered with by intergrowth with clinopyroxene. If 18 molecular percentage of $FeSiO_3$ is taken as the upper limit of enstatite then all the examined orthorhombic pyroxenes of the Skaergaard intrusion are hypersthene.

In the lower layered rocks orthopyroxene is not abundant, usually amounting to 5 per cent. or less, and it occurs as small ophitic crystals of a moderately late period of crystallisation. Inclusions of clinopyroxene consisting of fairly regularly orientated sheets, usually amount to 5 or 10 per cent. of the total bulk of the orthopyroxene. The orthopyroxene of 4087 has $\gamma = 1.708$ indicating that it is Fs_{41}, while that of 4077, 500 metres higher has $\gamma = 1.715$ and $-2V = 53°$ indicating Fs_{45}. In the higher layered rocks the orthopyroxene becomes more iron-rich. Thus in the analysed middle gabbro, 3662, the orthopyroxene has $\gamma = 1.732$ indicating Fs_{57} while in 3658, 200 metres higher, $\gamma = 1.735$ indicating Fs_{60}. There is much variation in the actual amounts of orthopyroxene present in these higher rocks, for example, there is 2 per cent. in 3662, while in the nearby rock, 3661, there is about 15 per cent. which occurs in small crystals with a somewhat similar habit to that of the clinopyroxene. In these rocks the orthopyroxenes are crowded with inclusions of clinopyroxene which probably amounts to 20 or 30 per cent. of the total. Orthopyroxene persists throughout the middle gabbro horizons but dies out just above. If the percentage of $FeSiO_3$ present in the orthopyroxenes is plotted against height in the intrusion and the curve extrapolated to the point where the mineral disappears, the composition of the highest orthopyroxenes is indicated as about Fs_{70}. In the Skaergaard layered rocks there is a long series of orthopyroxenes which might well repay more careful optical study than we have yet given them. The data so far collected are summarised in Table III.

The relationships between the orthorhombic pyroxene and the inclusions of clinopyroxene have been partly investigated. In favourable cases, e.g. in 1690 and 3661 it is easy to show the nature of the inclusions because they are optically continuous with large adjacent crystals of clinopyroxene. The form of the inclusions in the lower orthopyroxenes is that of irregular sheets which tend to break up into rows of flattened blebs. In the higher orthopyroxenes the sheets are more accurately parallel-sided and more continuous. In all cases there is a tendency for the sheets to have small projections in the direction of the optic plane (010). Some attempt has been made to determine the position of the sheets within the host. In the orthopyroxene of 1690 the sheets of clino-pyroxene are parallel to (112), or some plane near it, while determinations with universal stage suggest that in 4077 they are approximately parallel to (212). The pyramid planes (112) and (212) are not far removed from each other, and the sheets are not sufficiently regular in these two cases for greater precision to be expected. It is certain, however, that the sheets are not parallel to (001), (100), (010), or (110). The orientation of the clinopyroxene within the sheet is apparently variable and is often the same as that of adjacent large crystals of clinopyroxene. It seems probable that the association is due to intergrowth as a result of simultaneous precipitation of clino- and orthopyroxene in the later stages of crystallisation of the rock. The phenomenon is different, there-fore, from the fine-scale, even sheets of orthopyroxene in clinopyroxene which we have ascribed to exsolution. The intergrowth of clinopyroxene in orthopyroxene in the layered rocks does not resemble the fine-scale structure, parallel to the optic plane, which has frequently been observed but not yet explained, [see for instance comments in Bowen 1935, p.169] nor does it resemble the intergrowth of clinopyroxene described by Hess and Phillips [1938, pp. 450—56] which is parallel to the optic plane. It is, however, interesting to note that in the examples from the layered series the sheet inclusions of clinopyroxene which are parallel to a pyramid have projections which extend a little way in the direction of the optic plane of the orthopyroxene. In the more enstatite-rich orthopyroxene of the northern border group fine-scale structures parallel to the optic plane and simulating twinning are found. These are the unexplained structures mentioned above and are apparently not the result of inter-growth with other material.

5. Other minerals.

Iron ore is scarce in the lower layered rocks but is abundant in the middle gabbros and above. During the separation for analysis of the pyroxene from 3662 it was noticed that about half the opaque iron ore is strongly magnetic and is presumably magnetite, while the other half is only very feebly magnetic and is presumably ilmenite. In view of the interesting relationships discussed later between the amounts of TiO_2 and Fe_2O_3 in the varying Skaergaard magma and in the layered series, the amounts of magnetite and ilmenite in certain rocks of the layered series have been approximately determined. The iron ore was separated from the rest of the minerals with a suitably adjusted Clerici solution, and the ilmenite and magnetite were then separated with an electromagnet. This gave the ratio between ilmenite and magnetite in the rocks (Table XXXVIII). From these ratios and the total amount of iron ore in the mode (Table XXXXIII) the actual modal amounts of ilmenite and magnetite have been obtained (Table XXXVIII). These quantities are roughly comparable with the normative values (Table XXXXII).

A brown mica has been noted in 4296 in some quantity; it is also fairly common in the transitional layered rocks, and it occurs sporadically in many of the rocks of the main layered series. Amphiboles occur only occasionally as late stage alteration products of the pyroxene. Green and brown chlorite or serpentine from the hydrothermal alteration of the olivines are found and are sometimes well developed, for example in 1881. In iron-rich rocks of this type some of these minerals might repay detailed investigation but this has not so far been done.

(c) Specific Gravities.

The specific gravities of various layered rocks, usually determined in specimens weighing from 300—600 grams, are presented in order of height in the intrusion in Table IX. To obtain an idea of the range of variation in the specific gravity of the various types, several examples of each have been determined. In the first two columns is presented data for average rocks, that is rocks showing no obvious concentration of light or heavy constituents. Some irregular variation is to be observed, due no doubt to the effect of banding which has only been partly eliminated by selecting average rocks. Despite this there is to be clearly discerned a general rise in specific gravity with increasing height in the intrusion from a mean value of 3.06 for the hypersthene olivine gabbros to 3.36 for the purple band. Among the unlaminated layered rocks, there is a tendency for the specific gravity to fall with increasing height, and this is another difference between them and the main layered series.

The fact that the specific gravity of the layered rocks, excluding the unlaminated section, increases upwards is of interest because some authors have supposed that the density of the rocks should decrease upwards if gravity plays any part in the differentiation of an intrusion. For example, an attempt has been made, though it is not convincing, to show that the rocks of the Bushveld complex decrease in density upwards [cf. Hall, 1932, pp. 278—80]. Grout [1918B, p. 496] has given good reason for believing that such upward decrease in density in layered intrusions is not a necessary corollary from the gravitative

TABLE IX.

SPECIFIC GRAVITIES OF ROCKS OF THE LAYERED SERIES IN ORDER OF HEIGHT IN THE INTRUSION.

(Bracketed rocks are approximately the same height.)

		Average Rock		Average rock near border		Leucocratic bands		Melanocratic Bands	
		No.	S.G.	No.	S.G.	No.	S.G.	No.	S.G.
Unlaminated layered series	Hedenbergite-andesinite	{4136	2.82	..	..	..	..	..	..
		{1899	2.79	..	..	..	..	..	..
	Mean............	..	2.80	..	..	..	..	..	..
	Basic hedenbergite-granophyre	{4137	2.97	..	..	..	..	..	..
		{1905	2.84	..	..	..	..	..	..
		(1904	3.07[1])	..	..	..	..	..	..
	Mean............	..	2.91	..	..	..	..	..	..
	Fayalite-ferrogab-bro	4139	3.30	..	..	..	..	..	..
Ferrogabbros	Purple band	{4142	3.39	..	..	..	..	..	..
		{1881	3.18	..	..	..	..	..	..
		{1906	3.50	..	..	..	..	..	..
	Mean............	..	3.36	..	..	..	..	..	..
	Ferrohortonolite-ferrogabbro	4145	3.22	..	..	..	..	..	..
		4146	3.17	..	..	..	..	..	..
	Mean............	..	3.20	..	..	..	..	..	..
	Hortonolite-ferro-gabbro	..	..	(1732	3.03[2])	..	..	..	..
		..	..	1733	3.33	..	..	..	..
		..	..	1734	3.37	..	..	..	..
		1907	3.18	..	..	1925	2.77	1924	4.03
		2574	3.19	..	..	2568	2.75	2569	3.74
		2575	3.20	..	..	..	..	..	..
		..	..	1969	3.24	..	..	..	..
		..	..	4296	3.56	..	..	..	..
	Mean............	..	3.19	..	3.38	..	2.76	..	3.89
	Middle gabbro	3661	3.19	1815	3.29	..	..	..	..
		3662	3.01[3])	..	..	..	..	..	..
		1691	3.19	1919	3.56	..	..	..	..
		..	..	1920	3.41	..	..	..	..
	Mean............	..	3.10	..	3.42	..	..	..	..

[1]) Collected as a rather melanocratic type.
[2]) Petrographically also not typical.
[3]) Rather leucocratic see Table XI.

Table IX (continued).

	Average rock		Average rock near border		Leucocratic bands		Melanocratic bands	
	No.	S. G.	No.	S. G.	No.	S. G.	No.	S. G.
Hypersthene-olivine-gabbro	1690	3.06	2307	3.23	..	..	2308	4.05
	1687	3.06	..	..	..	..	..	..
	1689	3.14	..	..	..	..	..	..
	4077	3.04	..	..	..	..	..	..
	1684	3.00	..	..	..	..	..	..
	..	..	{ 4087	3.05	..	..	..	..
			1709	3.03	..	..	..	..
Mean............	..	3.06	..	3.10	..	..	..	4.05

hypothesis of their origin but comes about only if the earlier minerals have the greater specific gravity. In the case of the Skaergaard intrusion the early formed minerals are magnesium bearing and of less density than the later iron bearing minerals; thus the average specific gravity increases with height.

In the third and fourth columns the specific gravity of average rocks near the border is given for comparison with average rocks near the centre of the intrusion. With only one exception the former have a decidedly higher specific gravity than the corresponding average rock some distance from the margin. This is to be correlated with the field observations that a melanocratic facies is found adjacent to the inner contact. The one exception is the rock, 1732, from the south end of the Skaergaard Halvö. This rock has various other peculiarities which have not so far been explained but it does not significantly modify the generalisation that, on the average, more melanocratic and denser rocks occur near the inner contact.

The remaining columns give the specific gravity of certain extreme bands. In the case of the rocks from the hortonolite ferrogabbro horizon the examples are from the trough banding structures. The more felspathic bands have an average specific gravity of 2.76 which is very close to the specific gravity (2.75) of medium andesine. The specific gravities of the melanocratic bands are more variable and depend on which particular dark mineral predominates in the particular band. The figures in these two columns are admittedly extremes but they show how strong the variation in composition due to banding may be.

(d) Petrography of the Average Rocks.

There are two generalisations concerning the rocks of the layered series which greatly simplify their petrographical description:—

1. Rocks of the same horizon are similar except for differences in the relative abundance of the minerals; this has been called the rhythmic variation (see p. 64).

2. The differences between average rocks of different horizons, due to what has been called the major variation (see also p. 64), is a function of height in the intrusion, and this is gradual, except at a certain few horizons, where a particular mineral suddenly begins to be present or suddenly ceases to be present.

Rough field evidence for these generalisations has already been given and the more definite evidence from microscopic examination of the rocks will be found in the present section.

Ten rocks have been carefully selected which represent the more important types of the layered series; they have been analysed (Table XXXXI), their modes have been determined[1]) (Table XXXXII), the composition of the more significant mineral constituents have been found by analysis or optical methods (Table III) and their specific gravities have been determined (Table IX). As a result of the regularity of the major variation, these linear tabulations in order of height give a true picture of the layered rocks which is summarised in figures 18 and 25. In the succeeding sections detailed petrographic descriptions are only given for the analysed rocks or for types at, or near the horizons where a mineral phase appears or disappears. Other rocks will be mentioned to show the distribution of a particular type or to show intermediate stages in the evolution of special features. The position and height of the chief layered rocks described are given in plate 25.

1. Hypersthene-olivine-gabbros.

The lower 900 metres of the layered series are hypersthene-olivine-gabbros, but of this thickness only the upper 500 metres are typical layered rocks, the lower 400 metres being the transitional layered rocks which approach the border group in many features. One of the lowest rocks now exposed, 4087, has been analysed and will be described first. The rock was collected 150 metres from the northern contact, the actual position being shown in the lower part of plate 9, fig. 1. At this point the distinction between the layered series and the border group is indefinite

---

[1]) In some cases the modes are only approximate as the rocks are coarse grained and only a normal sized slice has been measured.

but 4087 is regarded as just within the transitional layered series. The rock does not look much finer-grained than higher layered rocks such as 4077, but its felspar crystals (see Table V) are about half the size of those in 4077, and the coarse appearance of the rock is due to the clinopyroxene being in rather large crystals. The plagioclase of 4087, An_{61}, is slightly more zoned than is usual in most of the layered series, and is subeuhedral (Fig. 24). The olivine, Fa_{34}, is a pale yellow colour; as a result of slight decomposition it shows irregular zones of magnetite granules and in some places a border of iron ore. It is usually idiomorphic and rounded and, although small well-shaped crystals of felspar occasionally occur in, or project into, the outer margin of the olivines, showing that both were early minerals, other relationships suggest that the olivine finished crystallising before the felspar. The clinopyroxene occurs in extensive ophitic crystals, and patches free from other minerals may equal or exceed the average size of the felspar and olivine crystals. The principal optical properties of the clinopyroxene are: $\alpha = 1.672$, $\gamma = 1.701$, $\gamma \wedge c = 39°$, $2V = 42°$. The plagioclase crystals embedded in the ophitic pyroxenes of this rock and others of this horizon (e.g. 1708) are small, and suggest that the pyroxene largely crystallised when only small crystals of plagioclase were available for incorporation. The clinopyroxene apparently finished crystallising after much of the plagioclase, and perhaps all the olivine, had ceased forming. Two directions of sheet inclusions of orthopyroxene occur in the clinopyroxene, and there are also inclusions which are presumably magnetite, and somewhat larger inclusions, perhaps of haematite, which transmit a brown colour. The hypersthene, Fs_{41} ($\gamma = 1.708$), is present only to the extent of 3 per cent. in the section measured but this may not be typical. It has the same ophitic habit as the clinopyroxene but occurs in much smaller crystals. It is fairly strongly pleochroic, and contains rather large inclusions of both magnetite and clinopyroxene. Iron ore is rare, and is often fringed by strongly pleochroic mica.

A rock, 1708, collected at the foot of the N.W. ridge of Peak 1300 metres of Gabbrofjaeld is similar to the analysed rock, 4087, and its estimated height in the layered series is also the same. Olivine in 1708 is about twice as abundant as the two pyroxenes together and this is perhaps a more common proportion in the low hypersthene-olivine-gabbros than the reverse proportion shown by the analysed rock. As elsewhere in the series, the relative proportion of the minerals is very variable, and rocks which were thought to represent average hypersthene-olivine-gabbro have turned out to be more variable than at higher horizons. Three average rocks from Uttentals Plateau, 4085, 4084, and 4067, collected at increasing distances from the margin, show a gradual change in texture towards the higher type of hypersthene-olivine-gabbro in which the clinopyroxene is not ophitic but is in independent grains. Other specimens, e.g. 1193 from the north ridge of Gabbrofjaeld (Peak 1300 metres) have clinopyroxene of a similar intermediate form, showing that this is a general feature of the horizon.

Another hypersthene-olivine-gabbro, 4077, at 500 metres in the layered series has been analysed. It was collected where Uttentals Plateau joins the W.N.W. ridge from Peak 1300 metres of Gabbrofjaeld, at a height of 350 metres above sea level, and 1400 metres from the nearest point of Uttentals Sund. It belongs to the main layered series which overlies the transitional group, and it shows moderate igneous lamination. The felspars reach 1 cm in length, and the olivine and pyroxene grains about 0.5 cm (Fig. 24 and Pl. 15, fig. 1). The plagioclase, An_{56}, is usually subeuhedral and is only slightly zoned at the extreme margin. The olivine, Fa_{37}, is idiomorphic and, although somewhat rounded, often shows relics of its crystal shape. The analyses and properties of the olivine are given in Table VI. The clinopyroxene crystals, whose composition and properties have also been given (Tables VII and VIII) contains fine-scale, sheet inclusions of orthopyroxene parallel to one direction only, and there are also opaque and brown inclusions as in 4087. The clinopyroxene in this rock is not ophitic but occurs in grains of about the same size as the olivine. The hypersthene, Fs_{45} ($\gamma = 1.715$), forms about a fifth of the total pyroxene; it occurs in small ophitic masses surrounding the olivine and clinopyroxene, and filling interstices between the plagioclase; it contains sheet inclusions of clinopyroxene. Iron ore is found in small grains in the clinopyroxene or as interstitial patches like the hypersthene. Apatite is uncommon, but when it occurs, it is associated with iron ore and biotite and is in the interstices of the felspar (Fig. 24).

An attempt has been made to estimate the probable sequence of the beginning and ending of crystallisation for the minerals of this rock. The olivine crystals of normal size occasionally have small felspars within them, especially near their borders; the felspars sometimes enclose small olivine and pyroxene crystals, and the clinopyroxenes often have small felspar crystals, and occasionally small olivine crystals, within them. These three minerals were crystallising simultaneously and each was therefore occasionally liable to incorporate one of the others. On the whole, small crystals of olivine and plagioclase are commonest in the pyroxene; this might be ascribed to slightly later growth of the pyroxene, but is more likely to be connected with the tendency, so frequently shown by pyroxene crystals, of forming ophitic masses. The hypersthene, the outer zones of the plagioclase and the bulk of the iron ore crystallised at a later stage, and tend to fill interstices. Apatite is also found only among the interstitial minerals.

The uppermost layers of hypersthene-olivine-gabbro show further changes in composition and texture. Thus 1690, from a height of 850 metres in the intrusion, contains plagioclase An_{53}, which is slightly zoned at the extreme margin in the way characteristic of the layered rocks. It occurs in thin plates parallel to 010 and these have sometimes been bent. A dusty appearance of the felspar is due to innumerable very fine, rod-shaped inclusions of an opaque mineral, presumably iron ore. This feature, which cannot be regarded as due to subsequent thermal metamorphism, is found in the plagioclase of the middle part of the layered series but, surprisingly, is absent from the uppermost, more iron-rich, layered rocks. The olivine of the rock has $\gamma = 1.760$, which corresponds with Fa_{44}. The clinopyroxene is also more iron-rich ($\gamma' = 1.733$). Hypersthene is present in small amounts surrounding olivine and clinopyroxene; it is interstitial to the bulk of the felspar. The olivine and clinopyroxene occur in grains of about equal size, reaching 0.5 cm in length. Both frequently contain small crystals of plagioclase, and they

---

Fig. 24. Average rocks of the layered series, magnification $\times$ 12.

4087. Hypersthene-olivine-gabbro, the lowest exposed rock of the transitional layered series, Uttentals Plateau. Hypersthene, showing similar ophitic habit to the clino-pyroxene, is present in small amount surrounding olivine at the top left, and in contact with clinopyroxene at the top right of the field. The remaining pyroxene is a single ophitic mass of clinopyroxene (p. 87).

4077. Hypersthene-olivine-gabbro, at 500 metres in the layered series, Uttentals Plateau. Clinopyroxene is in grains of about the same size as the olivine. Hyper-sthene (top centre) is seen as a single ophitic crystal surrounding olivine and clino-pyroxene. A small, well-shaped crystal of apatite is shown occurring between the felspar crystals at the top of the figure (p. 89).

3662. Middle gabbro, at 1200 metres in the layered series, Pukugaqryggen. A little hypersthene, not to be distinguished in ordinary light, occurs among clinopyroxene at the bottom right of figure; the remainder is clinopyroxene. Feathery reaction rims are developed sporadically between iron ore and felspar (p. 94).

1907. Hortonolite-ferrogabbro, at 1800 metres in the layered series, Main Base House. The pyroxene is wholly ferriferous clinopyroxene. Hortonolite occurs as independent crystals of the same size as the pyroxene and also as a reaction rim between pyroxene and iron ore (right hand side) or in close association with ore (p. 99).

4145. Ferrohortonolite-ferrogabbro, at 2200 metres in the layered series, West ridge of Basistoppen. All pyroxene is ferriferous clinopyroxene. Interstitial micropegmatite occurs. Apatite is abundant (p. 104. Figured also as Pl. 16, fig. 1).

4142. Fayalite-ferrogabbro from the purple band, 2500 metres in the layered series, west ridge of Basistoppen. The pyroxene over the whole field extinguishes approximately simultaneously. The part, usually marginal, showing good cleavages is brown, the remainder greenish and crowded with indefinite green inclusions. Quartz, forming about three per cent. of the total rock, occurs in small patches in the clinopyroxene and between the felspar, where, however, it is not easily distinguished in ordinary light (p. 108).

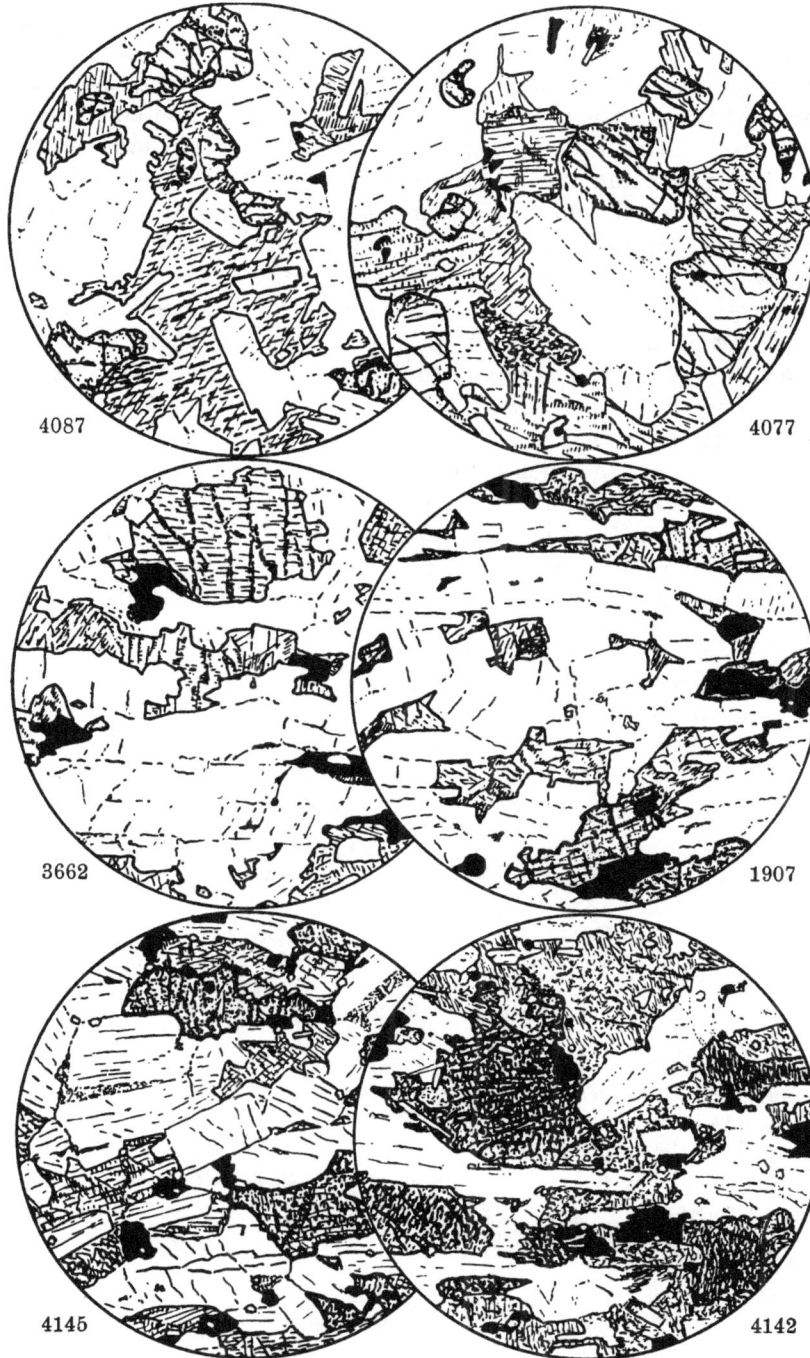

4087

4077

3662

1907

4145

4142

Fig. 24.

both may be found with an allotriomorphic relation to the plagioclase. These facts suggest that the period of crystallisation of both the olivine and clinopyroxene was more extended in the higher hypersthene olivine gabbros than in the lower. From the texture it may be inferred that the plagioclase, olivine, clinopyroxene and some of the iron ore were early crystallising minerals, but they probably did not finish early, as their outer fringes show evidence of simultaneous crystallisation with the hypersthene, towards the end of the solidification period.

Two rocks from Kraemers Ö, 2307 and 2308, are at about the same horizon as 1690. They contain similar minerals, but are somewhat finer-grained which is to be correlated with their nearness to the margin. In one, 2307, the compositions and proportions of the minerals are not very different from 1690 and the texture is similar. The rock 2308 from an adjacent melanocratic band has a greater proportion of olivine and pyroxene, and a much greater proportion of iron ore. In this rock the textures are somewhat different. Some of the ore is present as idio-morphic grains in the pyroxene and olivine, and the outer part of the plagioclase—no more zoned than usual—is frequently to be seen filling the interstices between the iron ore crystals. Both these features can be

TABLE X.

ANALYSES OF HYPERSTHENE-OLIVINE GABBROS.

	I	II	Norms				Modes (vol. %)		I	II
	(4087)	(4077)	I		II					
SiO_2	45.48	46.37	Or......	1.67		1.11	Plag.........		56	55
Al_2O_3	16.41	16.82	Ab.....	17.29		17.82	Clino-pyrox....		29	21
Fe_2O_3	2.09	1.52	An......	34.75		34.19	Ortho-pyrox...		3	5
FeO	9.29	10.44	Ne......	..		1.70	Oliv.........		11	17.5
MgO	11.65	9.61	Di.. 7.08	13.69	8.82	17.38	Ore..........		0.7	1.5
CaO	10.46	11.29		4.50	4.60		Apat..........		trace	trace
Na_2O	2.06	2.45		2.11	3.96					
K_2O	0.27	0.20	Hy. 0.30	0.43	..					
H_2O+	0.77	0.29		0.13			Ratios			
H_2O-	0.26	0.09	Ol. . 16.94	26.12	13.58	23.58				
TiO_5	0.94	0.79		9.18	10.00		$\dfrac{(FeO+Fe_2O_3)\times100}{MgO+FeO+Fe_2O_3}$		49	55
MnO	0.06	0.09	Ilm.....	1.67		1.52				
P_2O_2	0.05	0.06	Mt.	3.02		2.09				
	99.79	100.02	Ap.	0.17		0.34	$\dfrac{Fe_2O_3\times100}{FeO+Fe_2O_3}$ ...		18	18
			H_2O	1.03		0.38				
S.G.	3.00	3.04	Plag. $Ab_{33}An_{67}$		$Ab_{34}An_{66}$		$\dfrac{K_2O\times100}{Na_2O+K_2O}$ ...		12	8
			Diop. $Wo_{52}En_{33}Fs_{15}$		$Wo_{50}En_{27}Fs_{23}$					
			Hyp. $En_{70}Fs_{30}$		..					
			Oliv. $Fo_{65}Fa_{35}$		$Fo_{58}Fa_{42}$					

COMPARISONS.

	I	II	A	B	C	D	E	F
SiO_2	45.48	46.37	45.52	46.66	47.28	47.75	48.05	46.54
Al_2O_3	16.41	16.82	14.30	16.71	21.11	19.46	15.35	16.88
Fe_2O_3	2.09	1.52	3.43	2.69	3.52	2.31	1.86	3.20
FeO	9.29	10.44	9.00	5.87	3.91	6.28	7.53	7.41
MgO	11.65	9.61	10.65	12.36	8.06	7.90	12.53	9.77
CaO	10.46	11.29	9.54	12.57	13.42	11.32	11.02	9.54
Na_2O	2.06	2.45	2.21	1.16	1.52	2.46	1.26	3.14
K_2O	0.27	0.20	0.42	0.27	0.29	0.24	0.19	0.63
H_2O+	0.77	0.29	1.53	1.24	0.53	0.50	0.45	0.69
H_2O+	0.26	0.09	0.70	0.13	0.13	0.18	0.15	..
TiO_2	0.94	0.79	2.85	0.47	0.28	0.43	0.49	0.96
MnO	0.06	0.09	0.19	0.12	0.15	0.17	0.28	..
P_2O_5 . . . . ,.	0.05	0.06	0.23	0.13	trace	0.62	..	trace
CO_2	..	..	0.15	0.18	..	0.00	0.44	..
FeS_2	..	..	..	..	..	0.16	..	..
Cr_2O_3	..	..	..	..	0.02	0.05	0.14	..
	99.79	100.02	100.72	100.56	100.24[1]	99.83	100.10[2]	98.76

[1] + 0.02 V_2O_3.

[2] + 0.11 $(Ni,Co)O$; 0.05 CuO.

I. Hypersthene-olivine gabbro, 4087, 150 m from the northern margin, Uttentals Plateau.

II. Hypersthene-olivine gabbro, 4077, junction of Uttentals Plateau and W.N.W. ridge of Peak 1300 metres of Gabbrofjaeld.

A. Lava, basalt, E. side of Rudha Dearg, Morven, Mull [Bailey, Thomas etc. 1924]. Anal. F. R. Ennos.

B. Eucrite. S.E. of summit, Ben Buie, Mull [Bailey, etc. 1924] Anal. F. R. Ennos.

C. Olivine-gabbro. Floor of Coir'a'Mhadaidh Cuillins, Skye [Harker, 1904]. Anal. W. Pollard.

D. Hypersthene-gabbro. Ring-dyke. Beinn nan Codham, Ardnamurchan [Richey and Thomas, 1930]. Anal. B. E. Dixon.

E. Eucrite. N.E. of Summit of Allival, Rum, Inverness-shire [Harker, 1908]. Anal. W. Pollard.

F. Olivine dolerite from Krustorp, Breven, Sweden [Krokström, 1932]. Anal. K. Winge.

seen if carefully looked for in 1690 but were not noted before they had first been seen in this rock. They show quite clearly that iron ore, at this horizon, was capable of forming early in the crystallisation sequence, whereas in lower rocks, it seems to have been only a late mineral. In the melanocratic rock, 2308, hypersthene is rarer than in 2307 where it tends to build compact grains not much later than the clinopyroxene and olivine. Another melanocratic rock, 4063, from Kraemers Ö is from a horizon bordering on the olivine-free middle gabbros. Clinopyroxene and iron ore are abundant and hypersthene rare. Olivine in independent grains is also rare, but thin zones of a continuous crystal of olivine are frequently found enveloping the clinopyroxene and ore. This olivine and the outer part of the felspar crystals were clearly the last materials to crystallise. The structural inter-relations of the various minerals in this rock may differ from 1690 partly because it is more melanocratic, and partly because it lies on the actual dividing line between the hypersthene-olivine-gabbros and the overlying, olivine-free gabbros. Other rocks from the dividing line (e. g. 1920) show similar relations in their more melanocratic parts. The height in the intrusion at which olivine disappears is approximately nine hundred metres. Above this the layered rocks are olivine-free, and are classified as middle gabbros (Fig. 25).

The two analyses of hypersthene-olivine-gabbros are given in Table X, together with five somewhat similar rocks from the British Tertiary area and one from Sweden. The higher Greenland rock compared with the lower, shows slight increase in soda and iron, and decrease in magnesia—results which are to be expected from the changes in mineral composition just described. Comparison with the four plutonic rocks quoted shows that, in bulk composition, the lower rocks of the layered series are similar in many ways with other plutonic rocks of the province, the hypersthene-gabbro from Ardnamurchan (Anal D) being the closest. According to Thomas [Richey and Thomas, 1930, p. 224] this rock has similar iron-rich ferromagnesian minerals. None of the plutonic rocks selected for comparison have as much iron, particularly ferrous iron, as the Skaergaard rocks but it is possible to find them (see for example Table XII, Anal. E and Table XIII, Anal. F). On the whole the basic plutonic rocks, like the ones quoted, are relatively poor in iron, while basalts are frequently as rich, or even richer in iron than the hypersthene-gabbros of the Skaergaard intrusion; an example from Morven is quoted which belongs to the Plateau magma type (Anal. A). The analysis of olivine dolerite from Breven, Sweden, (Anal. F), is from a series of analyses of the Breven and Hällefors dykes described by Krokström [1932 and 1936]; many of these rocks have similarities with those of the Skaergaard intrusion and reference will often be made to them.

2. The Middle Gabbros.

Two middle gabbros have been analysed, — 3662, collected at the west foot of Pukugaqryggen about 25 metres above sea level and 3661, collected 50 metres higher. They come from about the middle of the middle gabbro series, their estimated height in the layered series being round 1200 metres. Our collection of middle gabbros from near the centre of the intrusion proved to be rather variable in the relative proportions of the different minerals, as can be seen from Table XI, which

shows the modes of five specimens collected at the west foot of Pukugaqryggen. The first rock which was analysed, 3662, is not representative
of the average composition of the middle gabbros, so that the adjacent,
3661, has also been analysed. The mean modal composition of these
two rocks is fairly close to the average for the five given in table XI,
and by averaging the two analyses a reasonable approach to the average
composition of the middle gabbro at this horizon is obtained.

A petrographic description of the lower analysed rock, 3662 (Fig. 24),
will first be given. The plagioclase is slightly zoned but the greater part
is An_{45}; it is subeuhedral and is more often idiomorphic than allotrio-

TABLE XI.

MODES OF MIDDLE GABBROS.

	3658	3660	3661	3662	1691	Average	Aver. of 3662, 3661
Plagioclase........	49	42	37	60	56	49	48
Pyroxene	44	49	51	35	27	41	43
Ore	7	9	12	5	16	10	8.5

morphic towards the pyroxene. The average size of the plagioclase
crystals is less than that of 1690 or 4077, and the dustiness due to opaque
inclusions is more conspicuous. Olivine is absent as independent grains
and also, in this particular rock, as a reaction rim. The clinopyroxene
has been separated and analysed, and its composition and optical
proportions are given in Tables VII and VIII. Under low magnification
rows of inclusions, apparently of hypersthene, can be seen as fine
sheets parallel to the basal plane, (001); these produce pale interference
colours in ordinary light (p. 81). Opaque rod inclusions of iron ore are
also abundant in the clinopyroxene, the best development being parallel
to (010). Hypersthene occurs only in small amount—about 2 per cent.
of the total rock—while there is 35 per cent. of clinopyroxene present.
From its refractive index ($\gamma = 1.732$), the hypersthene is determined
as Fs_{57}, but it is full of inclusions of clinopyroxene which makes the
determination only approximate. The hypersthene occurs, either surrounding the clinopyroxene, or in patches within the clinopyroxene
especially near the margin. It may have crystallised partly simultaneously with the outer part of the clinopyroxene or partly afterwards.
When the contact between the pyroxene and felspar is examined in
detail it is apparent that the felspar may fill in the interstices between
two pyroxene grains, or pyroxene may fill in the interstices between

the felspars. Iron ore also finished crystallising late as it usually has an interstitial habit. Sometimes a grain occurs well within the pyroxene which shows that it began crystallising fairly early. A feathery reaction rim occurs in places between iron ore and felspar. This effect is very sporadic, and although no conclusive evidence can be given we believe that it was probably produced at moderately low temperatures at points where volatile constituents happened to be concentrated.

The second middle gabbro which has been analysed, 3661, (Pl. 15, fig. 2) has apparently similar plagioclase, clinopyroxene and orthopyroxene. The pyroxene areas consist of clusters of grains of hypersthene and clinopyroxene in about equal amounts, and there does not seem to have been any significant difference in time of crystallisation between them. The clinopyroxene is crowded with sheet inclusions of orthopyroxene and the orthopyroxene is crowded with sheet inclusions of clinopyroxene, which are often seen following four different directions

TABLE XII.

ANALYSES OF MIDDLE GABBROS.

	III	IV	Norms				Modes (vol. %)		III	IV
	(3662)	(3661)	III		IV					
SiO_2	48.15	45.65	Or. 0.83		1.67		Plag.		60	37
Al_2O_3	18.02	15.08	Ab. 29.34		20.96		Clino-pyrox.		33	}51
Fe_2O_3	2.52	3.41	An. 32.94		29.19		Ortho-pyrox.		2	
FeO	9.50	14.86	Di. . . . 7.31	14.35	6.50	13.00	Ore		5	12
MgO	5.25	6.35	4.00		2.80		Apatite		tr.	tr.
CaO	10.17	9.18	3.04		3.70					
Na_2O	3.46	2.48	Hy. . . 2.40	4.25	5.20	12.02				
K_2O	0.14	0.28	1.85		6.82		Ratios			
H_2O+	0.20	0.22	Ol. . . . 4.69	9.28	5.60	13.15				
H_2O-	0.02	0.08	4.59		7.55					
TiO_2	2.64	2.59	Ilm. 5.02			5.02	$\dfrac{(FeO + Fe_2O_3) \times 100}{MgO + FeO + Fe_2O_3}$ . .		70	74
MnO	0.12	0.15	Mt. 3.71			4.87				
P_2O_5	0.05	0.08	Ap. 0.17			0.34				
CO_2	0.03	..	Cal. 0.05			..	$\dfrac{Fe_2O_3 \times 100}{FeO + Fe_2O_3}$		21	19
S	0.14	..	Pyr. 0.24			..				
ZrO_2	nil	..	H_2O 0.22			0.30				
SrO	0.07	..								
BaO	0.01	..					$\dfrac{K_2O \times 100}{Na_2O + K_2O}$		4	10
Cr_2O_3	nil	..	Plag. $Ab_{55}An_{45}$		$Ab_{40}An_{60}$					
CuO	0.006	..	Diop. $Wo_{50}En_{22}Fs_{28}$		$Wo_{50}Eb_{22}Fs_{28}$					
NiO	nil	..	Hyp. $En_{56}Fs_{44}$		$En_{43}Fs_{57}$					
	100.50	100.41	Oliv. $Fo_{51}Fa_{49}$		$Fo_{43}Fa_{57}$					
S. G.	3.01	3.19								

COMPARISONS.

	IVa	A	B	C	D	E	F
SiO$_2$	46.90	48.89	46.50	45.37	45.48	47.69	47.90
Al$_2$O$_3$	16.55	13.60	10.86	15.16	15.66	17.04	19.92
Fe$_2$O$_3$	2.96	4.72	2.70	3.38	3.64	3.77	4.92
FeO	12.18	9.65	12.77	11.58	10.56	9.39	9.78
MgO	5.80	4.54	6.27	6.72	6.99	4.56	4.55
CaO	9.67	10.09	11.04	8.11	8.24	9.17	8.56
Na$_2$O	2.97	2.74	2.69	2.90	2.68	2.89	2.75
K$_2$O	0.21	0.30	1.16	0.44	0.49	1.50	0.56
H$_2$O+	0.21	1.38	1.99	1.96	1.52	1.20	0.76
H$_2$O—	0.05	1.14	..	1.18	0.93	..	..
TiO$_2$	2.61	2.64	3.02	2.87	3.48	2.20	0.57
MnO	0.14	0.24	0.18	0.31	0.20	0.20	..
P$_2$O$_5$	0.06	0.39	0.64	0.29	0.26	0.20	..
CO$_2$	..	0.02	0.04	..	0.21	..	..
BaO	..	tr.	nil	nil	..	..	..
S	..	0.02	0.10	..	nil	..	..
	100.31	100.36	99.96	100.27	100.34	99.81	100.27

III. Middle gabbro, 3662, 25 m above sea-level W. foot of Pukugaqryggen.
IV. Middle gabbro, 3661, collected 50 m above 3662.
IVa. Average of 3662 and 3661.
A. Tholeiitic Basalt. Kap Dalton, E. Greenland [Wager, 1934]. Anal. H. F. Harwood.
B. Basalt, Kap Franklin, E. Greenland [Backlund and Malmquist 1932]. Anal. N. Sahlbom.
C. Lava, basalt. N.N.E. of Pennycross House, Mull [Bailey, Thomas etc. 1924]. Anal. E. G. Radley.
D. Lava, basalt. Cliff 200 yards W. of Lochaline Pier, Morven [Bailey, Thomas etc., 1924]. Anal. F. R. Ennos.
E. Glassy coarse-grained marginal dolerite, S.E. of Jacobsberg, Hallefors, Sweden. [Krokström 1936]. Anal. N. Sahlbom.
F. Diabase gabbro. East side of Birch Lake. Duluth gabbro complex [quoted from Grout 1918 p. 649]. Anal. A. N. Winchell.

corresponding with the four directions of faces of the general form. Where the abundant iron ore is deeply embedded in the pyroxene areas a wide reaction rim of olivine is found, but where it is on the outside of clusters of pyroxene, there is usually no reaction rim. Extremely local differences in composition of the residual liquid were apparently established towards the end of the crystallisation period causing reaction rims at certain places only. The olivine of the reaction rims in 3661 consists of many small crystals and not, as at lower horizons (e.g. 4063, 1920 and 1691), of a single crystal enveloping the greater part, or the whole of an iron ore grain. The refractive index of the olivine of the reaction rims has not been easy to measure, but the results indicate that the composition is approximately Fa_{48}. Feathery reaction products have developed between iron ore and felspar as in 3662. The absence of an olivine reaction rim in 3662 is no doubt to be correlated with the smaller amount of ore and pyroxene, the conditions corresponding with the outer parts of the pyroxene clusters of 3661.

A lower middle gabbro, 1691, from the shores of Uttentals Sund a little north of 3662 and about 100 metres below it, has hypersthene which is transitional in its manner of occurrence between that of the hypersthene olivine gabbros and that of the upper middle gabbros. It is a rock fairly rich in iron ore, and shows a good reaction rim of olivine between the pyroxene and iron ore. The olivine of the reaction rims is a single crystal for a distance several times the width of the rims, but not for such long distances as in the lower rocks 4063 and 1920 (pp. 93—94). A rock, 1738, from the west side of Uttentals Sund and at about the same horizon is very similar. Two middle gabbros, 3660 and 3658 from above the analysed examples on Pukugaqryggen, have areas of pyroxene consisting of grains of clinopyroxene and orthopyroxene of similar habit and apparently of the same period of crystallisation. Reaction rims of olivine are poorly developed in both. Middle gabbros collected from high on Gabbrofjaeld (1697, 1814 and 1815) are finer grained, in harmony with their nearness to the margin, but are otherwise very similar to the higher middle gabbros of Pukugaqryggen.

The average of the two analyses of middle gabbros (Table XII) is compared with two basalts from East Greenland (Anals A and B) and with two Plateau magma basalts from Mull (Anals C and D); the latter on the whole resemble the middle gabbros more closely. Comparable gabbros are not so common as comparable basalts but one is quoted from the Duluth complex (Anal F) and another is given in Table XIII (Anal D). Closely similar types exist in the Breven and Hällefors dykes and an example is quoted (Anal E).

3. Ferrogabbros—Types with Hortonolite.

The group of basic rocks to which we have given the name ferrogabbro are rich in iron which is largely present as ferrous iron in the olivines and pyroxenes. The term will be used for all the rocks of the main layered series above the middle gabbro, and also for the lowest member of the unlaminated layered series. These rocks contain olivine which varies from about Fa_{56} to almost pure fayalite and clinopyroxenes in which the percentage of $FeSiO_3$ varies from forty-five to seventy. The ferrogabbros, as so defined, are basic rocks with a silica percentage corresponding to the gabbros. Like normal gabbros they consist dominantly of plagioclase, olivine and pyroxene; they differ, however, in having ferromagnesian minerals which are iron-rich. In order to belong to the group we suggest that the olivine should have more than fifty molecular percentage of fayalite. If rocks, similar in ferrous iron content to the ferrogabbros as so defined, and containing iron-rich pyroxene but no olivine are some day found, it would probably be desirable to extend the definition given above so as to include them.

The lowest analysed ferrogabbro, 1907, at nineteen hundred metres in the layered series may be more precisely named hortonolite-ferro-gabbro from the composition of the olivine. Hypersthene, as a separate mineral, is absent, and iron ore is more abundant than in the lower hyper-sthene-olivine-gabbros; in other respects the proportion of the minerals is not very different. The rock shows moderate igneous lamination. Its specific gravity, 3.18, is fairly high, due partly to the ferriferous olivine and pyroxene and partly to the eight per cent. of iron ore. The plagio-clase is An_{40} from refractive index tests. Zoning is scarcely to be detected and needle-like inclusions of iron ore are not so conspicuous as in the middle gabbros. Since the pyroxenes and olivines have both been separated and analysed and the mode also determined the total amount of soda and alumina in these minerals can be estimated. The remainder of these oxides can with some certainty be ascribed to plagioclase whose com-position works out as An_{43}. The composition from refractive index deter-minations is An_{40} and the normative composition An_{50}. The presence of a little orthoclase in the plagioclase would lower the refractive index and may partly account for the lower value for anorthite obtained by the refractive index method. The olivine Fa_{61}, has been analysed and its properties recorded (Table VI). The texture of the rock indicates that the olivine probably began crystallising at about the same time as the plagioclase and pyroxene, and that, like them it continued crystallising late (Fig. 24). The pyroxene has also been analysed and carefully ex-amined optically (Tables VI and VII). It contains minute needle inclusions of iron ore, and sheet inclusions, which by analogy with the pyroxene of the middle gabbros are regarded as hypersthene. Iron ore is fairly abundant and is often idiomorphic towards pyroxene and olivine but the textures indicate that it also crystallised late. Although the olivine and pyroxene are particularly liable to exhibit obvious allotriomorphism (Fig. 24) we consider that all the minerals continued crystallising to a late stage. Reaction rims of olivine occur between the iron ore and pyroxene, and this olivine tends to be a continuous crystal for a con-siderable length of the rim. Feathery reaction rims are also sporadically developed between iron ore and felspar. In this rock quartz has not been found, either by itself, or as micropegmatite but it is present in small amounts in other rocks of this horizon. The olivine and pyroxene in

7*

the rocks containing free quartz have the same mean composition as in 1907. Late-stage reactions, such as marginal replacement of pyroxene by amphibole, are occasionally to be observed but are more conspicuous in the fayalite ferrogabbro from higher in the layered series.

Olivine, which is absent in the middle gabbros, is an important constituent of the ferrogabbros. The collection of rocks from Pukugaqryggen shows at what level in the intrusion, the olivine begins again to be an important mineral. Thus the rock 3658, at 1,300 metres in the intrusion, is a middle gabbro with olivine only present as thin reaction rims. In 3656, about 120 metres higher, a little altered olivine as separate grains is present, and in 3655, another 50 metres higher, olivine is abundant. The disappearance of olivine where the middle gabbros begin is also sudden. Thus in 1690, 4063, and others of this horizon, olivine is abundant while it is absent in 1691 and 4066. The estimated height in the intrusion at which olivine returns after its absence in the middle gabbros is 1,400 metres (Fig. 25). The rocks forming the coast north and northeast of the houses, lie on the border between the middle gabbros and the ferrogabbros; thus 1926 has no olivine; 1927 has a trace and 1254, at the northern house has abundant olivine. The boundary on the islet to the north of Ivnarmiut is also well defined as the specimen, 2270 from the centre of the island, has only a little independent olivine and must be about on the boundary.

Hypersthene is not present as a separate mineral in the analysed ferrogabbro 1907. It is present in 3656, the first olivine-bearing rock above the middle gabbro on Pukugaqryggen, and it is present in 3655, 60 metres higher. It is doubtful if it is present in 3654 and it is definitely lacking in 3653 and higher rocks. In 3655 the hypersthene forms 5 to 10 per cent. of the total pyroxene. Rocks collected a little inland from the coast, north and northeast of the Basishusene, also prove to be at about the limit of the hypersthene which is found to the extent of 10 per cent. of the total pyroxene in 1926 and 1927, and is absent in 1963, only 50 metres higher. The disappearance of the hypersthene is abrupt, like that of the olivine where the middle gabbros begin. Both these sudden changes are considered to be due to lower temperatures or a change in the composition of the magma causing the particular phase to be no longer precipitated. The height in the layered series at which hypersthene ceases to occur as a separate magmatic mineral is 1,600 metres, that is, about 200 metres above the reappearance of olivine after the olivine-free middle gabbros (Fig. 25). It should be noted that hypersthene as sheet inclusions in the clinopyroxene is probably present above this level but this hypersthene is ascribed to exsolution and not to direct precipitation from the magma.

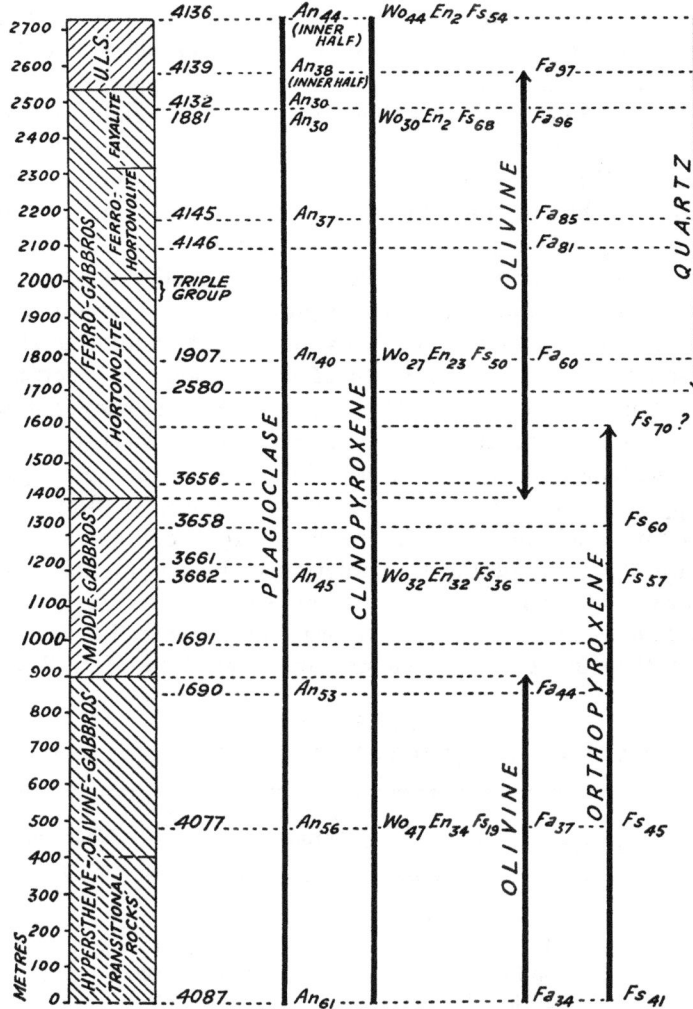

Fig. 25. Composition of the chief minerals of the layered series plotted against
height in the intrusion.

Quartz appears in certain rocks collected 100 metres above the level
of the disappearance of hypersthene (Fig. 25), but for 400 metres, it
occurs sporadically and only in small quantities—some rocks containing
it and others not. The lowest rock in which it has been found, 2580
is from the unbanded zone, one hundred metres below the analysed ferrogabbro 1907. It also occurs in 2568 and 1925 which are from leucocratic
bands of trough bands F and H, but it is not present in adjacent melanocratic bands, 2569 and 1924. Quartz has not been detected in the average
rock 1907, nor in 4265 collected some hundred metres higher but it is

found in the average rock 4272, and micropegmatite occurs in small amounts in rocks from the neck of the Skaergaards Halvö (e.g. 371, 1962 and 4299). On Pukugaqryggen quartz first appears as rare interstitial grains in 3651 which has the same estimated height in the layered series as 1907. It has not been found in 3650, sixty metres higher but is present in 3649, another sixty metres above. It is not to be seen in 3648 and 3647, but in these rocks some interstitial chlorite is found, in whose late stage formation a small amount of free quartz may have been absorbed. The incoming of quartz is a more gradual process than the reappearance of olivine after the middle gabbros. It is also a more gradual process than the disappearance of olivine at the beginning of the middle gabbro series, or the disappearance of hypersthene 200 metres higher. Although quartz first appears at 1,700 metres in the layered series, it only becomes a constant constituent above 2,200 metres.

The analysed hortonolite-ferrogabbro, 1907, contains just over twenty per cent. of total iron and it is difficult to find rocks with this amount which are not the result of iron ore concentration, in which case the value of Fe_2O_3 in relation to FeO is higher than in the Skaergaard rock. We have quoted two analyses of Greenland basalts and one from King Charles Land which approach the hortonolite-ferrogabbo in iron percentage and which have a similar low value for MgO (Table XIII,

TABLE XIII.
ANALYSIS OF HORTONOLITE-FERROGABBRO

V (1907)		Norm		Mode (vol. %)	
SiO_2	44.81	Or	1.67	Plag.	56
Al_2O_3	13.96	Ab.	22.53	Clinopyrox	20
Fe_2O_3	3.75	An.	22.24	Oliv.	16
FeO	16.66	Ne.	3.13	Ore.	8
MgO	5.54	Di	9.28 17.62	Apat.	tr.
CaO	8.53		2.80		
Na_2O	3.35		5.54		
K_2O	0.33	Ol.	7.98 22.06	Ratios	
H_2O+	0.34		14.08		
H_2O-	0.19	Ilm.	4.86	$\dfrac{(FeO + Fe_2O_3) \times 100}{MgO + FeO + Fe_2O_3}$	79
TiO_2	2.55	Mt.	5.34		
MnO	0.17	Ap.	0.34	$\dfrac{Fe_2O_3 \times 100}{FeO + Fe_2O_3}$	16
P_2O_5	0.08	H_2O	0.53		
	100.26				
S.G.	3.18	Plag.	$Ab_{50}An_{50}$	$\dfrac{K_2O \times 100}{Na_2O + K_2O}$	9
		Diop.	$Wo_{53}En_{16}Fs_{31}$		
		Oliv.	$Fo_{36}Fa_{64}$		

COMPARISONS

	V	A	B	C	D	E	F	G	H
SiO$_2$......	44.81	48.20	48.04	48.10	49.12	50.10	44.50	44.41	47.30
Al$_2$O$_3$.....	13.96	11.97	13.13	11.22	13.82	12.18	13.00	10.77	9.49
Fe$_2$O$_3$.....	3.75	4.52	6.89	6.94	6.76	4.35	8.25	5.51	5.54
FeO......	16.66	11.65	11.14	8.93	12.53	11.18	9.97	13.14	14.96
MgO	5.54	4.94	5.17	3.78	3.19	3.93	6.31	5.37	3.28
CaO......	8.53	10.21	10.87	8.09	8.70	8.85	11.10	9.98	7.60
Na$_2$O.....	3.35	2.39	2.83	2.56	2.49	3.06	1.88	2.60	2.76
K$_2$O......	0.33	0.79	0.06	1.55	1.26	0.96	0.25	1.08	2.69
H$_2$O+....	0.34	0.62	0.25	1.26	0.78	1.01	0.83	0.75	1.50
H$_2$O—....	0.19	..	0.98	0.96	..	0.53	0.29	..	..
TiO$_2$......	2.55	3.01	0.39	5.68	0.80	2.98	2.99	5.30	3.20
MnO	0.17	0.40	0.11	0.23	0.08	0.25	0.49	0.25	0.29
P$_2$O$_5$	0.08	0.92	0.07	0.60	..	0.17	0.06	0.76	1.44
CO$_2$	..	nil	0.79	nil	..	trace	0.14	..	..
BaO......	..	nil	..	0.03	..	0.05	nil	..	..
S	..	..	0.98	0.04	..	..	nil	..	..
FeS$_2$......	..	..	..	..	..	0.28	nil	..	..
Fe$_7$S$_6$.....	..	..	..	..	..	0.02	nil	..	..
	100.26	99.62	100.72	99.97	99.53	99.80	100.06	99.92	100.05

V. Hortonolite-ferrogabbro, 1907. Basishusene, Kangerdlugssuaq.

A. Dolerite. Clavering Island, E. Greenland [Backlund and Malmquist, 1932]. Anal. N. Sahlbom.

B. Basalt. Ovitak, West Greenland [quoted from Wolff 1931, Analysis No. 3 p. 932] Anal. G. Nauckhoff.

C. Basalt free from olivine. Hare Island, W. Greenland [Holmes, 1917]. Anal. H. F. Harwood.

D. Basalt, Kap Weissenfels. King Charles Land. Greenland [quoted from Wolff 1931. Analysis No. 12, p. 935]. Anal. N. Sahlbom.

E. Quartz-dolerite, Cone Sheet, Centre 2, Ardnamurchan [Richey, Thomas etc., 1930]. Anal. B. E. Dixon.

F. Fluxion Biotite-gabbro. Glendrian Ring-dyke, Centre 3, Ardnamurchan [Richey, Thomas etc., 1930]. Anal. E. G. Radley.

G. Glassy Halleförs dolerite. S. of Lake Skogssjön, Sweden [Krokström, 1936]. Anal. N. Sahlbom.

H. Halleförs Dolerite, Ölmstorp [Krokström, 1936]. Anal. N. Sahlbom.

Anals. B, C and D). A quartz-dolerite from Ardnamurchan (Anal. E) is rather similar but has a much higher silica percentage, and a gabbro (Anal. F) from the same locality has a similar percentage of total iron but is richer in Fe$_2$O$_3$ than the Skaergaard rock. The closest comparison we have found is with dolerites from the Halleförs and Breven dykes (Anals. G and H).

4. Ferrogabbros—Types with Ferrohortonolite.

A rock, 4145, collected 250 metres above the Basishusene on the west ridge of Basistoppen, is representative of the ferrohortonolite-ferrogabbros (Fig. 24 and Pl. 16, fig. 1). Its estimated height in the layered series is 2,175 metres. It is moderately laminated and has a specific gravity of 3.22. The amount of plagioclase in this rock has fallen to 45 per cent. while the amount of pyroxene has increased compared with average rocks from lower horizons. No new minerals have appeared but there is now always a little micropegmatite occupying interstices, and apatite has become an important constituent. The plagioclase, An_{37}, is not zoned except for a narrow border adjacent to the micropegmatite and here the zoning is strong. On the whole it is subeuhedral, and the crystals attain a length of 0.6 centimetres. The plagioclase does not contain the needle-shaped, opaque inclusions present at lower horizons. The olivine is crowded with ore inclusions while that from 4146, 100 metres lower, contains far fewer, and was therefore separated for analysis. The analysed olivine from 4146 is Fa_{81} (Table VI), while from a determination of 2V (= 53°) the olivine in 4145 is Fa_{85}. The clinopyroxene has a fairly strong brown colour but is only feebly pleochroic. It contains only a few iron ore inclusions and no inclusions of hypersthene are visible. Iron ore is present to the extent of about 6.5 per cent. and is sometimes idiomorphic towards the pyroxene, olivine and plagioclase. Pyroxene, olivine and iron ore may occupy some of the interstices between the felspar but others are occupied by micropegmatite which at this horizon is a constant constituent. Apatite is also a significant constituent amounting to 4 per cent. of the total rock. It occurs in euhedral crystals embedded in the olivine, pyroxene, iron ore and plagioclase but it tends to be in the outer part of these minerals. It is not particularly concentrated in the micropegmatite and is always idiomorphic towards it. It is thus one of the early minerals to crystallise.

The incoming of abundant apatite is abrupt and takes place at a height of 1,850 metres in the layered series (Fig. 18). Thus in the rock 3650 of Pukugaqryggen no apatite is seen while in 3649, 60 metres above, it is abundant, forming about 2 per cent. of the total rock. In 3648, 40 metres higher, it is abundant and present as rather big crystals up to 1 millimetre across. On the west face of Basistoppen and in the Skaergaarden region apatite is found to be rare in the lower ferrogabbros, eg. 1907 and 4299, but it is present abundantly in the higher, eg. 4265, 371 and 1962. Apatite is not absent from the lower rocks but is decidedly rare and the amount of P_2O_5 determined by analysis confirms this. It is significant that apatite appears only as an interstitial mineral of late crystallisation in the lower members of the layered series (cf. drawing of 4077, Fig. 24) while it occurs as a mineral of early crystallisation where it becomes abundant.

The disappearance of visible sheets of hypersthene inclusions in the clinopyroxene takes place at about 2,100 metres. Very fine sheets are to be detected in 4147 and in lower specimens but are not found above. The increase in abundance of the opaque inclusion of ore in the olivine is a more gradual process and more erratic in the height in the layered series at which it becomes important. Thus inclusions increase rapidly in amount in the rocks of Pukugaqryggen between 3650 and 3647, the latter having a dense mass of small opaque inclusions. Low on the west ridge of Basistoppen the inclusions in the olivine are coarse and not numerous but they become suddenly abundant in the analysed rock 4145, that is two hundred and fifty metres higher in the series than the point at which they become abundant on Pukugaqryggen. Some of the changes such as the disappearance or appearance of a phase happen at a constant horizon, but others such as the incoming of a dense mass of inclusions in the olivine, though a conspicuous feature in thin sections, appears at different heights in different parts of the intrusion and seems to be less directly associated with the steady differentiation processes which give rise to the cryptic layering.

The analysis of the ferrohortonolite-ferrogabbro (Tabel XIV) is compared with a basic epidolerite from the Breven dyke (Anal C). This analysis is the only one of which we are aware that is at all close to the Skaergaard rock, and it is a somewhat doubtful comparison since the Breven rock is obviously altered. Although certain oxides may have been modified in amount by metasomatic effects the analysis appears to represent a similar type of original rock, and it is interesting that the values for TiO_2 and MnO and the exceptionally high value of P_2O_5 are comparable in both. We have also added for comparison two iron-rich and magnesia-poor basalts which like the Skaergaard rock are low in potash (Anals A & B). Their silica percentage, however, is considerably higher than that of the ferrohortonolite-ferrogabbro. Nockolds [1937] has classified the fields of association of common rock forming minerals and the analysis of the ferrohortonolite-ferrogabbro 4145, (MgO : FeO = 6 : 94) belongs to section X of set I (rocks without appreciable potash and without hornblende). The analyses of the fayalite-ferrogabbros, 4142, 1881, and of the unlaminated layered rocks, 4139, 4136 and 4137 also belong to this section for which Nockolds could find no examples when he wrote his paper. In each case the plot of the analyses lies within the proper field for the mineral assemblages of these iron rich rocks.

TABLE

ANALYSES OF FERROHORTONOLITE-

	VI (4145)	VII (4142)	VIII (1881)	Norms	VI		VII	
SiO_2	44.61	44.13	48.27	Qu.		..		..
Al_2O_3	11.70	7.88	8.58	Or.		2.22		2.78
Fe_2O_3	2.05	4.05	4.06	Ab.		25.15		18.34
FeO	22.68	26.63	22.89	An.		17.51		10.29
MgO	1.71	0.25	1.21	Di.	6.03	12.67	12.06	26.38
CaO	8.71	10.03	7.42			0.70	0.20	
Na_2O	2.95	2.15	2.65			5.94	14.12	
K_2O	0.35	0.47	0.34	Hy.	1.90	17.87	0.30	24.85
H_2O+	0.22	0.30	1.13			15.97	24.55	
H_2O-	0.20	0.19	0.37	Ol.	1.12	11.94	0.07	2.83
TiO_2	2.43	2.48	2.20			10.82	2.76	
MnO	0.21	0.48	0.26	Ilm.		4.56		4.71
P_2O_5	1.85	1.61	0.65	Mt.		3.02		6.03
CO_2	0.04	..	..	Ap.		4.37		3.70
S	0.31	..	..	Calc.		0.10		..
ZrO_2	0.01	..	..	Pyr.		0.54		..
SrO	0.08	..	..	H_2O		0.42		0.49
BaO	0.02	..	..					
Cr_2O_3	trace	..	..					
CuO	0.016	..	..					
NiO	trace	..	..					
	100.15	100.65	100.03	Plag.	$Ab_{59}An_{41}$		$An_{65}An_{35}$	
				Diop.	$Wo_{48}En_5Fs_{47}$		$Wo_{45}En_1Fs_{54}$	
				Hyp.	$En_{11}Fs_{89}$		En_1Fs_{99}	
S.G.	3.22	3.39	3.18	Oliv.	$Fo_{10}Fa_{90}$		Fo_2Fa_{98}	

COMPARISONS

	VI	A	B	C	VII
SiO_2	44.61	52.52	50.36	40.76	44.13
Al_2O_3	11.70	12.28	13.80	10.27	7.88
Fe_2O_3	2.05	} 21.55	19.22	12.28	4.05
FeO	22.68			14.57	26.63
MgO	1.71	1.26	1.16	3.30	0.25
CaO	8.71	8.67	10.22	8.47	10.03
Na_2O	2.95	2.71	1.92	0.61	2.15
K_2O	0.35	0.29	0.07	0.97	0.47
H_2O+	0.22	..	..	4.06	0.30
H_2O-	0.20	..	..	1.15	0.19
TiO_2	2.43	..	..	2.20	2.48
MnO	0.21	..	..	0.55	0.48
P_2O_5	1.85	..	..	1.68	1.61
CO_2	0.04	..	..	..	..
S	0.31	..	..	..	..
ZrO_2	0.01	..	..	..	..
SrO	0.08	..	..	..	..
BaO	0.02	..	..	..	..
Cr_2O_3	trace	..	..	..	..
CuO	0.016	G.-V.	G.-V.	..	..
NiO	trace	1.00	0.31	..	..
	100.15	100.28	97.07	100.87	100.65

XIV.
AND FAYALITE-GABBROS

	VIII		Modes (vol. %)	VI	VII	VIII
		4.44	Qu.	1	6[1])	11[2])
		1.67	Plag.	45	24	29
		22.53	Clino-pyrox.	28	37	34
		10.56	Oliv.	17	21	17
	9.28	19.58	Ore	6.5	9	8
	0.80		Apat.	2.5	3	1
	9.50					
	2.20	28.07				
	25.87					
		..	Ratios			
		..				
		4.26	$\dfrac{(FeO + Fe_2O_3) \times 100}{MgO + FeO + Fe_2O_3}$	94	99	96
		6.03				
		1.68				
		..				
		..	$\dfrac{Fe_2O_3 \times 100}{FeO + Fe_2O_3}$	8	13	15
		1.50				
			$\dfrac{K_2O \times 100}{Na_2O + K_2O}$	11	22	9
	$Ab_{68}An_{32}$					
	$Wo_{47}En_4Fs_{49}$					
	En_8Fs_{92}					
	..					

[1]) Includes some intergrowth of quartz and felspar.
[2]) Includes marginal felspar, probably perthite.

VI. Ferrohortonolite-ferrogabbro, 4145. W. ridge of Basistoppen. 250 m. above Basishusene.

VII. Purple Band, fayalite-ferrogabbro, 4142. At 550 m west face of Basistoppen.

VIII. Purple Band, fayalite-ferrogabbro, 1881. 10 m from top of more easterly nunatak to the south of Forbindelsesgl.

A. Basalt, Hvammur unter Baula, Iceland [quoted from Wolff, 1931, Analysis No. 4 p. 945]. Anal. Kjerulf.

B. Dolerite, Keflavik, Reykjanes [quoted from Wolff, Analysis No. 29, p. 948], Anal. Kjerulf.

C. Basic Epidolerite. W.N.W. Högsater, Breven, Sweden [Krokström, 1932]. Anal. N. Sahlbom.

5. Fayalite-Ferrogabbros, excluding those from the Unlaminated Layered Series.

The purple band horizon occurs within the fayalite-ferrogabbro series. On the west face of Basistoppen it is about 150 metres thick and its estimated height in the layered series is 2,350—2,500 metres. This band grades into the other rocks of the layered series, especially downwards, yet it was mapped in the field with much greater precision than the other subdivisions and, in working out the relative heights of many of the rocks in the upper ferrogabbros, the purple band has been used as a datum line.

Two examples of the purple band have been analysed—4142 from the west face of Basistoppen at an estimated height of 2,375 metres in the layered series, and 1881 from about the same height in the layered series and collected 10 metres below the highest exposure on the more easterly nunatak south of Forbindelsegletschen (Nunatak II). The localities of these two rocks are two and a half kilometres apart but they are strikingly similar in chemical and mineralogical composition. The pyroxene of 1881 is almost free from inclusions while they are abundant in the pyroxene of 4142, and for this reason the pyroxene of the former rock has been separated and analysed (Tables VII and VIII).

The more typical example of the purple band, 4142, will be described first. Like all the examples from this horizon it is feebly laminated and extremely tough. Its density is 3.39. Micrometric analysis shows that it contains forty per cent. of leucocratic constituents—felspar and quartz—but it would be estimated from the hand specimen to be far more melanocratic than this. The moderately euhedral plagioclase (Fig. 24) has been determined by refractive index determinations as An_{30}. Outside the narrow border zone there is a felspar which is not repeatedly twinned and may be perthite. Orientated inclusions of iron ore are lacking in the felspar. Quartz, which is not abundant, occurs in small irregular patches within the pyroxene and also in interstitial patches among the felspar, where it may be intergrown with felspar to give micropegmatite. The olivine crystals are often larger than the plagioclase. They are crowded with orientated ore inclusions which are sometimes less numerous near the margin and allow the clear yellow colour of the olivine to be seen. The olivine is probably of the same composition as that in 1881 which from optical tests is Fa_{96}. The form of the olivine suggests that it had the same period of crystallisation as the other minerals. The clinopyroxene has a habit which may be described as poikilitic and which is therefore very different from that in the lower layered rocks. It occurs in extensive, irregular masses, all parts of which have approximately the same orientation. In 4142 these are often as much as one centimetre across; they contain olivine and felspar crystals of the size usually found in the ferrogabbros and also many small pieces of quartz. In the nearby rock, 1906, such poikilitic crystal are as much as two centimetres in length, while the felspars only reach 0.6 centimetres. Parts of these masses, usually away from the margins, are greenish in colour and are cloudy with indefinite, greenish inclusions of irregular orientation. Other parts, which are usually marginal, are brownish in colour and free from inclusions. Cleavage cracks which are conspicuous in the brown parts persist in the green but are not easily seen. The clinopyroxene which was separated from 1881 and analysed, proved to be an almost pure hedenbergite-clinoferrosilite solid solution. The refractive index is 1.772 and since γ' for the pyroxene of 4142 is 1.770 we may safely conclude that the latter pyroxene is also a heden-bergite-clinoferrosilite solid solution. Iron ore is fairly abundant and is usually one of the more idiomorphic minerals. Apatite also occurs in abundance as well shaped crystals enclosed in all the other minerals.

The second analysed example of the purple band, 1881 (Pl. 16, fig. 2), is on the whole very similar to 4142. The plagioclase is for the most part An_{30} but outside this is a zone, usually much wider than in 4142, of acid felspar which is free from repeated twinning and cloudy with slight decomposition products. The rock contains eleven per cent. of interstitial quartz and, unlike that of 4142, it has no tendency to form a micrographic intergrowth with felspar. The olivine has $\gamma = 1.875$ and $2V = 48°$ which indicates that its composition is Fa_{96}. It has a good pinacoidal cleavage and occurs in clear, yellow, scarcely pleochroic, crystals which are free from ore inclusions. It is partly replaced by chlorite and iron ore, these minerals being developed in relatively large crystals. The pyroxene, for the most part a clear green colour but here and there a light purple brown, has been separated and analysed (Tables VII and VIII). Excluding the sesquioxides, which are not abundant, the pyroxene is an almost pure member of the hedenbergite clinoferrosilite solid solution series. In habit this pyroxene is essentially the same as that in 4142; it occurs in poikilitic masses or in ragged branching crystals appearing in the slice as isolated areas with the same orientation. In detail the pyroxene is often seen to be subeuhedral towards the quartz but it is usually conspicuously allotriomorphic towards the plagioclase. Like the olivine this pyroxene is free from orientated iron ore inclusions, and it is also free from the minute indefinite inclusions which make the pyroxene of 4142 unsatisfactory for analysis or optical determinations. Iron ore is fairly abundant in the rock as an idiomorphic, early con-stituent and it has also developed during the late stage replacement of the fayalite. Apatite is fairly abundant and occurs in well-shaped crystals embedded in all the other minerals. There has been a considerable amount of late stage reaction—about a fifth of the olivine having been replaced by brownish serpentine or chlorite—while a greener chlorite occupies spaces among the pyroxene crystals and in some cases seems to replace felspar. The analysis shows 1.50 per cent. of water while 4142 and the analysed rocks from lower in the series usually have a water content of about 0.5 per cent. The higher water content of 1881 can not be regarded as the result of weathering but must be water originally present in the magma.

A rock, 1884, from the same horizon as 1881 but 100 metres to the east, is very similar except that the fayalite and pyroxene occur in smaller crystals and the fayalite is about twice as abundant as pyroxene while in 1881 the ratio is reversed. About half of the pyroxene is brown. Another rock, 1885, from about 50 metres below 1881 was regarded as belonging to the purple band when mapping. It proved on sectioning to be very similar and to contain green and brown pyroxene in moderate sized poikilitic masses. A rock, 1883, lying a further 100 metres below, contains brown non-poikilitic pyroxene as in the ferro-hortonolite bearing rock 4145.

On the west face of Basistoppen below 4142 the strong purple colour of the weathered rocks decreases but 4143, collected about 80 metres below 4142 has been included within the purple band division. In the size of the felspar crystals and their abundance the rock is similar to 4142 although in the hand specimen it appears much less melanocratic. The plagioclase, from refractive index determinations, is An_{31}, a value which cannot be regarded as significantly different from that of 4142. As usual the plagioclase is only zoned close to the margin, and in this rock there is very little felspar present which is not repeatedly twinned. Quartz is also not abundant. The olivine and its inclusions are similar to those of 4142. The pyroxene, however, shows a remarkable new feature: it occurs in patches of about the same size and distribution as in 4142 and in the lower ferrohortonolite ferrogabbros such as 4145 but these patches consist of small, interlocking and disoriented grains of pale green colour, with here and there a narrow border of brown pyroxene. Two other rocks, 1713 and 1974 collected from about this horizon but one and three-quarter kilometres south-west, contain similar pyroxene in the form of interlocking grains (Pl. 16, figs. 1 and 2). Another rock, 4132, from the purple band on the nunatak immediately east of Basistoppen (Nunatak I) shows similar features except that the pyroxene grains are smaller and not so obviously interlocking. The pyroxene which occurs characteristically as small interlocking grains is difficult to examine optically but it can be shown to have oblique extinction, and approximately the same + ve optic axial angle as that of 1881. Several determinations of γ' gave an average value of 1.769. These properties are sufficient to show that this clinopyroxene has essentially the same composition as that of 1881.

The remarkable structure shown by the pyroxene of rocks low in the purple band, e. g. 4143, 1713, 1974 and 4132, implies that the pyroxene now present has formed by inversion from some other mineral. From the experimental work of Bowen, Schaïrer and Posnjak [1933] it is known that the material to separate from melts of a wide range of composition in the system CaO-FeO-SiO_2 is β-wollastonite of the solid solution series $CaSiO_3$—$FeSiO_3$, and that this phase inverts at a lower temperature to a member of the hedenbergite-clinoferrosilite solid solution series. Combining the evidence from the rocks and the experimental work, we suggest that the solid phase, which has inverted to give the small, interlocking grains of pyroxene of the hedenbergite-clinoferrosilite solid solution series, was an iron rich β-wollastonite. In 1713 and similar rocks the inversion apparently took place from very many centres scattered through the β-wollastonite, and thus a large number of interlocking pyroxene crystals were formed. In rocks such as 4142 which come from slightly higher horizons, only a few centres of recrystallisation must have been formed, and the inversion, spreading from these, gave clinopyroxene in extensive poikilitic masses.

The poikilitic pyroxene formed by inversion is usually a pale green colour, and it is cloudy with inclusions which are so small and indefinite that their nature has not been made out. Associated with this pyroxene, as small internal patches or as a discontinuous border, is a brownish pyroxene in optical continuity with the green, and of approximately the same refractive index. The brownish pyroxene is free from inclusions and has not been found as small interlocking grains, so that there is no evidence to show that it has inverted from β-wollastonite. It seems likely that it was formed at a slightly later stage when the temperature was sufficiently low for clinopyroxene to be directly precipitated instead of β-wollastonite. If this is so then the β-wollastonite must have inverted to the greenish pyroxene at the appropriate temperature and with no appreciable lag since the brown pyroxene, which seems to have formed only a little later, was able to crystallise in optical continuity with it. The analysed pyroxene from 1881 is largely the green type with perhaps 5 or 10 per cent. of the brown. The relative freedom from inclusions of the green part of this particular pyroxene from 1881 may be due to the greater amount of water contained in the rock which acted as a flux for the inversion, causing a purer and better crystallised product. Although lacking the vague cloudiness due to inclusions which is found in the other pyroxene formed by inversion, there seems no reasonable doubt that the analysed pyroxene was also formed by inversion as it is present in extensive poikilitic masses.

On the west face of Basistoppen a hundred metres below 4143, which contains the pyroxene in small interlocking grains, a rock, 4144, was collected in which the pyroxene is in large brown crystals as in the ferrohortonolite-ferrogabbro 4145. This pyroxene was clearly not formed by inversion but was precipitated directly from the magma. Similarly the pyroxene of 1883 which lies only 150 metres below the analysed purple band, 1881, is the usually brown type which was not formed by inversion. There was apparently a definite horizon at which iron-rich β-wollastonite was deposited from the magma instead of clinopyroxene and this lies at approximately 2,350 metres on the scale of heights. There is only a narrow, and probably discontinuous horizon at which the pyroxene is entirely present as areas of interlocking grains. Traced upwards the grains increase in size; by 2,450 metres they have been replaced by the extensive poikilitic plates found in 4142, 1881, 1906, etc.

The analyses of 4142 and 1881 are given in Table XIV. The differences between them are not great but the analysis VII of 4142 must be regarded as the more representative of the horizon. Some of the differences may be due to the high water content of 1881 and the late stage decomposition which has taken place. The only analysis which we have found comparable with either of them is that from the Breven dyke which has already been compared with the ferrohortonolite-ferrogabbro.

6. The Unlaminated Layered Rocks.

On the west face of Basistoppen there are about 200 metres of unlaminated layered rocks between the purple band and the base of the raft-like inclusion of spotted gabbro. The lower third consists of unlaminated fayalite-ferrogabbro, and a specimen, 4139, from this horizon has been analysed. The fayalite-ferrogabbro is succeeded fairly abruptly by a light weathering and finer-grained, intermediate rock without olivine and with an abundant micropegmatitic mesostasis. This type, of which 4137 is an analysed example, contains a pyroxene close to hedenbergite and we shall call it basic hedenbergite-granophyre. By the gradual incoming of large, strongly-zoned, plagioclase crystals, ranging from basic andesine to albite, the granophyre passes into a hedenbergite-andesinite which forms a thin discontinuous layer beneath the spotted gabbro raft or injects it in narrow veins. Roughly the same sequence, but telescoped to about 50 metres, is found on the east face of Basistoppen and on Nunatak I. The north face of Basistoppen is difficult of access but it appears that there the unlaminated layered rocks are represented by a thin zone of coarse ferrogabbro with large fayalite crystals and rather abundant micropegmatite. Between the purple band on Nunatak II and the upper border rocks of Kilen there is not room for more than a few metres of rock. Exposures are poor but a coarse fayalite-ferrogabbro, similar to that from the north face of Basistoppen, was collected here. The thickness of the unlaminated layered rocks is variable and they probably only extended over a small part of the intrusion. They are well developed on Basistoppen, and this may be connected with the presence of the raft of sill gabbro.

The analysed, unlaminated ferrogabbro, 4139, at a height of about 2,600 metres in the layered series, contains only 35 per cent. of leucocratic minerals. Olivine is conspicuous in hand specimens as it forms large crystals up to 1 centimetre long. The plagioclase is less euhedral than at lower horizons. It is dominantly An_{30} but has a narrow and strongly zoned border. Interstitial micropegmatite mounts to 9 per cent. of the rock. The olivine which has been separated and analysed (Table VI) contains an appreciable amount of the larnite molecule; excluding this it is Fa_{97}. It has abundant orientated ore inclusions which partly mask its yellow colour and give it the appearance of being pleochroic. Inclusions of plagioclase, iron ore and apatite are abundant in it. The pyroxene is green except occasionally at the margin where the brown type is developed. The green part is crowded with indefinite greenish-brown inclusions, and some iron ore inclusions also occur. The pyroxene tends to form extensive poikilitic areas, and it is likely that the green part was formed by inversion. Textural evidence suggests that plagioclase, olivine, iron ore and apatite had a somewhat earlier period of crystallisation than much of the pyroxene or iron-rich β wollastonite which seems to have crystallised only slightly earlier than the micropegmatite; this, however, may be a misleading appearance due to the inversion. In mineralogical and chemical composition the fayalite-ferrogabbro from the unlaminated layered series is very similar to that from the purple band. The main differences are textural, and they probably indicate a slightly different manner of origin.

The light-coloured, basic hedenbergite-granophyre, represented by 4137 and 1905, seems to be the dominant material for 100 metres above the fayalite-ferrogabbro just described, but exposures of rock in situ are rare due to scree cover. In the analysed rock, 4137, the bulk of the plagioclase occurs as thin laths up to 1 centimetre long, but a few larger crystals are 1.5 centimetres long, thus resembling those of the andesinite. The inner half of the large, plagioclase crystals is An_{40}, and the outer half shows strong zoning, an appreciable amount being as acid as An_{25}. About 30 per cent. of the rock is plagioclase. Neither olivine, nor the products of its decomposition are present. The pyroxene occurs in ragged, small crystals of green colour which separated after the early plagioclase and simultaneously with the later. The pyroxene shows no evidence that it formed by inversion; it is apparently similar to that in 4136,

TABLE

ANALYSES OF UNLAMIN-

	IX (4139)	X (4137)	XI (4136)	Norms			
				IX		X	
SiO₂	45.19	52.13	55.30	Qu.	0.48		6.84
Al₂O₃	9.37	15.87	18.52	Or.	2.78		8.34
Fe₂O₃	5.78	5.61	2.18	Ab.	20.44		30.92
FeO	23.77	11.17	7.47	An.	13.34		22.80
MgO	0.43	1.11	0.21	Di.	11.14 23.72	0.81	1.67
CaO	9.05	5.80	8.20		0.30	0.20	
Na₂O	2.43	3.63	6.01		12.28	0.66	
K₂O	0.49	1.38	0.78	Hy.	0.80 25.22	2.60	16.59
H₂O+	0.57	0.86	0.41		24.42	13.99	
H₂O−	0.31	0.25	0.06	Ol.	..		..
TiO₂	1.67	1.14	0.94				
MnO	0.32	0.30	0.09	Ilm.	3.19		2.13
P₂O₅	0.91	0.70	0.42	Mt.	8.35		8.12
	100.29	99.95	100.59	Ap.	2.02		1.68
				H₂O	0.88		1.11
S.G.	3.30	2.97	2.82				
				Plag.	Ab₆₀An₄₀	Ab₅₈An₄₂	
				Diop.	Wo₄₇En₁Fs₅₂	Wo₅₀En₁₂Fs₃₈	
				Hyp.	En₃Fs₉₇	En₁₆Fs₈₄	
				Oliv.	..	..	

COMPARISONS.

	X	A	XI	B	C	D	E
SiO₂	52.13	52.16	55.30	57.98	55.76	56.64	57.34
Al₂O₃	15.87	11.95	18.52	13.58	16.55	12.84	24.90
Fe₂O₃	5.61	4.86	2.18	3.11	3.10	4.06	1.10
FeO	11.17	9.92	7.47	8.68	6.02	8.43	0.94
MgO	1.11	3.77	0.21	2.87	1.08	2.01	0.25
CaO	5.80	7.14	8.20	2.01	3.23	4.08	7.99
Na₂O	3.63	2.36	6.01	3.56	6.28	3.74	5.37
K₂O	1.38	1.74	0.78	3.44	3.87	3.17	1.23
H₂O+	0.86	1.95	0.41	} 2.47 {	0.95	0.90	0.33
H₂O−	0.25	0.56	0.06		0.80	0.52	..
TiO₂	1.14	3.25	0.94	1.75	1.78	1.88	..
MnO	0.30	0.18	0.09	0.13	0.22	0.15	..
P₂O₅	0.70	0.24	0.42	0.29	0.40	0.89	tr.
CO₂	..	0.18	..	..	0.03	0.11	..
S	..	0.18	..	0.04	..	0.24	0.40
BaO	..	..	..	..	0.07	0.10	..
	99.95	100.44	100.59	99.91	100.14	99.76	100.25

XV.

ATED LAYERED ROCKS.

XI			Modes	IX	X	XI
	..		Qu...........................	9[1])	39[2])	26[2])
	4.45		Plag...........................	26	28	56
	50.83		Clino-pyrox...................	33	22[3])	15
	21.13		Oliv...........................	18	..	..
7.08	15.10		Ore...........................	12	10	2
0.10			Apatite.......................	2	1	1
7.92						
0.20	1.39					
1.19						
0.14	1.16		Ratios			
1.02						
	1.82		$\dfrac{(FeO + Fe_2O_3) \times 100}{MgO + FeO + Fe_2O_3}$	99	94	98
	3.25					
	1.01					
	0.47		$\dfrac{Fe_2O_3 \times 100}{FeO + Fe_2O_3}$	19	33	23 ·
	..					
$Ab_{71}An_{29}$						
$Wo_{47}En_1Fs_{52}$			$\dfrac{K_2O \times 100}{Na_2O + K_2O}$	17	28	12
$En_{14}Fs_{86}$						
$Fo_{12}Fa_{88}$						

[1]) Includes some ore.

[2]) This figure is for material which is interstitial to the early well shaped plagioclase; it includes an outer border of felspar in graphic intergrowth with quartz and also clear quartz and some apatite and chlorite.

[3]) Includes little chlorite.

IX. Fayalite-ferrogabbro, 4139, 100 m above the "Purple Band" W. face of Basistoppen, Kangerdlugssuaq.

X. Basic hedenbergite-granophyre, 4137. W. face of Basistoppen, 200 m above "Purple Band", Kangerdlugssuaq.

XI. Hedenbergite-andesinite, 4136. W. face of Basistoppen, 250 m above "Purple Band", Kangerdlugssuaq.

A. Quartz-dolerite, cone-sheet, Cruachan Dearg, Mull [Bailey Thomas etc., 1924]. Anal. F. R. Ennos.

B. "Intermediate rock", selected to show the type of rock between gabbro and "red rock". Pigeon Point (Anal. W. F. Hillebrand).

C. Lava, mugearite, 290 yards E. of Kinloch Hotel, Mull [Bailey, Thomas etc., 1924]. Anal. E. G. Radley.

D. Fayalite-monzonite, Korsudden, Nordingra, Sweden [Washington, 1919, p. 352, No. 26].

E. Andesine rock. Alvaerstrommen near Bergen, Norway [Washington 1919, p. 301, No. 7].

which has been analysed and proves to be close to hedenbergite in composition (Table VII). Iron ore is abundant as small crystals embedded in the pyroxene or micropegmatite but not in the central part of the plagioclase. The micropegmatite which makes up 30 per cent. of the rook is composed of quartz (in unusual acicular crystals built up into skeletal groups) and cloudy felspar, probably perthite. The rock contains 1.38 per cent. K_2O which is equivalent to 8.4 per cent. of orthoclase; this is the highest amount present in any of the layered rocks. Apatite is not abundant but occurs as inclusions in all the other minerals. The later-crystallising, leucocratic part of the unlaminated fayalite-ferrogabbro is rather similar to the basic hedenbergite-granophyre, and addition of fayalite and pyroxene to the basic hedenbergite-granophyre would give a rock very close to the fayalite-ferrogabbro. We have temporarily used the name basic hedenbergite-granophyre in describing this rock in order to avoid making a new name. It is much poorer in orthoclase and quartz than a typical granophyre and richer in ferrous iron and lime. A very similar rock occurs on the east face of Basistoppen and on Nunatak I.

The analysed hedenbergite-andesinite from the west side of Basistoppen immediately below the raft of sill gabbro, contains about sixty per cent. of large plagioclase crystals up to 1.5 centimetres long, and these are set in a base which is essentially the same as the basic hedenbergite-granophyre, 4137. There is a gradual passage from basic granophyre into andesinite due to the increase in the number of large plagioclase crystals, a few of which were found in the analysed basic hedenbergite-granophyre. The central half of the large plagioclase crystal is An_{40} and the crystals are strongly zoned. Surrounding the plagioclase is a micropegmatite often with the acicular type of quartz described from 4137. The pyroxene which has been analysed and described (Table VII and p. 81) is a greenish brown variety close to hedenbergite. It shows no evidence of having been formed by inversion from wollastonite but occurs sporadically in large crystals and as small granular masses widely scattered in the mesostasis. Iron ore occurs as euhedral grains but is not abundant. Apatite is not so abundant as in the ferrohortonolite-ferrogabbros. Hedenbergite-andesinite is also developed to a small extent on the east face of Basistoppen and on Nunatak I. It has also been found veining rocks of the Basistoppen spotted gabbro raft on the south side of Basisgletschen.

Analyses of unlaminated layered rocks are given in Table XV. The basic hedenbergite-granophyre is roughly comparable with a quartz dolerite from Mull which is regarded as belonging to the Nonporphyritic Central magma. It is also fairly similar to a rock intermediate between the Duluth gabbro and the Red Rock. The hedenbergite-andesinite resembles in some ways the mugearites (Anal. C), the main difference, as would be expected, being more lime and less potash in the andesinite. There are similar differences between the analyses of the Skaergaard andesinite and a fayalite-monzonite from Sweden (Anal. D). An andesine rock from Norway (Anal. E) is fairly similar in lime and alkalies but has much less iron.

(e) Extreme Rocks of the Layered Series.

Rhythmic layering is present in all the rocks of the layered series except the fayalite-ferrogabbros and unlaminated layered rocks but it varies much in conspicuousness. Since the average rocks described in the previous section are simply parts of the layered series where, so far as can be estimated by eye, the light and heavy constituents are evenly mixed, it is considered safe to assume that these is no essential difference in manner of origin between the average rocks and the banded rocks and certain conditions of formation which can be deduced for the latter, will be extended to the whole series.

The mean density of three average hortonolite-ferrogabbros is 3.19 (see Table IX). Two extreme leucocratic bands from the same horizon (1925 and 2568) have densities of 2.77 and 2.75 which is not far removed from the specific gravity of andesine, 2.67. The specific gravities of two melanocratic bands (1924 and 2569) adjacent to these two leucocratic bands are 4.03 and 3.74 respectively. These examples are sufficient to show how very different in composition, adjacent extreme bands may be. The leucocratic layers are always due to concentration of plagioclase: the melanocratic layers are mainly due to the abundance of iron-rich pyroxene and olivine but there is usually also some extra concentration of iron ore. Melanocratic bands composed dominantly of iron ore, though common in other layered complexes, are not a normal feature of the Skaergaard intrusion and were only found at a few places on Kraemers Ö in association with inclusions. The rock, 2308, from the hypersthene-olivine-gabbro horizon which has a density of 4.05 (Table IX) is an example of a rock rich in iron ore.

Rhythmic layering, which is conspicuous in extensive rock faces (Pl. 6), becomes inconspicuous in hand specimens and not readily appreciable in thin sections, except in the case of the extreme bands at about the horizon of the Base Houses (Figs. 26 and 27). Only examples of strongly differentiated bands will be described, since these alone show more than quantitative differences from the average rocks. A melanocratic band, 1924, from the trough band F, consists of grains of olivine and pyroxene which are present in about equal amounts and together form three quarters of the rock (Fig. 27). As seen in thin section most of

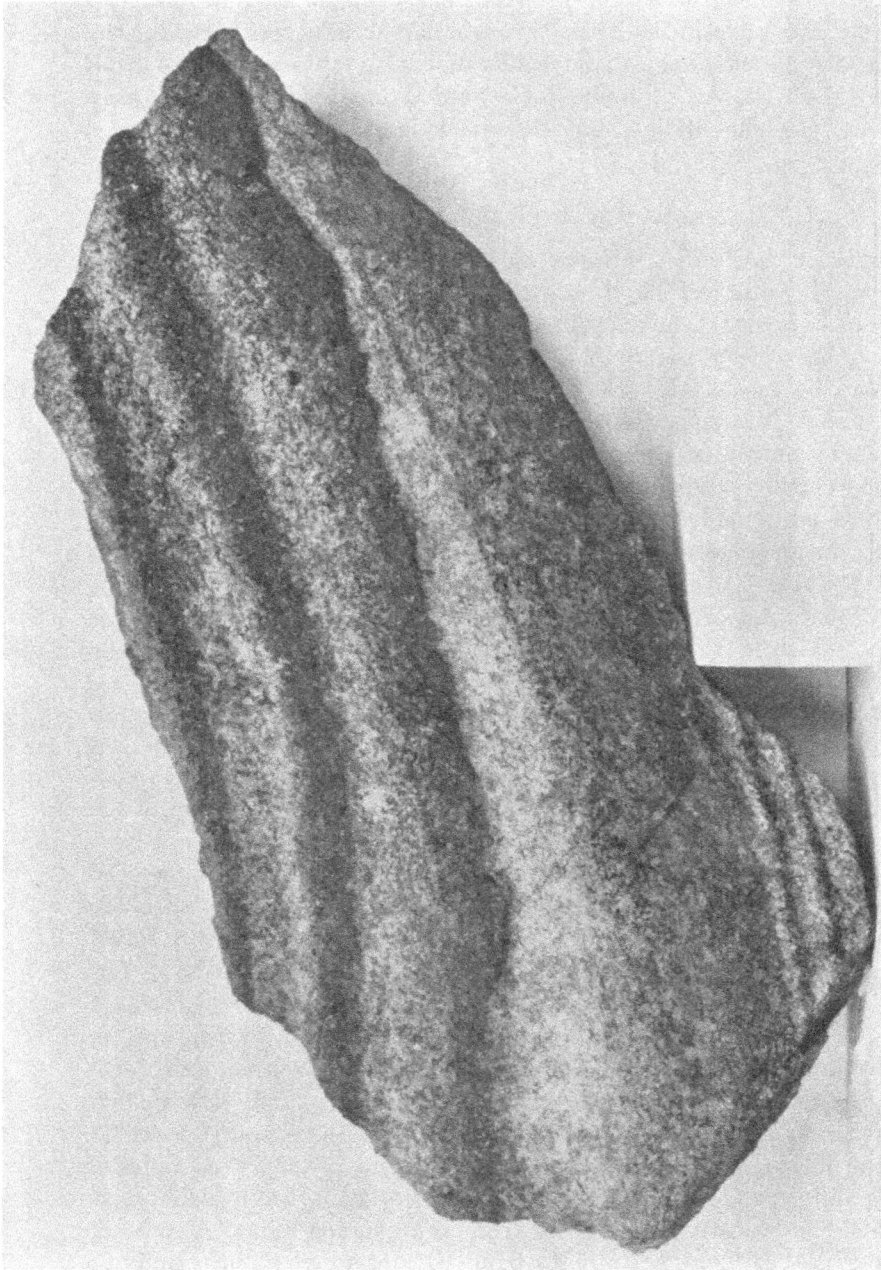

Fig. 26. Specimen showing extreme banding (arranged as in original position). Height of specimen 56 centimetres. Collected near Main House from trough banding G.

the olivine and pyroxene crystals have the form of idiomorphic grains but some are extended to fill interstices between the neighbouring grains (e. g. the top right hand olivine in the drawing of 1924, fig. 27). Felspar and ore which together make up the remainder of the rock consist partly of crystals of about the same size as the pyroxene and olivine, and partly of smaller, interstitial masses. An example of the latter mode of occurrence is the small patch of interstitial plagioclase surrounding

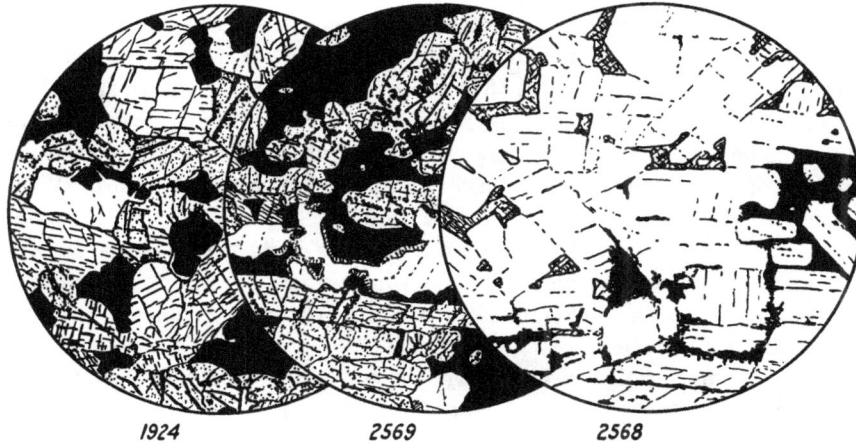

1924   2569   2568

Fig. 27. Extreme bands at the hortonolite ferrogabbro horizon, Magnification × 12.

1924. Melanocratic band from trough banding F near the Main Base House. Clino-pyroxene and hortonolite occur in about equal amounts. Rare prismatic crystal of plagioclase are present (e. g. left centre), elsewhere plagioclase is interstitial pp. 118—19.

2569. Melanocratic band from trough banding H near the Main Base House. Clino-pyroxene, subordinate to hortonolite, is only present at left centre. Fibrous reaction rims are sporadically developed between iron ore and plagioclase (p. 119).

2568. Leucocratic band adjacent to 2569, trough banding H near the Main Base House. Ophitic pyroxene is developed interstitially—all shown being optically continuous. Iron ore and dark-green, almost opaque, decomposition products also fill interstices (p. 119—20).

the ore grains at the right hand centre of drawing (1924, fig. 27). The larger masses of ore and felspar are allotriomorphic, but they are to be interpreted as idiomorphic grains extended by further material which occupies the interstices. The same structure is shown by another melanocratic band, 2569, but in this case the heavy constituents are mainly olivine and ore, there being only 5 per cent. of pyroxene (Fig. 27). Reaction rims between iron ore and felspar are conspicuous in this rock.

A leucocratic band, 2568, immediately above the melanocratic band, 2569, is a tightly compacted collection of prismatic felspar crystals

(Fig. 27) with the usual sporadic fringe of strong zoning. The interstices of the plagioclase crystals are partly occupied by ophitic pyroxene, partly by iron ore and partly by dark green, almost opaque, late-stage minerals. The pyroxene which occurs as ophitic crystals in this rock is essentially the same as that occurring in grains in the melanocratic bands, 2569; at first sight it is surprising to find it occurring as ophitic crystals in the one, and only as separate grains in the other. The same feature is seen in other, strongly leucocratic bands (e. g. 1925 and 1493).

Since the grading in composition of the bands is always from melanocratic below to leucocratic above and never the reverse, it has already been pointed out (pp. 39—40) that gravity must have been a controlling factor in the production of the banding. The gravitative action would occur during crystallisation, and would be due to the different rates of sinking or rising of crystals having different specific gravities. Without dealing with the wider aspects of the process, which involved gentle horizontal flow of the magma and its included crystals as will be shown later, we may here consider whether the microscopic structure of the banded rocks provides any evidence of the action of gravity during their formation.

When sand grains are loosely thrown down there is forty-eight per cent. of interstitial space, and after shaking to make the packing tighter there is still about twenty-six per cent. Crystals separating from a magma and accumulating as a precipitate under the influence of gravity would resemble the loosely packed sand and have about 40 per cent. of magma in the interstices. As the pile accumulated settling of the crystals and expulsion of some of the interstitial liquid would be expected to take place. For the layered rocks we believe that the amount of interstitial liquid which finally crystallised round the mineral precipitate lay between ten and thirty per cent, and that the amount was variable within these limits. It does not seem possible to estimate the amount from measurements in thin section for reasons which will be apparent later. From some of the analyses there is evidence suggesting a rather small amount of interstitial liquid (p. 228). The amount would depend on the pressure exerted by the overlying mass of crystals, and on whether the shape of the crystals allowed close packing. For the sake of simplifying later discussions, it will be assumed that on the average, there was twenty per cent. of interstitial magma surrounding the original crystal precipitate. This considerable quantity of interstitial magma must have crystallised either to the same minerals as were already precipitated, or partly to the same minerals and partly, to new minerals. Material crystallising from the interstitial magma which belongs to the same solid solution series as occur among the original precipitate will not have exactly the same composition as the original crystals. Consider for instance the case of the plagioclase: it is known from the experimentally determined thermal diagram for the plagioclase felspars, that the composition of the early precipitate of plagioclase crystals would be more basic than the composition of the plagioclase remaining in the interstitial liquid. During solidification of the interstitial twenty per cent. of magma the plagioclase already precipitated should either be made over continuously to more acid plagioclase, or be surrounded by zones of increasingly acid material. The olivine and pyroxene would be expected to behave similarly.

Whether these expectations are realised in the layered rocks may now be considered. At first sight thin sections of the melanocratic rocks have the appearance of a collection of granular crystals with here and there interstitial iron ore and felspar. Looked at more carefully, olivine and pyroxene are also found to fill interstices. The original precipitate appears to have been mainly olivine and pyroxene with an occasional grain, of felspar and iron ore, and the interstitial material from the twenty per cent. of magma which originally surrounded the precipitate appears to consist of all these minerals. The plagioclase grains, if examined carefully, usually reveal a narrow and discontinuous fringe of strong zoning. In these rocks, which must have cooled slowly, it appears that reaction did not take place to any significant extent between the interstitial liquid and the original plagioclase crystals, but instead, zones of increasingly acid composition were formed[1]. In the same way some of the olivine crystals can be shown to have narrow outer zones of more ferriferous and therefore lower temperature olivine[2]. Thus the textures exhibited by the melanocratic bands may be interpreted as fulfilling the conditions to be expected if the rocks were formed by the accumulation of a pile of discrete crystal surrounded by a small amount of magma which crystallised later.

In the extreme, leucocratic bands, felspar crystals were collected together in some way, to the virtual exclusion of augite and olivine grains. Apparently this collection of felspars was surrounded by twenty

[1]) It is often believed that the absence of zoning in crystals of some plutonic rock is due to slow cooling allowing equilibrium to be maintained. Vogt held this view [1923, pp. 247—52] and has given his reasons. The same idea is implicit in Bowen's papers, although it is also accepted that other methods such as crystal sorting will give unzoned crystals. The absence of zoning, except at the extreme margin, in the minerals of the layered rocks is due to the crystals having formed from a magma which was effectively stirred by convection currents. The zoning of the crystals at the margin indicates that equilibrium was not established even during slow cooling of this large intrusion. It may be that absence of zoning should be regarded as evidence that a rock has formed by crystal fractionation rather than slow cooling.

[2]) We wish to thank Dr. S. I. Tomkeiff for confirming our observations on this matter by the methods he has recently described [1939].

per cent. of liquid, which crystallised to pyroxene, olivine, iron ore and
felspar. The narrow discontinuous zone of acid felspar round the original
crystals may be interpreted as material crystallised from the interstitial
liquid. The interstitial pyroxene, iron ore and the decomposition products
(probably after olivine) represent other material from the interstitial
magma. It is interesting that the pyroxene and iron ore occur as extensive
poikilitic crystals, a form not found in the melanocratic bands or average
rocks. The reason for this appears to be that the original precipitate
of the extreme, leucocratic bands consisted only of felspar and therefore
there were no pyroxene, olivine and iron ore crystals to form nuclei
on which the interprecipitate liquid could deposit crystals of these
minerals. Deposition therefore started about new centres, and, apparently,
there were but few of them—conditions which resulted in the interstitial pyroxene and iron ore having a poikilitic form.

Details of the microscopic structure of the strongly banded rocks
may be satisfactorily explained on the supposition that they are the
result of the accumulation of a mass of discrete crystals, surrounded
by interstitial liquid which later crystallised either on, or round the
grains of the original accumulation; we shall call the former the
primary precipitate and the latter the interprecipitate material. Rhythmic
layering of varying degrees of conspicuousness is found in practically
all the rocks of the layered series, and therefore general conclusions
about the origin of the strongly layered rocks should apply to the whole
layered series. The microstructures of the average rocks of the layered
series will now be considered to find if they are compatible with their
formation in the way deduced for the strongly banded layers.

The rock, 4077 (Fig. 24) and others of this horizon may be interpreted as a collection of olivine, clinopyroxene and plagioclase crystals
surrounded by small outgrowths of these minerals and of orthopyroxene
crystals. The large olivine, pyroxene and felspar crystals often contain
smaller crystals of these minerals embedded in them and are composite
masses. The same features are found in most of the layered rocks, and
in the drawing of 1907 (Fig. 24), two small plagioclase crystals are
shown enclosed in one of the olivines which thus begins to take on
a poikilitic relation to the felspar. These structures may be explained
as the result of small crystals becoming attached to larger different
crystals as they grew in the free liquid. In some cases it appears that
the discrete units which formed the primary precipitate were composite
masses of more than one mineral which had grown in contact when
still suspended in the free liquid. The incorporation of one crystal in
another during growth in the free magma must be postulated to have
frequently taken place if the rocks are to be interpreted as originating
by the collecting together of discrete units. The rock, 4087, from the

transitional layered series (Fig. 24) has extensive, ophitic pyroxene masses which may possibly be explained in this way but the structures are more likely to be due to the rock being transitional to the border group which had a somewhat different origin. The ophitic habit of the pyroxene in the purple band is also not the result of the primary precipitate being complex units, but is due to the pyroxene having formed by inversion from iron wollastonite (p. 111). Because of the incorporation of one crystal in another during growth in the free magma, and because of the crystallisation of the interstitial magma round the primary precipitate, the texture of the layered rocks is very much that of ordinary plutonic rocks and, without evidence from the extreme layers, a special mode of origin would probably not have been postulated from their micro-structures alone. There is, however, nothing in the microstructure of the average rocks incompatible with the view that they were formed, like the extreme rocks, from two parts—a primary precipitate of discrete units, either single crystals or small complexes, and an interstitial liquid, amounting to about twenty per cent. of the total rock, which crystallised about the primary precipitate.

(f) Layered Rocks showing Increase in Heavy Constituents near the Margin.

The increase in heavy, melanocratic constituents near the margin of the layered series is obvious in the field, and it is found to take place where the even layering changes to a more irregular, false-bedded type of banding. In this part of the layered series the bands consist of a lower, melanocratic part which usually grades upwards into a medium type of rock and only rarely into leucocratic material. The increase in heavy constituents near the margin is confirmed by specific gravity measurement. Thus a large specimen, 4296, 3 metres from the inner contact at the south-east corner of Ivnarmiut (the position is shown on Pl. 5, fig. 1, directly above the stern of the boat) is of sufficient size to give a fair estimate of the average specific gravity, the value being 3.56. Another rock, 1969, collected 15 metres from the inner contact on the islet to the south, has a specific gravity of 3.24. These rocks are from the hortonolite-ferrogabbro horizon where the average specific gravity away from the margin is 3.19. Two middle gabbros, 1919 and 1920, collected near the margin on Kraemers Ö have specific gravities of 3.56 and 3.41, and another rock, 1815, from near the margin on the summit of Gabbrofjaeld, has a specific gravity of 3.29, while the average value for the middle gabbros away from the margin is about 3.10. Other examples are given in Table IX, and also one exception which, however, cannot be regarded as significant (see p. 86).

The melanocratic rock, 4296, three metres from the inner contact at the south-east corner of Ivnarmiut, has been studied in detail, and it has been found that the minerals are the same as at this horizon further from the margin. The estimated height of 4296 in the layered series is 1,500 metres. This height is partly deduced from dips and partly from the observation that olivine returns after its absence in the middle gabbros at about the centre of the small island to the north of Ivnarmiut. The plagioclase of 4296 is An_{41}, the value to be expected for a rock low in the hortonolite-ferrogabbro series (Table III). It occurs partly as large crystals up to a centimetre in length, and partly as small crystals interstitial to most of the other minerals. The large crystals, which show a little zoning near the margin, are larger than any present in the rocks of this horizon away from the contact, a point already mentioned (p. 71). Olivine occurs abundantly in large grains up to 0.7 centimetres across, this being three times the size attained by the larger olivines of the analysed hortonolite ferrogabbro, 1907, and other rocks of this horizon away from the margin. The composition of the olivine is Fa_{60} ($\gamma = 1.797$), and this is the type found in the lower ferrogabbros. The clinopyroxene is usually in smaller crystals than the olivine. In general appearance and refraction index ($\gamma' = 1.739$) it fits well with others of the same horizon. Some of the iron ore occurs in idiomorphic grains and some as large well shaped skeletal masses with infillings of felspar or pyroxene. Orthopyroxene is rare and a brown mica is rather abundant. Alteration of olivine to coarse crystals of serpentine, and of pyroxene to various types of amphibole, is more common than is usual in the lower ferrogabbros. This rock is clearly made up of the minerals appropriate to its horizon in the layered series, but the proportions are not those of the average rock, there being a much lower proportion of felspar and a correspondingly higher proportion of olivine, pyroxene and iron ore. A mode has not been determined since the rock is banded, but the high density, 3.56, of a large fragment indicates the high proportion of heavy constituents. Although 4296 is essentially a hortonolite-ferrogabbro showing a high concentration of heavy constituents it also shows certain unusual features such as an abundance of mica, extreme rareness of orthopyroxene, considerable late stage alteration and abnormally large felspar and olivine crystals.

Another example, 1969, collected fifteen metres from the inner contact on the islet south of Ivnarmiut shows the typical uneven banding and some concentration of heavy constituents. The plagioclase, olivine and pyroxene, are the types appropriate to the horizon, yet the rock is not very similar to 4296, to which it is near. The olivine and plagioclase crystals are not especially large as they are in 4296, and orthorhombic pyroxene is abundant instead of being rare. A similarity with 4296 is the considerable alteration of the pyroxenes to brown and green chlorite. The difference between 4296 and 1969 may be partly due to the former occurring very close to the margin and the latter some fifteen metres away.

Rocks from near the inner contact at the middle gabbro horizon, e. g. 1919, 1920 show increase in the heavy constituents at the expense of the light, and at the same time the composition and character of the minerals are those appropriate to the horizon. We have not sufficient material to trace in detail the changes in the layered series as the inner contact is approached, but it is clear from field and laboratory examination that there is a decrease in light constituents and corresponding increase in heavy, and at the same time the minerals present and their composition are similar to those in rocks from the same horizon further from the margin. Some peculiarities of these rocks, such as the large olivine and plagioclase crystals of 4296, may be features of especial significance, while others such as the abundance of brown mica may not be.

(g) Preliminary Deductions concerning the Origin of the Layered Series.

1. The Layered Series due to Fractional Crystallisation.

Early experimental work on silicate melts has shown that the plagioclase felspars form a series of solid solutions of Roozeboom's type I. More recently Bowen and Schairer (1935, pp. 161—3) have shown that the olivines form a similar series of solid solutions ranging from the high temperature component, forsterite, to the relatively low temperature component, fayalite. In the $MgSiO_3$—$FeSiO_3$ series of clinopyroxenes the more iron-rich solid solutions separate out at lower temperatures than the more magnesium-rich (Bowen and Schairer, 1935, pp. 163—67). The full thermal behaviour of pyroxenes involving calcium, iron and magnesium is complex and not yet fully known, but the available evidence (Bowen, Schairer and Posnjak 1933) clearly suggests that increasing concentration of iron lowers the temperature of formation of the pyroxene. In the case of pyroxenes poor in sesquioxides such as are present in the layered series, it is reasonable to conclude that the more iron-rich will have a lower temperature of crystallisation than the more magnesium-rich.

The variation found in the layered series has been distinguished into two kinds. There is the minor variation, due to changes in the relative amount of the heavy and light constituents and this gives the rhythmic layering. There is also the major variation causing the cryptic layering. The major variation is the more fundamental and is the one considered in the present section. An explanation of the conspicuous rhythmic layering will be attempted later as an extension of our theory explaining the major variation. The most significant features of the major variation in the layered series in passing upwards are:—

1. Gradual change in the plagioclase composition from An_{60} to An_{30}, except for a partial reversal of the trend in the uppermost unlaminated rocks.

2. Gradual change in the composition of the olivine from Fa_{34} to Fa_{97}, except at the middle gabbro horizon, where olivine is absent.

3. Gradual change in the rhombic pyroxene from En_{40} to about En_{70} after which it ceases to form.

4. Gradual change in the clinopyroxene from a type having 19 per cent. Fs. to a type having 68 per cent. Fs. while there is only slight variation in the amount of the wollastonite molecule.

The changes in these minerals which are the ones making up the bulk of the layered rocks, are from relatively high temperature solid solutions at the base of the series to progressively lower temperature solid solutions as the series is ascended. This means that at the temperature at which the lower layered rocks were crystallising the higher rocks must have been liquid, and it must be concluded that the order of formation of the layered series was from below upwards.

If the crystals slowly forming in a cooling magma are repeatedly collected together by some process, the successive batches of crystals will vary gradually from high temperature to low temperature assemblages. For this petrogenetic process Becker, in 1897, introduced the chemists' term fractional crystallisation [1897B]. Since then Harker [1905] and many other petrologists have considered the process to be of primary importance in producing the diversity among igneous rocks, and Bowen [1928 and many other papers] has greatly elaborated the idea by appeal to the detailed experimental results now available. Since the highly regular layered series of the Skaergaard intrusion is composed of assemblages of minerals which, from below upwards, have a steadily decreasing temperature of crystallisation, it is evident that the rocks are essentially successive crystal fractions and that the variation in their composition is the result of what Becker called differentiation by fractional crystallisation.

In the fractional crystallisation of a magma various mechanical processes may reasonably be postulated by which separation of crystals and residual liquid may take place. In the average layered series of the Skaergaard intrusion, materials which must be regarded as successive fractions from the crystallising magm, occur as sheets resting one above

the other. Such a disposition strongly suggests that the crystals separating
from the magma collected at the bottom of the intrusion under the action
of gravity due to their density being greater than that of the liquid.
Crystals accumulating as a precipitate in this way would gradually
build upwards into a layered mass, having a major variation like that
actually found in the layered series. The general nature of the rhythmic
layering shown by the layered series has also led to the view that the
rocks were produced as a precipitate of discrete crystals collected under
the influence of gravity (pp. 120—23): it has indeed, been possible to
distinguish from the microstructures between the primary precipitate
collected together in this way and the interstitial material which
crystallised from the magma filling the spaces in the pile of crystals.
Convergence of these lines of evidence indicates that gravity was probably the factor which produced the crystal fractionation, and we shall
adopt this view as a working hypothesis. Detailed consideration of
the hypothesis and some of its difficulties will be considered in a
later section when all the different units which make up the Skaergaard intrusion have been described.

2. The Primary Phases and their Relation to the Interpre-cipitate Material.

The above general considerations support the view, also deduced
from the microstructures, that the material forming the rocks of the
layered series can be divided into two parts:—

1. The primary precipitate consisting of discrete crystals or small
glomeroporphyritic groups which separated from the overlying magma
and formed about 80 per cent.[1]) of the final rock.

2. The interprecipitate material, amounting to about 20 per cent.
of the total rock, which crystallised from the magma surrounding the
primary precipitate.

The magma which at one time occupied the space between the
primary precipitate, crystallised partly as the same minerals as the
primary precipitate and partly as different minerals. It is not always
possible from an examination of thin sections to decide definitely what
material belongs to the primary precipitate and what to the interprecipitate material. That part of the interprecipitate material which crystallised as the same minerals as the primary crystals, is found as outer
zones of increasingly lower temperature solid solutions, and such outer

[1]) The relative amount of the primary crystals and the interprecipitate material
is not known precisely (p. 120) and no doubt the amount varied in different cases.
We have adopted the figures of 80 and 20 per cent. as a first approximation, and in
later discussions we shall use these values without further qualification.

zones often occupy interstices. Neglecting the outer fringe, the primary precipitate may be found to have an idiomorphic relation to a certain mineral, but taking into consideration the outer fringe, the crystals would be put down as allotriomorphic to the same mineral. In describing the rocks we have seldom found it satisfactory to use these terms without qualification. Where the interprecipitate material consists of different minerals from the primary precipitate it can usually be distinguished by its allotriomorphic or interstitial habit. As a further help in distinguishing between the primary precipitate and the interstitial material the relative proportions may be used: on general grounds (p. 120) it seems that the amounts of these two units should approximate to 80 per cent. of the former and 20 per cent. of the latter.

In the analysed hypersthene-olivine-gabbro 4077, from the main layered series and in other rocks from this horizon, olivine, felspar and clinopyroxene are the primary phases, while orthopyroxene and apatite are new phases from the interprecipitate magma. The rare apatite crystals, one of which is shown in the drawing of 4087 (Fig. 24) are euhedral but as already stated (p. 89) they are confined to the interstitial areas and the outer zones of the plagioclases. In the rocks of the layered series below the ferrohortonolite-ferrogabbros apatite is rare, but no doubt could always be found if searched for since there is a fairly steady 0.05 per cent. of P_2O_5 in the rocks. It is believed that apatite is a constant interprecipitate mineral in the lower rocks of the layered series, although it has not always been observed. In descending from the well layered hypersthene-olivine-gabbros into the transitional layered rocks, the distinction between primary crystals and interprecipitate material becomes less clear and by the time the horizon of 4087 is reached a distinction is not possible (cf. Fig. 24). In ascending from the horizon of 4077, iron ore soon becomes a definite early mineral; then it must be regarded as a primary precipitate although it is only in certain melanocratic bands that it approaches the size and abundance of the other primary crystals.

In the middle gabbros plagioclase, clinopyroxene and iron ore are always primary phases. Orthopyroxene is an interprecipitate mineral in the lower middle gabbros, but seems to become a primary phase in the upper middle gabbros and in the lower hortonolite-ferrogabbros. The olivine which appears as a reaction rim in some of the middle gabbros may be regarded as an interprecipitate mineral. After the olivine-free middle gabbros fractionation of the Skaergaard magma resulted in olivine being again precipitated. If, as seems to have happened, the interprecipitate magma crystallised with partial fractionation due to zoning then, at the middle gabbro horizon, it might be carried sufficiently far in the direction later taken by the main body of magma for olivine to be formed. This seems to be a partial explanation of the olivine reaction rims but other factors must be involved since they are usually localized between pyroxene and iron ore, and since reaction rims occur in the lower ferrogabbros where olivine is also a primary phase. In the lower ferrogabbros the reaction rims seem to be the more usual type due to late stage solutions which produced reaction between iron ore and pyroxene at a relatively low temperature

In the lowest hortonolite-ferrogabbros the primary phases are plagioclase, olivine, clinopyroxene, orthopyroxene and iron ore. Traced upwards the orthopyroxene[1]) soon ceases to form either as a primary phase or as an interprecipitate mineral. The interprecipitate material of the low ferrogabbros consists of the outer zones of the primary phases with, here and there, apatite and quartz. When olivine reappears above the olivine-free middle gabbros it immediately becomes an abundant primary phase. The incoming of quartz is gradual and its occurrence sporadic because it is a mineral crystallising only as the interprecipitate magma is further fractionated. It has been stated during the petrological description of these rocks that the quartz must be regarded as in equilibrium with olivine because it occurs crystallised beside that mineral. It is not, however, a primary phase and did not crystallise from the same magma as the primary phases, but formed at a lower temperature from the interprecipitate magma when this had been changed in composition by further fractionation. The quartz was in equilibrium with an outer zone of the olivine which was crystallising from the interprecipitate magma, and this olivine is richer in fayalite than the primary olivine. The olivine whose composition has been estimated by refraction index measurement is the more abundant primary olivine and not the thin outer zone. The bulk of the olivine of 2580, 3651, 1907, which are all approximately at the horizon where quartz first appears, is Fa_{60}, while the outer part of the olivine with which the quartz is in equilibrium is richer in iron than this.

At approximately the horizon where the ferrohortonolite-ferrogabbros begin the primary phases include all those of the upper hortonolite ferrogabbros together with apatite. In the lower rocks apatite is only present to the extent of about 0.3 per cent.; in these upper rocks apatite amounts to over 4 per cent. and it occurs embedded in the heart of pyroxene, olivine, and plagioclase crystals so that it is here definitely a primary phase. Vogt [1921, p. 324] has suggested that

---

[1]) For five hundred metres above this, submicroscopic sheets of some mineral, believed to be orthopyroxene, occur in the clinopyroxene as the result of exsolution but these apparently disappear completely at higher horizons. Orthopyroxene as a result of exsolution is neither a primary phase nor interprecipitate material and therefore does not concern us in this place.

TABLE XVI.

PRIMARY PHASES OF THE LAYERED SERIES AND ADDITIONAL MINERALS OF THE INTERPRECIPITATE MATERIAL.

	Height in layered series	Primary phases	Minerals additional to the primary phases occurring as interprecipitate material
Ferro-gabbros	2350—2500 m	Plagioclase, olivine, iron-wollastonite, iron ore, apatite	Quartz
	1850—2350 m	Plagioclase, olivine, clino-pyroxene, iron ore, apatite	Quartz
	1700—1850 m	Plagioclase, olivine, clino-pyroxene, iron ore	Apatite, quartz
	1600—1700 m	Plagioclase, olivine, clino-pyroxene, iron ore	Apatite
	1400—1600 m	Plagioclase, olivine, clino-pyroxene, orthopyroxene, iron ore	Apatite
Middle gabbros	Ca. 1200—1400 m	Plagioclase, clinopyroxene, orthopyroxene, iron ore	Olivine, apatite
	900—ca. 1200 m	Plagioclase, clino pyroxene, iron ore	Orthopyroxene, olivine, apatite
Hyper-sthene-Olivine-gabbros	Ca. 700—900 m	Plagioclase, olivine, clino-pyroxene, iron ore.	Orthopyroxene, apatite.
	Ca. 400—ca. 700	Plagioclase, olivine, clino-pyroxene	Orthopyroxene, iron ore, apatite

apatite separates early from magmas owing to its relative insolubility in silicate melts. When, on a later page an estimate has been made of the composition of the successive residual magmas from which the primary minerals of the Skaergaard intrusion were precipitated, an estimate will also be made of the threshold value of P_2O_5 below which apatite was not precipitated. The primary phases of the upper ferrogabbros are plagioclase, olivine, iron ore, apatite and the mineral, believed to be iron-rich β-wollastonite, from which the pyroxene later formed by inversion. Quartz still seems to be a phase only developed from the interprecipitate magma but its occurrence in small masses in the pyroxene of the unlaminated fayalite-ferrogabbro, 4139, suggests that it may there be a primary phase. However, the unlaminated layered rocks do not appear to have been formed by the same process of fractionation as the other members of the layered series, and the distinction between primary phases and interprecipitate minerals, though still to be made, has not quite the same significance (see p. 282). In the unlaminated fayalite-ferrogabbro, olivine, iron wollastonite, iron ore, plagioclase, and perhaps quartz are to be regarded as primary phases while in the basic hedenbergite-granophyre no certain distinctions into primary phases and interstitial material can be made.

The primary minerals occurring in the layered series at different horizons are given in Table XVI. The interprecipitate minerals are the same as the primary, but additional minerals may also occur and these are listed in the third column of the table. The sequence in which the primary phases appear and disappear has a significant relationship to the new phases appearing in the interprecipitate magma. Thus the appearance of a new primary phase is preceded by its appearance as an interprecipitate mineral. This is in harmony with the view that the layered series is the result of crystal fractionation. The primary phases are fractions separated from the magma at a particular temperature while the interprecipitate material is the result of further fractionation of the same magma which was entangled in the interstices of the primary crystals. The further fractionation which takes place in the interprecipitate material due to zoning, should be in the same direction as later took place in the main mass of the magma as the later layered rocks were formed. Thus it would be expected that the fractionation of the interstitial material due to zoning should forecast the trend later taken by the whole layered series. When a primary phase disappears it would be expected on theoretical grounds that it would also disappear from among the interprecipitate minerals, and this is true of the ortho-pyroxene. If the reaction rims of olivine in the middle gabbros are inter-precipitate material, then at this horizon the disappearance of olivine as a primary phase is not accompanied by its disappearance as an interprecipitate mineral; this is to be correlated with the reappearance of olivine among the primary phases after further fractionation.

The temporary disappearance of olivine as a primary phase during the formation of the middle gabbros is remarkable since there is no break in the olivine solid solution series. However, for certain compositions of the melt system MgO—FeO—SiO$_2$, recently described by Bowen and Schairer [1935, pp. 209—13], strong fractionation results in a sequence of phases, as follows:— 1) olivine, 2) clinopyroxene, 3) clino-pyroxene and tridymite, 4) olivines rich in iron and tridymite. Despite the fact that the Skaergaard magma is much more complex than this three-component system, the effect of fractionation resembles the

9*

simpler system, as olivine is first precipitated, then ceases to form, and finally is precipitated again.

The most natural subdivision of the layered series is by means of the primary phases, and since these are usually large crystals, present in considerable amount, it is also a practical method which we have followed in the three major divisions, the hypersthene-olivine-gabbros, the middle gabbros and the ferrogabbros. Subdivision on the basis of the interprecipitate minerals is not so satisfactory as the latter are not directly related to the composition of the magma giving the bulk of the rock, but to the slowly changing composition of the interprecipitate liquid. Moreover they are only present in small amount and sporadically. However, the incoming of quartz as an interprecipitate material gives an approximate method of subdividing the ferrogabbros and we have put this boundary down on the map. Subdivision of the ferrogabbros can also be based on the composition of one of the important primary phases, which is a varying solid solution. For this purpose we found it convenient to use the composition of the olivine. Although it is necessary to determine the composition of the olivine from refractive index tests or from the optic axial angles, this requires little more time than searching for the presence or absence of quartz. The composition of the pyroxene could have been used but not so satisfactorily as it is a more complex mineral. The disappearance of orthopyroxene and the appearance of apatite as primary minerals could also have been satisfactorily used and the latter could have been determined with a pocket lens in the field; had we realised this at the time, we should have used it as a datum line for the field mapping of the upper part of the layered series.

3. Variation Diagram of the Layered Series.

The general variation in composition of the rocks of the layered series is shown graphically in figure 28 by plotting the percentages of the various oxides against height in the intrusion. In the case of the two analysed middle gabbros of nearly the same horizon, the mean of the analyses has been used. It is not practicable to plot a variation diagram against the silica percentages of the rocks, since there is scarcely any variation in this value, but by plotting against height in the intrusion, a diagram is obtained which shows variation in the composition of the rocks developed at successive time stages of the differentiation, and this is what is usually roughly shown by plotting against the silica percentages. In a general way, therefore, figure 28 is comparable with the ordinary variation diagram. The most valuable variation diagrams, as Bowen has insisted, show the composition of glassy rocks which generally represent the composition of a once

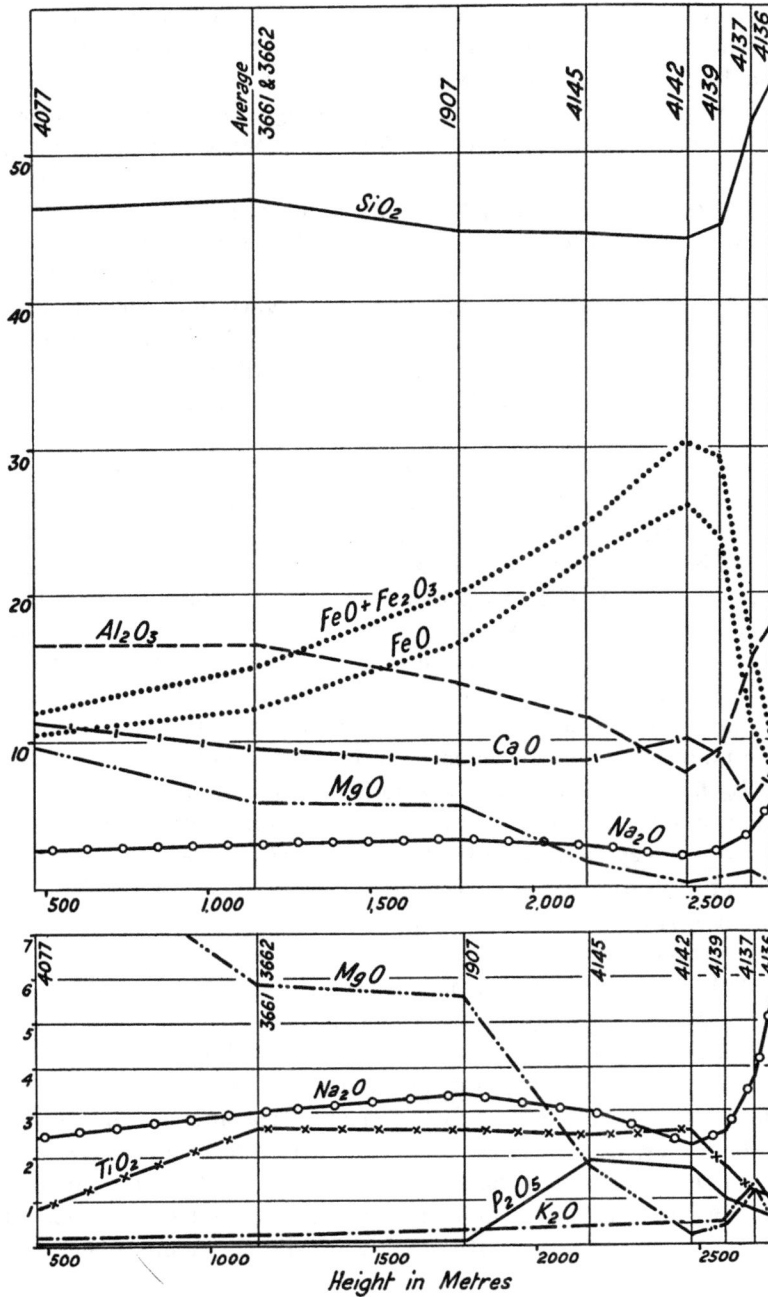

Fig. 28. Variation diagram of the analysed layered rocks plotted against height in the intrusion.

completely liquid magma. The diagram given in figure 28 represents the variation in the composition of a series of rocks, and from it, without further data, it is not possible to deduce the composition of the magma from which the rocks were forming. The diagram shows an approach to the composition of the fraction separating from the magma at a particular time, but it does not show this exactly, as the rocks consist of the primary precipitate plus about 20 per cent. of the magma which was then in existence. However, the interprecipitate magma apparently had a considerable similarity in composition to the total composition of the primary phases, and the variation diagram may be taken as giving the approximate composition of the fractions separating at particular moments during the magmatic differentiation.

The rocks of the layered series all show banding due to variation in the relative amounts of the primary crystals, and the rocks analysed were only selected by eye to represent as nearly as possible the average material. The fact that the analyses, excluding the unlaminated layered rocks, fall close to even curves suggests that the selection of average rocks has been fairly successful, and that we are right in assuming that the layered series is the result of an ordered process of differentiation. The changes in the composition of some of the minerals making up the layered series, e.g. plagioclase and clinopyroxene, are gradual, and the relative proportion of these minerals separating also changes gradually. The variation diagram for the oxides which are present only in these minerals should be smooth, and this is found to be the case with Na_2O, K_2O, and Al_2O_3 which are largely confined to the felspars. In variation diagrams showing the changing composition of a residual magma during crystallisation there are, theoretically at any rate, sudden changes in the slope of the curves when new phases separate (cf. Fenner quoted in Bowen 1928, p. 97). In the case of a variation diagram of a series of rocks consisting of successive crystal fractions, there should be theoretically, not only changes of slope, but also discontinuities in the curves where primary phases suddenly appear or disappear. Except in the case of P_2O_5 we have not sufficient analyses and our method of estimating average rocks is not sufficiently precise, to show definitely the points where changes of direction or discontinuities of the curves occur. However, in this particular case, with the incoming of apatite as a primary phase there is a sudden increase in the amount of P_2O_5 to thirteen times as much as was previously present in the rock series, and thus the curves of variation for P_2O_5 show a strong discontinuity at this point[1]). The marked change in direction of most of the curves

---

[1]) The discontinuity in the P_2O_5 curve should occur somewhere between the points for the analyses of 1907 and 4145. In order to keep figure 28 as much like a normal variation diagram as possible the discontinuity has been smoothed.

which is shown with the beginning of the unlaminated layered rocks distinguishes them sharply from the rest of the layered series. This is probably due to a difference in the manner of origin between the well laminated and unlaminated layered rocks.

In a later section when the border rocks have been considered, an estimate will be made of the composition of the successive residual magmas which were formed during the solidification of the Skaergaard intrusion. The variation diagram of the magma will then be compared with the variation diagram of other magma series. Then also the variation diagram of the layered series which essentially shows the variation in composition of successive crystal fractions, will be compared with that of the liquid from which the successive fractions were forming. In view of the fuller treatment later it is sufficient here to point out that the variation diagram of the layered series can be divided into two parts:— first, there is the part representing the laminated layered rocks in which the variation shown by most of the oxides is small, except that ferrous oxide increases rapidly and magnesia falls rapidly to almost vanishing point; secondly there is the part due to the unlaminated layered rock where the curves for iron fall rapidly and those for silica, soda, potash and alumina rise rapidly. The amount of the unlaminated layered rocks is relatively small but they form an interesting stage in the history of the Skaergaard complex.

4. Summary of position reached.

The deductions so far made concerning the origin of the layered series may be summarised briefly as follows:—

1. The material of any rock of the layered series may be divided:— first, into primary crystals which separated simultaneously from the magma, and collecting at the bottom of the liquid as an aggregate of loose grains; and secondly, into interprecipitate material which crystallised from the twenty per cent. of interstitial magma, surrounding the primary precipitate.

2. The observed variation in composition of the primary phases which are solid solutions, combined with what is already known from experimental work on silicate melts, shows that the primary phases are crystal fractions which separated successively from the magma.

3. The successive fractions are found lying one above the other which suggests that the minerals composing them were collected under the influence of gravity at the bottom of the magma.

4. The rhythmic layering, due to variation in the relative amounts of the same few minerals also shows the influence of gravity since

melanocratic layers, consisting of heavy constituents, grade upwards, and never downwards, into leucocratic layers consisting of light constituents.

5. The interprecipitate magma, amounting to about twenty per cent. of the rock, crystallised, partly as the same minerals as the primary crystals, and partly as new minerals. Where plagioclase, olivine and pyroxene occur as interprecipitate material, they form zones of increasingly lower melting point material which surround the primary phases. Despite the slow cooling of the rock, reaction between the interprecipitate magma and the primary phases has not produced homogeneous crystals. As a result of zoning the interprecipitate magma underwent further fractionation during crystallisation.

6. Bulk chemical analyses of the average rocks provide data showing roughly the trend in composition of the fractions separating at successive stages.

Points connected with the origin of the layered series which must await consideration until the border rocks have been described are:—

1. The extent to which the primary crystals sank vertically through the liquid and the extent to which they were involved in flow movements of the magma.

2. The significance of the saucer form of the layers, the trough banding, the increase in melanocratic constituents near the inner contact and the variation in size of the plagioclase.

3. The original composition of the magma and the changes it underwent during fractionation.

4. The cause of the trend of differentiation which has given rise to highly iron-rich rocks, so far not found elsewhere.

VII. PETROLOGY OF THE BORDER GROUP

(a) The Chilled Margin.

On all sides of the intrusion near the margin, moderately fine-grained rocks are encountered which are taken to be the chilled and undifferentiated, original Skaergaard magma. Two of these rocks have been analysed and will be described first, then certain other examples which show the constancy in mineralogical composition and structure found in all the chilled border rocks.

One of the chilled gabbros, 1825, which has been analysed, was collected 3 metres from the contact with the basalts at the head of Uoloberen. In the hand specimen it is conspicuously patchy, about three-quarters being fine grained, with felspars from 1—3 millimetres in length, and the remainder coarser, with felspars up to 1 centimetre in length. The coarser material has roughly the same composition as the finer and the whole is regarded as the result of solidification of the undifferentiated magma. Under the microscope the rock (Pl. 20, fig. 1) is seen to be an olivine gabbro containing both rhombic and monoclinic pyroxene. The plagioclase is moderately strongly zoned and is thus in marked contrast with the plagioclase of the layered series. The average composition is about An_{60} with central parts An_{65} and margins fairly acid. The mode has only been roughly determined because of the heterogeneity, but it is sufficient to indicate that there is between 60 and 70 per cent. of plagioclase. The texture suggests that plagioclase was among the first minerals to begin crystallising as small crystals project into, or lie completely embedded in olivine. The olivine, with margins and cleavage cracks darkened by magnetite due to incipient alteration, occurs as somewhat rounded, zoned grains, usually less than 1 millimetre across. It appears to be a common type of olivine (chrysolite or hyalosiderite), and this is also to be inferred from the bulk analysis of the rock. Both rhombic and monoclinic pyroxenes occur, but in this particular rock the latter is about ten times as abundant as the former. Both types occur in ophitic masses enclosing plagioclase and olivine and they are apparently very similar to the types occurring in the lower rocks of the layered series such as 4077. The monoclinic pyroxene has $\gamma' = 1.714$. The rhombic pyroxene has the orientated, blebby inclusions of clino-pyroxene which is such a constant feature of the rhombic pyroxenes of the layered series. Iron ore is usually interstitial and decidedly rare. The early minerals were plagioclase and olivine and probably pyroxene. There seems to have been no significant difference in the time at which crystallisation of rhombic and monoclinic pyroxenes ended and both apparently continued to crystallise long after the olivine had ceased. Some of the iron ore crystallised earlier than he rhombic pyroxenes and some later; the greater part may be considered to have had an intermediate period of crystallisation.

The other analysed chilled gabbro, 1724, was collected 25 metres from the contact with the basalts at the coast on the east side of Skaergaardsbugt. Like 1825, the rock shows patchy variation in grain size but both the fine and coarse grained parts are coarser than in 1825 which is in harmony with its greater distance from the margin. Except for the grain size there are practically no differences between the two rocks. In 1724 the plagioclase is strongly and deeply zoned but averages An_{60}. The olivine, with black margins and cleavage cracks is in somewhat larger crystals with a more definite ophitic relationship to the plagioclase, but it seems to have about the same composition. The rhombic pyroxene is a little more abundant relative to the mono-clinic than in 1825 but is otherwise similar. The monoclinic pyroxene has $\gamma' = 1.714$ and is the same therefore as in 1825. Both pyroxenes are usually allotriomorphic towards olivine but in some cases the relation-ships suggest simultaneous crystallisation. In some parts of the rock the olivine as well as the two pyroxenes have an ophitic relation towards the plagioclase. Iron ore is again rare, and the evidence indicates that it had mainly an intermediate period of crystallisation. In this particular rock some alteration of olivine and rhombic pyroxene has occurred.

A ten inch apophysis from the Skaergaard intrusion, 1721, col-lected near 1724 and 5 metres from the outer margin of the intrusion, was intended for analysis but proved on sectioning to be too altered. It is intermediate in texture between the fine-grained parts of 1825 and 1724; the plagioclase is of similar composition and similarly zoned; the monoclinic pyroxene is fresh, while the rhombic is completely replaced by a bastite pseudomorph which contains the usual unaltered orientated, blebby inclusions of clinopyroxene. Other areas of irregularly orientated serpentine are presumably replacing olivine. There is no doubt that this apophysis was once essentially the same as the analysed chilled rocks.

A rock, 1922, collected one and a half metres from the contact and 20 metres above sea level at the southern part of the outcrop on

Kraemers Ö is similar to the rocks already described. It is patchy in grain size; the plagioclase is strongly and deeply zoned but in composition is about the same as that of 1825; rhombic and monoclinic pyroxenes are present in about equal amounts and show the usual features; olivine is less abundant than in the analysed rocks while iron ore is more abundant; the textural relations of the minerals are the same.

The last chilled contact rock which will be described, 4093, comes from the northern contact; it was collected 800 metres north of Strömstedet and at a height of about 175 metres above sea level. A band of gabbro about 10 metres wide is found between the outer contact and the large inclusion of gabbro-picrite which is here present in the border group. To the south of this point, the contact and the chilled gabbro is hidden by scree and to the north the marginal chilled gabbro is less conspicuous owing to the confusion produced by inclusions of sediment. The rock 4093, which was collected about 3 metres from the contact, has the usual variable texture. The plagioclase, which is moderately zoned, averages about An_{60}. Rhombic and monoclinic pyroxenes are about equally abundant and are similar to those already described, except that the rhombic pyroxene has no visible inclusions of monoclinic pyroxene. The olivine has the usual appearance but is without conspicuous dark borders. The textures are the same as in the analysed rocks and suggest a similar order of crystallisation, except that the rare iron ore is more idiomorphic towards the pyroxenes, and is often surrounded by biotite with a fibrous reaction rim between it and surrounding felspar.

The chilled contact rocks from all sides of the intrusion have a striking similarity in mineral composition and texture, and the two analysed examples are almost identical in chemical composition (Table XVII). It is safe to conclude that all the chilled marginal gabbro was produced by the solidification of the same magma, but may it be assumed that this magma is representative of that from which all the Skaergaard rocks were differentiated? In many sills and dykes, and in some plutonic intrusions such as the outermost ring dyke of Centre II in Ardnamurchan [Richey, Thomas etc. 1930, pp. 217—21], the marginal rocks are believed to have been derived from a different magma from that responsible for the central part, and this is considered to be due to a change in the nature of the magma passing along the dyke or sill. Because of its large dimensions it is unlikely that the whole space now occupied by the Skaergaard intrusion was occupied by a magma which was subsequently replaced by a second and different magma. It might, however, be argued that an early cone sheet preceded the main Skaergaard intrusion and gave rise to the chilled marginal rocks. Such a hypothesis would seem to be reasonable on tectonic grounds. Two facts are, however, opposed to this view; first, the texture of the chilled marginal gabbro is not that of a normal, basic, hypabyssal rock and second, the chilled material is of similar composition and character round the whole margin. Even if the postulated cone sheet had been originally complete it would scarcely be likely to have survived in its entirety the subsequent formation of the main intrusion. The more probable hypothesis, which is the one here adopted, is that the fine-grained marginal gabbro was produced by chilling of the outside part of the original Skaergaard magma before differentiation had had time to modify it.

The two analyses of the chilled border gabbro (Table XVII) are so similar that, for most purposes, the average will be used. It is natural to compare the original Skaergaard magma with other rocks of the

TABLE XVII.
ANALYSES OF CHILLED MARGINAL GABBROS.

	XII (1825)	XIII (1724)	Norms				Modes (vol. %)	
			XII		XIII		XII	XIII
SiO_2	48.01	47.83	Or........	1.11		1.11	Plag............. 65.5	57.2
Al_2O_3	19.11	18.62	Ab........	19.91		21.48	Clinopyrox....... 21.2	25.5[2])
Fe_2O_3	1.20	1.16	An........	40.87		38.64	Orthopyrox. ..... 1.7[1])	..
FeO	8.44	8.87	Di.... 4.18	8.13	5.80	11.34	Oliv............. 11.1	16.3[3])
MgO	7.72	7.92	2.50		3.30		Ore. 0.5	1.0
CaO	10.33	10.59	1.45		2.24			
Na_2O	2.34	2.54	Hy. .. 9.60	14.88	5.40	8.96		
K_2O	0.17	0.20	5.28		3.56			
H_2O+	0.55	0.27	Ol. ... 5.04	8.92	7.70	12.80		
H_2O-	0.05	0.16	3.88		5.10		$\dfrac{(FeO + Fe_2O_3) \times 100}{MgO + FeO + Fe_2O_3}$ 7	7
TiO_2	1.51	1.29	Ilm.......	2.89		2.43		
MnO	0.12	0.09	Mt........	1.86		1.86		
P_2O_5	0.07	0.06	Ap.	0.34		0.17		
CO_2	0.11	0.01	Cal.	0.20		..	$\dfrac{Fe_2O_3 \times 100}{FeO + Fe_2O_3}$ 12	12
S	0.29	0.25	Pyr.......	0.54		0.48		
ZrO_2	nil.	nil.	H_2O	0.60		0.43		
SrO	0.21	0.19						
BaO	0.02	0.02	Plag. $Ab_{32}An_{68}$		$Ab_{35}An_{65}$		$\dfrac{K_2O}{Na_2O + K_2O}$ 56	56
Cr_2O_3	tr.	nil.	Diop. $Wo_{53}En_{30}Fs_{17}$		$Wo_{51}En_{30}Fs_{19}$			
CuO	0.006	0.008	Hyp. $En_{64}Fs_{36}$		$En_{60}Fs_{40}$			
NiO	nil.	nil.	Oliv. $Fo_{57}Fa_{43}$		$Fo_{60}Fa_{40}$			
	100.26	100.07						
S. G.	2.98	2.95						

[1]) Perhaps a little low as doubtful material was assumed to be clinopyroxene.
[2]) Includes some uralite.
[3]) Includes some serpentine.

COMPARISONS.

	XIIIa	A	B	C	D	E	F	G	H
SiO$_2$	47.92	48	47	47.6	46.40	47.24	48.28	47.75	48.20
Al$_2$O$_3$	18.87	20	14	18.0	16.30	18.55	20.38	19.46	18.07
Fe$_2$O$_3$	1.18	} 9	13	} 10.3 {	3.60	6.02	1.78	2.31	1.76
FeO	8.65				7.17	4.06	6.70	6.28	9.16
MgO	7.82	7.75	6.5	7.3	6.00	5.24	7.93	7.90	5.94
CaO	10.46	12	10.0	11.3	11.04	11.72	11.80	11.32	9.93
Na$_2$O	2.44	2	2.6	2.2	2.14	2.42	1.75	2.46	2.25
K$_2$O	0.19	0.5	0.7	0.6	0.29	0.15	0.14	0.24	1.46
H$_2$O+	0.41	..	..	..	1.10	2.24	0.76	0.50	0.94
H$_2$O—	0.10	..	..	..	2.40	0.21	0.09	0.18	0.28
TiO$_2$	1.40	..	..	..	3.05	1.46	0.23	0.43	1.54
MnO	0.11	..	..	..	0.23	0.31	0.28	0.17	0.15
P$_2$O$_5$	0.07	..	..	..	0.23	0.26	0.02	0.62	0.38
CO$_2$	..	..	..	..	nil.	0.19	0.03	tr.	..
BaO	..	..	..	..	..	nil.	..	..	..
(Co,Ni)O	..	..	..	..	..	0.05	..	..	..
FeS$_2$	..	..	..	..	..	nil.	0.04	0.16	..
Cr$_2$O$_3$	..	..	..	..	..	..	..	0.05	..
	99.62	..	..	..	99.95	100.12	100.21	99.83	100.06

XII. Marginal olivine gabbro. 1825. 3 m from contact, head of Uolöberen, Kangerdlugssuaq.

XIII. Marginal olivine gabbro. 1724. 25 m from contact E. side Skaergaardsbugt Kangerdlugssuaq.

XIIIa. Average of marginal olivine gabbro. 1825, 1724.

    A. Composition of Porphyritic Central Magma-Type (SiO$_2$ = 48 %) [Richey, Thomas etc. 1930, p. 87].

    B. Composition of Non-porphyritic Central Magma-Type (SiO$_2$ = 47 %) [Bailey, Thomas etc. 1924, p. 14].

    C. Average of A and B in ratio 2:1.

    D. Basalt, with porphyritic felspars, Faeroes [Simpson, 1928]. Anal. W. H. Herdsman.

    E. Lava, Basalt, Porphyritic Central Type. Half a mile S.S.W. Derrynaculan, Mull. [Bailey, Thomas etc. 1924]. Anal. E. G. Radley.

    F. Gabbro variant of Great Eucrite Ring-Dyke Centre 3, Ardnamurchan. [Richey, Thomas etc. 1930]. Anal. E. G. Radley.

    G. Hypersthene-gabbro. Ring-dyke, centre 2. Ardnamurchan [Richey, Thomas etc., 1930]. Anal. B. E. Dixon.

    H. Olivine bearing facies of the central dolerite. Johannesberg, Breven [Krokström, 1932]. Anal. N. Sahlbom.

north Atlantic Tertiary igneous province and among these there are types which are closely similar in their general chemical composition. Thus the Porphyritic Central magma type, first distinguished in Mull [Bailey, Thomas etc. 1924, pp. 23—24], and later recognised as dominant among the basic plutonic intrusions of Ardnamurchan [Richey, Thomas, etc. 1930, pp. 86—90] has given rise to rocks varying from 45—50 per cent. of silica and, from the variation diagram, the average composition of a rock belonging to this magma series and having 48 per cent. SiO_2 may be estimated; this comparable Porphyritic Central material (Table XVII, A) corresponds closely with the original Skaergaard magma. In the Mull memoir three analyses of Porphyritic Central types of basalts are given and the percentages of SiO_2, Al_2O_3, (Fe_2O_3 + FeO), Na_2O, K_2O for the Skaergaard magma lie within the range shown by these Mull basalts. The percentage of MgO in the Skaergaard magma is greater and of CaO less than in the Porphyritic Central types of basalt, but gabbros and dolerites from Ardnamurchan which are ascribed to the same magma type, include within their range the percentages of MgO and CaO found in the Skaergaard rock. In one respect the composition of the Skaergaard magma lies outside the range shown by rocks ascribed to the Porphyritic Central type by Richey and Thomas and that is in the low state of oxidation of the iron. The ratio Fe_2O_3/Fe_2O_3 + FeO for all the analyses of Porphyritic Central types given in the Mull and Ardnamurchan memoirs lies between 0.22 and 0.60, the average being 0.28. The average values for Plateau magma and Non-porphyritic Central magma are 0.28 and 0.31 respectively. The values for the two analysed, chilled gabbros of the Skaergaard intrusion are both 0.12, and there is a similarly low ratio for most of the analysed rocks of the layered series.

The composition of the original Skaergaard magma, as shown by the analyses of the chilled marginal gabbro, was also fairly close to the average Mull Normal magma which has a silica percentage of 47 (Table XVII, B). While the average Porphyritic Central magma of silica percentage 48 is a little too rich in alumina and lime the Normal magma is too poor in these and the closest approach is obtained by taking two parts of Porphyritic Central magma (SiO_2 48 per cent.) and 1 part of Normal magma (SiO_2 47 per cent.) (Table XVII, C). The Skaergaard magma may be regarded as being close to the Porphyritic Central magma type as defined in the Mull and Ardnamurchan Memoirs, but it also resembles to some extent the Normal magma series and the closest approach is obtained by striking the weighted average. In two respects, however, these magma types and the weighted average differ from the Skaergaard magma: first, the iron is in a much less reduced state; and second, the amount of potash is decidedly greater.

Besides similarity in chemical composition there are also close mineralogical and textural similarities between the chilled Skaergaard magma and some of the products of the Porphyritic Central magma of the British area. The closest comparison is with the hypersthene gabbro of the first ring dyke of Centre II, Ardnamurchan Point, whose analysis is quoted in Table XVII, G. Comparison cannot be made with the chilled marginal material of this ring dyke as it was apparently derived from a slightly different magma. The average rock of the central part, however, contains labradorite, ferriferous olivine, hypersthene, monoclinic pyroxene and but little iron ore; the textures are also similar [Richey, Thomas etc. 1930, pp. 223—25]. In general remarks on the rocks formed from the Porphyritic Central magma Thomas states [p. 86] that the special characteristic of the magma is early separation of plagioclase, a feature also clearly exhibited by the chilled marginal Skaergaard magma.

Several analyses of East Greenland basalts have been made but these are certainly not representative of the huge bulk of basalts lying between Kangerdlugssuaq and Scoresby Sund. None of the basalts so far analysed are close to the Porphyritic Central type but this is certainly due to lack of analyses rather than to the absence of these basalts. Two analyses of non-porphyritic basalts from the Blossville Coast [Wager 1934, Table I. Analyses II, III] may be taken as indicating something of the composition of the non-porphyritic basalts, but there are at present no analyses of the basalts containing conspicuous porphyritic plagioclase crystals which are as abundant as the non-porphyritic types. From microscopic examination, combined with the analysis of the non-porphyritic basalts already made, it may be surmised that these porphyritic basalts will have a composition closely related to that of the Skaergaard magma.

The average analysis of chilled Skaergaard gabbro is also compared with a basalt having porphyritic felspars from the Faeroe Islands and with a Porphyritic Central basalt from Mull (Table XVII, D and E). The Faeroes basalt has a general similarity of composition with the chilled Skaergaard gabbro, and the Mull basalt has a similarly low content of K_2O. Two gabbros from Ardnamurchan (Table XVII, F and G) which are considered to belong to the Porphyritic Central magma type are similar to the original Skaergaard magma in many respects, and the similarity extends even to the amount of Fe_2O_3, FeO, and K_2O. An analysis of a rock from the Breven Dyke (Table XVII, H) shows that similar rocks also occur there.

Although the unique iron-rich rocks of the upper layered series have been produced by strong fractionation of the Skaergaard magma, the composition of the undifferentiated magma appears to have been closely similar to that of the Porphyritic Central magma of the British area. Before the analyses were made we expected that the chilled marginal gabbro would be iron-rich but it proves to contain less iron than average basalts of Plateau, and Non-porphyritic Central types and less than

Daly's average for all basalts. Initial richness in iron cannot have been the cause of the unusual trend of the differentiation. The relatively reduced condition of the iron in the Skaergaard magma may be a factor but it is also possible that the cause was not initial composition but unusual environmental conditions, and the matter will be left for detailed consideration until later.

(b) The Perpendicular Felspar Rock.

The rock having much elongated felspar crystals set at right angles to the margin of the intrusion (Pl. III, fig. 1) is found developed sporadically in the border group along the western, northern and eastern borders between 15 and 30 metres from the outer contact (pp. 26—27, 29, 31). Conditions for the formation of this unusual rock were widespread during certain early stages in the formation of the border group, and its origin is therefore worth considering in detail. The perpendicular felspar rock is an olivine-gabbro somewhat similar to the chilled gabbro except for the lath-shaped felspars 3—5 centimetres long and averaging 0.1×0.3 centimetre in cross section. These form gently-curved, and sometimes branching, crystals which are set with their length roughly perpendicular to the nearby wall of the intrusion, and there are from 5 to 10 such crystals in a cross section of 1 square centimetre (Fig. 30). Between the elongated felspars the rock is fine grained except for occasional poikilitic pyroxenes. The average size of the felspar and olivine crystals in the fine-grained part between the elongated plagioclase crystals is actually somewhat less than in the fine-grained parts of the analysed, chilled marginal rock, 1825 (cf. Pl. 20, figs. 1 and 2). The perpendicular felspar rock forms reefs, 10—100 centimetres wide, set parallel to the margin of the intrusion; in a single traverse from the margin inwards, two or three such reefs may be found between which olivine-gabbro is developed which has apparently the same composition but is without the elongated felspars. No field evidence was found to show that the perpendicular felspar rock was a later injection into the normal olivine-gabbro. On the contrary it is apparently an integral part of the border group which, however, has developed unusual structures due to the temporary occurrence of especial conditions during the solidification process.

The elongated plagioclase of the analysed example of the perpendicular felspar rock, 1851, shows some zoning but 90 per cent. of it is An_{65}. The zoning differs from that in the chilled marginal rocks in being oscillatory; the plagioclase becomes steadily more acid in composition and then it reverts abruptly to a more basic type which in turn gradually becomes more acid. Two or three oscillations of this

Fig. 29. Perpendicular felspar rock, Mellemö. Specimen 13 centimetres wide.

kind are frequently seen in the thin sections of 1851 and neighbouring
rocks, and the elongated felspars of another example of perpendicular
felspar rock from the eastern border group, 1768, show similar features
and have the same composition. The elongated felspars are evidently
more basic than the average value for the zoned felspars of the chilled
marginal gabbro, a point which is also suggested by a comparison of
the analyses (Tables XVII and XVIII). The elongation of the felspars
is parallel to the X crystallographic axis and the flattening of the laths
is parallel to (010). It has already been mentioned (p. 70) that the
felspars of the upper layered series have the same habit but without
the extreme elongation in the X direction. The fine-grained material
between the elongated felspars consists of small, zoned, plagioclase
crystals rarely 1 millimetre long, olivine grains between 0.5 and 1 millimetre across and pyroxene. The bulk of the pyroxene is monoclinic;
it occurs in poikilitic masses up to 4 millimetres across and encloses
small plagioclase and olivine crystals. The olivine, from refractive index
tests, is Fa_{30}. Some rhombic pyroxene occurs but it contains none of
the orientated inclusion of monoclinic pyroxene which are so abundant
in other rocks of the intrusion. The rhombic pyroxene can only be
distinguished with certainty from the +ve monoclinic pyroxene with
moderate 2V in those sections in which its —ve sign and large 2V

can be proved. Probably about twenty per cent. of the pyroxene is rhombic but this estimate is only approximate. The monoclinic pyroxene $\gamma' = 1.702$) has a weaker pleochroism than the rhombic, and it often includes orientated inclusions of iron ore. Iron ore, as an ordinary magmatic product, is decidedly rare and where present it shows relationships suggesting that it finished crystallising before the rhombic pyroxene which frequently surrounds it.

The chemical composition of the perpendicular felspar rock, 1851 (Table XVIII) is not far removed from that of the chilled marginal gabbro but such differences as exist are significant. The alkalies are decidedly less abundant and the normative plagioclase is An_{78} compared with about An_{65} for the chilled marginal gabbro. From the norm it appears that the perpendicular felspar rock differs from the chilled marginal gabbro in having more anorthite and a little more of the magnesium silicates relative to the iron. It is compared with the com-

TABLE XVIII.

ANALYSIS OF PERPENDICULAR FELSPAR ROCK.

XV (1851)		Norm		Mode (vol. %)	
SiO_2	46.15				
Al_2O_3	21.34	Or.	0.56	Plag.	54.2
Fe_2O_3	0.79	Ab.	14.15	Clino-pyrox.	22.7
FeO	8.21	An.	49.76	Oliv.	22.8
MgO	9.97	Cor.	0.20	Ore	0.3
CaO	10.24	Hy. 9.40	14.68		
Na_2O	1.72	5.28			
K_2O	0.12	Ol. 10.92	16.84		
H_2O+	0.27	5.92			
H_2O-	0.10	Ilm.	1.06	Ratios	
TiO_2	0.55	Mt.	1.16		
MnO	0.15	Ap.	0.17		
P_2O_5	0.05	Calc.	0.20	$\dfrac{(FeO + Fe_2O_3) \times 100}{MgO + FeO + Fe_2O_3}$	47
CO_2	0.13	Pyr.	0.20		
S	0.12	H_2O	0.37		
ZrO_2	nil.				
SrO	0.07	Plag. $Ab_{22}An_{78}$		$\dfrac{Fe_2O_3 \times 100}{FeO + Fe_2O_3}$	8
BaO	0.03	Hyp. $En_{68}Fs_{32}$			
Cr_2O_3	0.04	Oliv. $Fo_{54}Fa_{46}$			
CuO	0.007				
NiO	nil.			$\dfrac{K_2O \times 100}{Na_2O + K_2O}$	7
	100.06				
S.G.	3.01				

COMPARISONS.

	XV	XIIIa	A	B
SiO_2...................	46.15	47.92	47.28	48.28
Al_2O_3.................	21.34	18.87	21.11	20.38
Fe_2O_3.................	0.79	1.18	3.52	1.78
FeO....................	8.21	8.65	3.91	6.70
MgO	9.97	7.82	8.06	7.93
CaO...................	10.24	10.46	13.42	11.80
Na_2O..................	1.72	2.44	1.52	1.75
K_2O...................	0.12	0.19	0.29	0.14
H_2O+..................	0.27	0.41	0.53	0.76
H_2O-..................	0.10	0.10	0.13	0.09
TiO_2...................	0.55	1.40	0.28	0.23
MnO	0.15	0.11	0.15	0.28
P_2O_5	0.05	0.07	tr.	0.02
CO_2	0.13	0.06	..	0.03
S	0.12	0.27	..	0.04
ZrO_2	nil.	nil.	..	..
SrO	0.07	0.20	..	..
BaO...................	0.03	0.02	..	..
Cr_2O_3.................	0.04	nil.	..	..
CuO...................	0.007	0.007	..	..
NiO...................	nil.	nil.	..	..
	100.06	100.18	100.20	100.21

XV. Perpendicular felspar rock, 1851. W. side Mellemö. Kangerdlugssuaq.

XIIIa. Average of the chilled marginal gabbros (1825, 1724).

    A. Olivine-gabbro, floor of Coir'a'Mhadaidh, Cuillins, Skye [Harker, 1904].
       Anal. W. Pollard.

    B. Gabbro-variant of Great Eucrite Ring-dyke, Centre 3, Ardnamurchan
       [Richey, Thomas etc. 1930]. Anal. E. G. Radley.

position of other marginal border rocks and with the layered rocks in the variation diagram, Fig. 31. The plagioclase-rich gabbro from the Cuillin Hills of Skye (A) and a gabbro variant of the Great Eucrite of Ardnamurchan (B) are two rocks of similar composition from the British Tertiary area.

The elongated felspars cannot have been orientated by magmatic flow during crystallisation as that would have placed them parallel, and not perpendicular, to the walls of the intrusion: they can only have assumed their present disposition by inward growth of appropriately orientated felspars attached to the already solidified, outer material. Well-shaped crystals, growing inwards from the boundary walls towards the free liquid, are not usually developed in plutonic intrusions

10*

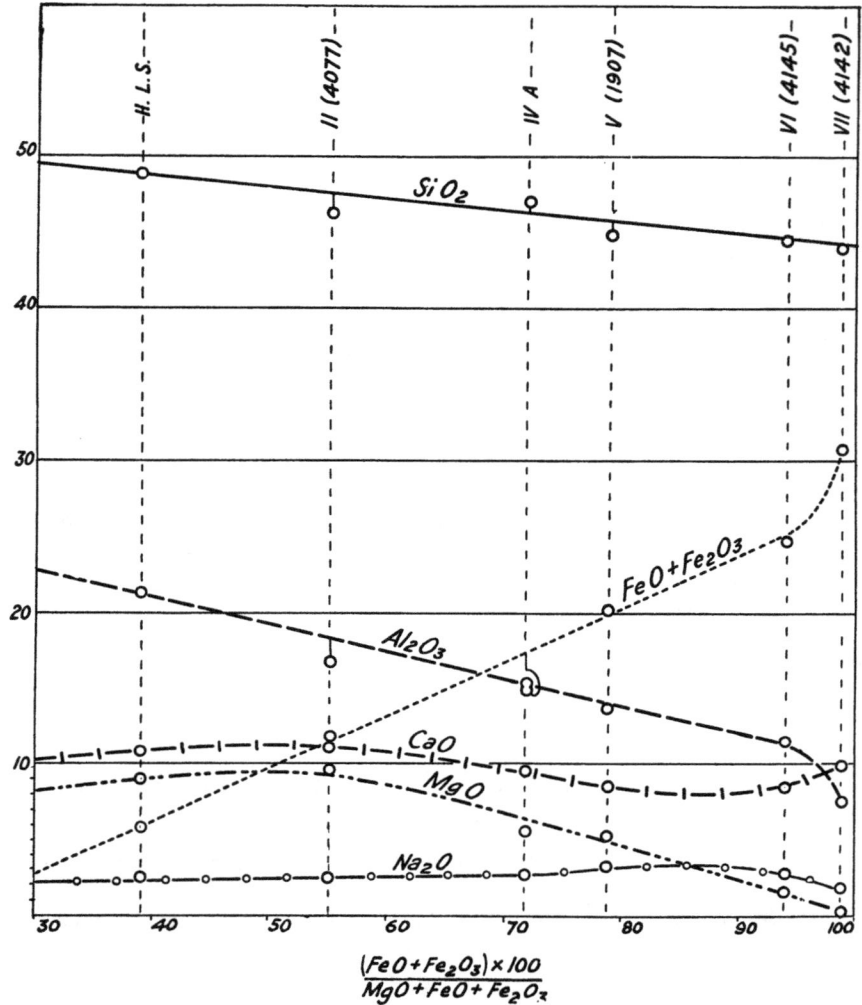

Fig. 30. Variation diagram of the layered series plotted against iron-magnesium ratios.

but they are found in some pegmatites and in geodes, drusies and mineral veins. In order that large acicular crystals of this kind may grow, diffusion, aided perhaps by circulation of the liquid, must keep the growing crystals supplied with material appropriate to their composition, and this free diffusion and circulation must take place so effectively that new centres of crystallisation are not started in the liquid ahead of the growing crystals. The liquids from which pegmatites and mineral veins are deposited differ from normal silicate magmas in having a lower viscosity which allows easier flow and probably freer diffusion; as a result

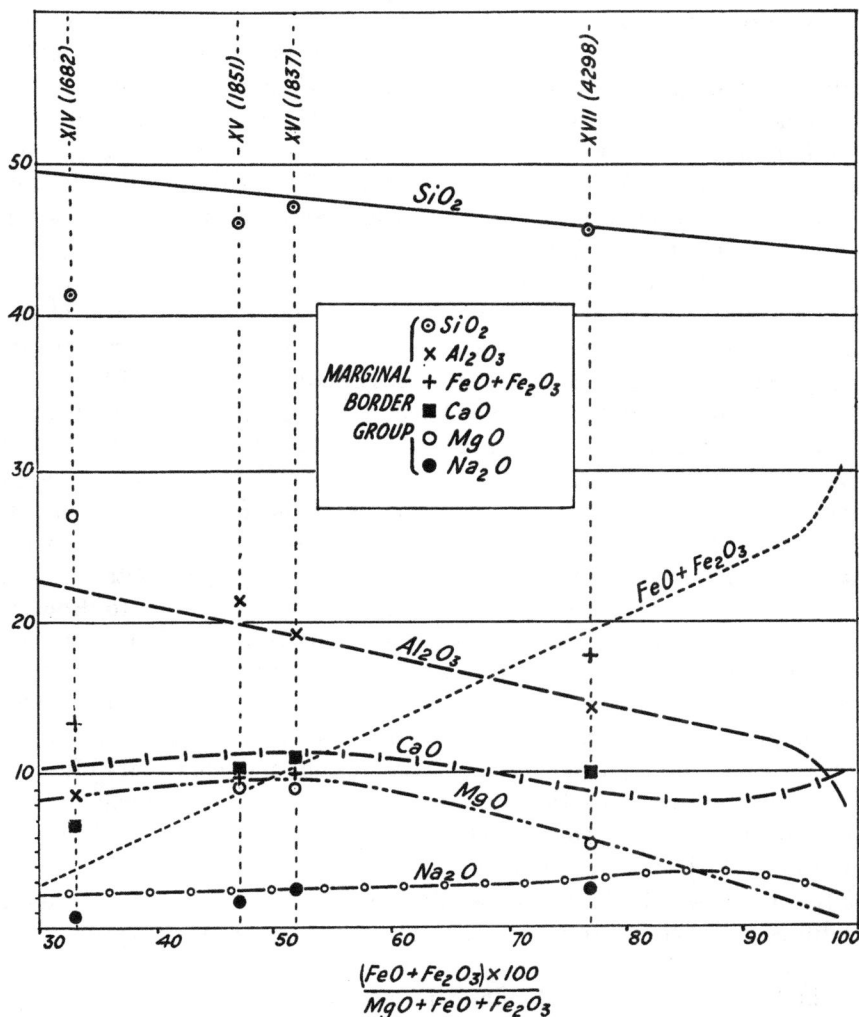

Fig. 31. Variation diagram of the layered series (taken from Fig. 30) with plots of the analysed marginal border rocks.

relatively large crystals form by inward growth from the walls of the pegmatite or vein. It seems likely that the same factors, namely lower viscosity and freer diffusion than is usual in basic magmas, were responsible for the production of the perpendicular felspar rock.

Localisation of crystallization along the cooling walls has frequently been invoked to explain the gradual variation, shown by many igneous masses, from rocks rich in early constituents near the margin to types rich in the later constituents near the centre. In order to bring this about it is necessary that the marginal impoverishment of the liquid in the early crystallizing substances shall be made good either by free diffusion from all parts of the liquid, or by convection currents which continually bring fresh magma to the marginal parts. Even among those who hold that one or other of these processes has produced certain cases of differentiation, there is a difference of opinion as to which process has been dominantly responsible. Becker [1897 B] maintained that convection currents in the magma were the cause of the differentiation while Harker [1894, p. 328 and 1908, pp. 317—20] considered that the variation in the Carrock Fell gabbro, which seemed to be an example of this kind of differentiation, was to be explained by diffusion of the early crystallizing material towards and the late crystallizing material away from the margin. Bowen [1915, p. 12] considered that diffusion in silicate magmas could not have been as free as would be needed for effective differentiation of large bodies of rock since "when centres of crystallization are once established, diffusion as free as that postulated should lead only to the continued growth of the crystals so initiated. There is no reason why new centres should be established. The result should be the growth inward from the contact of huge crystals of the minerals of early separation". The rock which, according to Bowen, should result from a high degree of free diffusion is similar to the perpendicular felspar rock. Bowen, however, goes on to say that convection currents should give a result similar to that produced by diffusion. In 1921 Bowen presented experimental data on the rate of diffusion of molten diopside into molten plagioclase and from this he concluded that diffusion, set up by the crystallisation of material along the cooling walls of an intrusion, could only produce relatively small differentiation effects in the time which would be available on any reasonable assumptions [1921, pp. 312—16]. While this would seem to dispose of the likelihood of diffusion having produced gradual variation in rock types over distances amounting to hundreds of metres, it does not preclude the possibility that the perpendicular felspar rock was formed in this way, since it only occurs in reefs 10—100 centimetres in width.

The above discussion suggests that the perpendicular felspar rock may have formed either as a result of particularly free diffusion or as a result of currents in the magma, and so far our observations on the rock itself have not provided a means of distinguishing between these two possibilities. However, evidence will be given later (pp. 262—66) for the existence of convection currents in the Skaergaard magma, and on these grounds it seems likely that circulating currents in the magma were the cause of the perpendicular felspar rock of the border group.

(c) The Major Unit of the Marginal Border Group.

1. Western and Eastern Margins.

On Mellemö and Ivnarmiut and indeed along most of the western side of the intrusion, the rocks forming the main part of the border series, are so variable that it is difficult to collect representative material. Although they stand up above the adjacent gneisses on one side and the layered series on the other, they are friable and break down easily to a gravel of coarse crystals which enhances the difficulty of collecting satisfactory specimens. For 20—100 metres from the inner contact the main rock is a massive, moderately coarse, olivine-hypersthene-gabbro, usually with conspicuous large patches of pyroxene enclosing felspar and occasionally olivine. The large pyroxenes sometimes extend into wavy masses set roughly at right angles to the wall of the intrusion (p. 26 and Pl. 3, fig. 2) and these are thought to have an origin similar to that of the perpendicular felspar crystals. The change from the chilled marginal olivine-gabbro to the coarser type is gradual, except for the interruption of the perpendicular felspar rock. The remaining part of the border group on Mellemö, that is from 100 to 250 metres from the outer contact consists of variable and coarse olivine-gabbros which give place near the inner contact to olivine-free types. In the outer 100 metres of the border group there are rounded inclusions of granophyre, surrounded by a narrow zone of hybrid material. In the rest of the border group irregular patches and streaks of granophyre, associated with ten or more times as much coarse hybrid material, are abundant and it is not easy to collect rocks which can with certainty be regarded as uncontaminated.

Rocks from Mellemö, 315, 320 and the analysed 1837 (Table XXI) and from Ivnarmiut, 4289 (Pl. 21, fig. 1), all of which were collected between thirty and sixty metres from the outer contact, are representative of the outer part of the major unit of the border group. Similar rocks occur in a similar position on Kraemers Ö (2292) and on the south-southeast ridge of Gabbrofjaeld (1766, 1767, 1819). On the whole these rocks tend to be coarser than the layered rocks and in this respect 4289 (Pl. 21, fig. 1) is typical. Their plagiocalse is moderately strongly and deeply zoned; the bulk has a composition between An_{60} and An_{66} and is thus about the same as that in the perpendicular felspar rock. There is no orientation of the felspars or other minerals except the large, wavy pyroxenes which are only found sporadically. Olivine occurs abundantly in grains up to five millimetres across and small felspars are occasionally embedded in it or indent the margin. The olivines of 315 and 320 have a negative 2V of about 80° indicating Fa_{36} while the olivines of 1837 and 4289 have $\gamma = 1.746$, indicating Fa_{37}. Both rhombic and monoclinic pyroxene

ANALYSES, OPTICAL PROPERTIES AND FORMULAE OF
GROUP AND TRANSITIONAL

	VI (from 4289 pp.151 and 52)	VII (from 1222 pp.172-73)	VIII (from 4087 pp. 76 and 152)	VIIIa[1])	VI	
					Mol. ratios	Atomic ratio to 6 (0,0 H)
SiO_2.............	50.39	49.68	50.58	49.98	0.8387	1.836 } 1.99
Al_2O_3............	3.54	4.99	4.06	4.49	0.0347	0.152 }
TiO_2.............	0.87	1.17	0.55	0.61	0.0109	0.024
Fe_2O_3............	1.95	1.12	1.39	1.53	0.0122	0.053
FeO.............	8.47	11.76	11.07	9.81	0.1179	0.258
MnO	0.19	0.16	0.25	0.28	0.0027	0.006
MgO	15.82	12.79	22.07	21.76	0.3923	0.859 } 1.98
CaO.............	18.41	17.97	10.01	11.07	0.3283	0.719
Na_2O............	0.70	0.56	0.16	0.18	0.0113	0.049
K_2O.............	0.14	0.13	0.01	0.01	0.0015	0.008 }
H_2O+...........	0.11	0.18	0.16	0.18	..	..
H_2O-...........	0.04	0.15	0.09	0.10	..	..
	100.63	100.66	100.40	100.00	..	..

	2V[2])(+ve)	$\gamma \wedge c$[2])	Refractive Indices[3]			Pleochroism
			α	β	γ	
VI.............	46	39	1.684	1.692	1.712	weak
VII.............	42-52	38	1.690	1.697	1.716	weak
VIII.............	42°	39°	1.672	1.678	1.701	weak

[1]) After subtracting 9.6 % hypersthene (Fs_{41}) and recalculating to 100.00 %. It is doubtful if this estimated composition of the clinopyroxene is reliable (see text p. 153).

[2]) ± 2°.

[3]) ± 0.002.

[4]) Neglecting other constituents and recalculating to 100 %.

occur, the former having the usual orientated inclusions of monoclinic pyroxene and the latter the usually orientated iron ore inclusions. The monoclinic pyroxenes of 1837 and 4289 have a γ' of 1.708 and 1.711 respectively. It was thought that the clinopyroxenes of the border group might have a rather different composition from those of the layered series, and the large poikilitic pyroxenes of the border gabbro, 4289, from Ivnarmiut have been analysed (Table XIX) as it was fairly easy, using a drill, to obtain material almost free from contamination with the other minerals, before making a final separation with heavy liquids. The pyroxene proves to be low in sesquioxides and rather similar to that from 4077 (Table VII) but not like that from 4087 as we expected. It seems possible that the analysed material from the transitional border

XIX.

CLINOPYROXENES FROM THE ROCKS OF THE BORDER LAYERED SERIES.

	VII				VIII		
Mol. ratios	Atomic ratio to 6 (O,OH)			Mol. ratios	Atomic ratio to 6 (O,OH)		
0.8271	1.862	0.138	2.00	0.8322	1.835	0.165	2.00
0.0490	0.221	0.083		0.0441	0.194	0.029	
0.0146	0.033			0.0076	0.017		
0.0070	0.039			0.0097	0.043		
0.1637	0.369			0.1366	0.301		
0.0022	0.005		2.01	0.0039	0.009		2.04
0.3172	0.714			0.5391	1.190		
0.3204	0.721			0.1973	0.435		
0.0090	0.041			0.0029	0.013		
0.0014	0.006			0.0001	..		
..	..			..	..		
..	..			..	..		
..	..			..	..		

	Dispersion	Spec. grav.	Weight %[4])		
			Wo	En	Fs
VI..............	r > v	3.35	41	42	17
VII..............	r > v	3.37	41	35	24
VIII..............	r > v	3.34	24	57	19

rock, 4087, contained more orthopyroxene than has been allowed for and we doubt if the estimated composition of the clinopyroxene is reliable. Iron ore in the outer border rocks is rare, amounting to no more than two per cent. of the whole; it has roughly the same period of crystallization as the pyroxene. The monoclinic pyroxenes which form large poikilitic crystals, three centimetres or more long, contain abundant plagioclase and only a little olivine, which is always near the periphery. It is these large poikilitic pyroxenes that develop in places into the wavy masses set approximately at right angles to the margin. Away from the large pyroxene crystals the rock is a troctolite with approximately thirty per cent. of olivine, sixty per cent. of labradorite and the remainder rhombic and monoclinic pyroxene and ore. The name troctolite was used for these rocks in the field, but taking

COMPARISON OF MINERALS FROM THE WESTERⲚ

Border Group.

	Plag. Mol. % An	Oliv. % Fa	Clino-pyrox.
Olivine-free gabbro, 4298. Ivnarmiut	ca. 48	none	$\gamma' = 1.735$
Olivine-hypersthene-gabbro, 4289. Ivnarmiut..	ca. 62	Fa_{37} ($\gamma = 1.746$)	$\gamma' = 1.711$
Olivine-hypersthene-gabbro, 1837. Mellemö ...	ca. 66	Fa_{37} ($\gamma = 1.746$)	$\gamma' = 1.708$
Perpendicular felspar rock, 1851. Mellemö....	ca. 65	Fa_{30} ($\gamma = 1.732$)	$\gamma' = 1.702$

into consideration, the large scattered pyroxene crystals, the rock as a whole, is better described as olivine-gabbro. It is remarkable that the large pyroxenes enclose abundant plagioclase but seldom olivine; we can give no satisfactory explanation of this.

The olivine-hypersthene-gabbros from the border group are similar in many ways to the olivine-hypersthene-gabbros from the lower layered rocks, and data showing this similarity are given in Table XX. The analysis of the hypersthene-olivine-gabbro, 1837, from the border group, is not far removed from that of the perpendicular felspar rock, 1851, and it also resembles fairly closely both the analysed hypersthene-olivine-gabbro from the layered series and the average chilled marginal gabbro (Table XXI). Its composition is compared with that of the layered series in the variation diagram, Fig. 31.

A specimen, 4298, from the inner part of the marginal border group collected at the south-east corner of Ivnarmiut, about three hundred and fifty metres from the outer contact and only six metres from the inner contact with the layered series, may be taken as representing the uncontaminated rock of the innermost part of the border series at this place (Pl. 21, fig. 2). The plagioclase which forms about fifty-five per cent. of the rock, is not so strongly or deeply zoned as it is in the olivine-gabbros of the outer border group, but it is much more zoned than in the layered series. Its composition, from refraction index tests, approximates to An_{48}. Neither olivine nor quartz is present. Brown monoclinic pyroxene, with $\gamma' = 1.735$ indicating a decidedly ferriferous type, is more abundant than rhombic, and it has the usual orientated iron ore inclusions. The orthopyroxene has continuous, parallel sheets of monoclinic pyroxene inclusions exactly

BORDER GROUP AND THE LAYERED SERIES.

Layered Series.

	Plag. Mol. % An	Oliv. % Fa	Clino-pyrox.	Height in metres
Hortonolite ferrogabbro, 1907	40	60 ($\gamma = 1.795$)	$\gamma = 1.749$	1900
Hortonolite ferrogabbro, 4290	41	60 ($\gamma = 1.797$)	$\gamma' = 1.739$	1500
Middle gabbro, 3662	45	none	$\gamma = 1.735$	1200
Olivine-hypersthene-gabbro, 4077........	56	37 ($\gamma = 1.748$)	$\gamma = 1.722$	500
Olivine-hypersthene-gabbro, 4087........	61	34 ($\gamma = 1.739$)	$\gamma = 1.701$	0

as in the rocks of the middle gabbros of the layered series. Iron ore forms approximately 10 per cent. of the rock and has a fairly early period of crystallisation. Rocks resembling this olivine-free gabbro probably occur in a belt which runs south from the south-east corner of Ivnarmiut and gradually withdraws from the inner contact. Thus 1731 from the south coast of the Skaergaard Peninsula and 300 metres from the inner contact, is olivine-free and in other ways is also similar. Crossing over the inner contact near the locality of 4298, the layered rocks are olivine bearing and have plagioclase which is An_{41} (see for example 4296, which was collected only 9 metres from 4298 and which is described on page 124).

The olivine-free gabbro, 4298, of the inner border group is strikingly similar in average mineral composition, and also in the detailed character of the minerals, to the middle gabbros of the layered series (Table XX). Analysis shows that its composition is almost a mean between that of the average middle gabbro and the ferrogabbro, 1907 (Table XXI); its composition must correspond closely to that of the highest middle gabbros or lowest ferrogabbros of the layered series (Fig. 31). Since this unusual rock type has been developed in the border group as well as in the layered series, it must be concluded that the lines along which the border group and layered series developed were essentially the same, despite considerable differences in the structures and situation of the rocks.

The change from the coarse olivine-gabbros of the outer border group to the olivine-free gabbro just described is gradual if we exclude the complications due to hybridisation with the granophyre inclusions. Contamination has produced obvious hybrids of coarse quartz gabbro, and it has also produced olivine-free gabbros which are less obviously hybrid in origin. Thus a rock, 1856, collected about the centre of Mellemö, is free from olivine and a little quartz is present. It has strongly and deeply zoned plagioclase from An_{60} to An_{40}. There is a little rhombic pyroxene with the usual inclusions, but the bulk of the pyroxene is monoclinic with $\gamma' = 1.710$. In the olivine-free layered rocks the monoclinic pyroxenes have γ' ranging from 1.734—1.738, so that the pyroxene of this olivine-free border rock is markedly dissimilar. This rock which is believed to be a hybrid, is adjacent to a thoroughly coarse and obvious hybrid rock, 1858, consisting of bladed pyroxenes up to 10 centimetres long, strongly zoned plagioclase and abundant quartz. Another rock 1249, which forms a resistant band on the upper side of a granophyre inclusion 150 metres from the outer contact on Mellemö has only two small remains of olivine and no rhombic pyroxene, in the thin section examined. It is otherwise rather similar to 1856 and it is also regarded as a hybrid. The border rocks which in the field seemed to be uncontaminated, such as those considered in Table XX, have a mineralogical

TABLE XXI.

ANALYSES OF GABBROS FROM THE MAYOR UNIT OF THE MARGINAL BORDER GROUP.

	XVI (1837)	XVII (4298)	Norms				Modes		XVI	XVII
			XVI		XVII				XVI	XVII
SiO_2	47.01	45.51	Or. 1.67			2.22	Plag.		64	56
Al_2O_3	19.12	14.36	Ab. 19.91			20.44	Clino-pyrox.		8[1])	35[1])
Fe_2O_3	0.71	5.04	An. 40.59			27.24	Ortho-pyrox.		2	
FeO	9.12	12.65	Di. . . . 5.34	10.39	8.47	16.76	Olivine.		25[2])	. .
MgO	8.98	5.38		3.20		4.20	Ore.		1	9
CaO	10.88	9.91		1.85		4.09				
Na_2O	2.33	2.43	Hy. . . 1.60	2.52	8.90	17.48				
K_2O	0.28	0.35		0.92	8.58					
H_2O+	0.24	0.12	Ol. . . 12.46	20.82	0.28	0.69	Ratios			
H_2O-	0.09	0.10		8.36	0.41					
TiO_2	1.47	3.96	Ilm.	2.89		7.45	$\dfrac{(FeO+Fe_2O_3) \times 100}{MgO+FeO+Fe_2O_3}$		52	77
MnO	0.08	0.20	Mt.	0.93		7.19				
P_2O_5	0.06	0.32	Ap.	0.34		0.67				
	100.37	100.33	H_2O	0.33		0.22	$\dfrac{Fe_2O_3 \times 100}{FeO + Fe_2O_3}$		7	28
S. G.	3.01	3.22	Plag. $Ab_{33}An_{67}$		$Ab_{43}An_{57}$					
			Diop. $Wo_{51}En_{31}Fs_{18}$		$Wo_{51}En_{25}Fs_{24}$		$\dfrac{K_2O \times 100}{Na_2O + K_2O}$		11	12
			Hyp. $En_{64}Fs_{36}$		$En_{51}Fs_{49}$					
			Oliv. $Fo_{60}Fa_{40}$		$Fo_{41}Fa_{59}$					

[1]) Includes a little chlorite.
[2]) Includes a little talc.

COMPARISONS.

	XVI	I	XIIIa	XVII	IVa	V
SiO_2...............	47.01	45.48	47.92	45.51	46.90	44.81
Al_2O_3..............	19.12	16.41	18.87	14.36	16.55	13.96
Fe_2O_3..............	0.17	2.09	1.18	5.04	2.96	3.75
FeO..............	9.12	9.29	8.65	12.65	12.18	16.66
MgO	8.98	11.65	7.82	5.38	5.80	5.54
CaO..............	10.88	10.46	10.46	9.91	9.67	8.53
Na_2O..............	2.33	2.06	2.44	2.43	2.97	3.35
K_2O..............	0.28	0.27	0.19	0.35	0.21	0.33
H_2O+.............	0.24	0.77	0.41	0.12	0.21	0.34
H_2O-.............	0.09	0.26	0.10	0.10	0.05	0.19
TiO_2..............	1.47	0.94	1.40	3.96	2.61	2.55
MnO	0.08	0.06	0.11	0.20	0.14	0.17
P_2O_5	0.06	0.05	0.07	0.32	0.06	0.08
	100.37	99.79	99.62	100.33	100.31	100.26

XVI. Olivine-gabbro, 1837, Border Group. 30 m. from contact, Mellemö, Skaer-
    gaarden. Kangerdlugssuaq.

  I. Hypersthene-olivine-gabbro (Table X, p. 92).

XIIIa. Average of chilled marginal gabbros (Table XVII, p. 140).

XVII. Olivine-free gabbro, 4298, Border Group. 6 m. from inner contact with the
    layered series, S.E. corner of Ivnarmiut, Kangerdlugssuaq.

  IVa. Average of Middle Gabbros 3662 and 3661 (Table XII, p. 97).

   V. Hortonolite ferrogabbro (Table XIV, p. 106—07).

and chemical composition which can be closely paralleled by rocks
from the layered series. Those such as 1856 and 1249 are not similar to
any rocks of the layered series; they are regarded as hybrids, related
to other, obviously hybrid rocks of the border group which are described
in section VIII.

2. Northern Margin.

The chilled olivine-gabbro, 4093, occurring along the outer contact
on Uttentals Plateau, and already described (p. 139), is succeeded inwards
by coarser olivine-gabbros which enclose many inclusions of gabbro-
picrite and also several of norite, metamorphosed gneiss from the metamorphic complex, and metamorphosed sediment. The olivine-gabbro of
the northern border may have large wavy pyroxenes and one case of
a reef of perpendicular felspar rock was found. Some of the border
gabbro weathers a strong brown colour due to the abundance of olivine
and thus approaches the gabbro-picrite in appearance and composition.
Neither in the field nor in the laboratory is it possible to distinguish

sharply between the olivine-gabbros of the border group and the overlying rocks which have been classified as the transitional layered series.
For 100—150 metres adjacent to the contact, the olivine-gabbro shows
practically no banding and contains abundant inclusions especially of
gabbro-picrite; this we have considered to be the border group proper.
Inwards, banding of the false-bedded kind develops and rocks such as
4087, the lowest member of the transitional layered series, are found.
The change from the false-bedded type of banding to more and more
continuous banding is gradual and it is not until some one and a half
kilometres from the margin that the banding is sufficiently extensive
and even, to be called layering; the analysed rock, 4077, belongs here.
There is not only a gradual change in the nature of the banding, but
also in the mineral composition and microstructures of the rock. Whereas
on the east and west borders the division between the border group
and layered series takes place in about a metre, along the northern
border there is a transition taking place in five or ten metres to poorly
layered rocks—the transitional layered series—which still have some
features of the border group.

Two rocks will be described to show approximately the range in
the olivine-gabbros of the border group on Uttentals Plateau. The first,
4088, is a brown weathering, rather olivine-rich gabbro collected about
30 metres from the margin and about 20 metres below and to the north
of the lowest rock of the transitional layered series, 4087. Under the
microscope it is seen to consist of about 50 per cent. of plagioclase, the
rest being equal amounts of olivine and pyroxene (including both
rhombic and monoclinic) with a little iron ore. The plagioclase is deeply
zoned varying in composition from about 72—60 per cent. of anorthite.
The olivine occurs in rounded masses up to 2 millimetres across and it
sometimes contains small felspar crystals. Both the monoclinic and
rhombic pyroxenes occur in small ophitic masses enclosing felspar and
olivine. The monoclinic pyroxene, a little more abundant than the
rhombic, had a period of crystallization beginning slightly earlier than
the rhombic; the usual orientated inclusions of iron ore are present.
The rhombic pyroxene contains coarse, poorly-orientated, blebby
inclusions of monoclinic pyroxene. Iron ore occurs in small, scattered,
ophitic patches enclosing olivine and felspar; it apparently crystallized
mainly before, or during the early stages of, the pyroxene crystallization
period. A little biotite is usually associated with the ore. The second
rock, 4089, collected about the same distance from the margin and
200 metres west of 4088, contains less olivine and is therefore less brown
weathering; it has large ophitic pyroxenes as other border gabbros,
and these sometimes extend into wavy sheets. Under the microscope
this rock is seen to be very similar to 4088. It contains a little more

TABLE XXII.
MINERAL COMPOSITION OF NORTHERN BORDER AND RELATED ROCKS.

	Plag. Mol. % An	Oliv. % Fa	Clino-pyrox. γ or γ'	Ortho-pyrox. % Fs
Hypersthene-olivine gabbro, 4077. Main layered series.	56	37 ($\gamma = 1.748$)	$\gamma = 1.722$	45
Hypersthene-olivine gabbro, 4087. Transitional layered series.	61	34 ($\gamma = 1.739$)	$\gamma = 1.701$	41
Hypersthene-olivine gabbro, 4088. Northern border group.	ca. 65 (zoned	33 ($\gamma = 1.737$)	$\gamma' = 1.716$	not determined
Perpendicular felspar rock, 1851. Mellemö. Western border group.	ca. 65	31 ($\gamma = 1.732$)	$\gamma' = 1.702$	not determined
Gabbro-picrite, 1682, inclusion in northern border group.	65—45	19 ($\gamma = 1.709$)	$\gamma' = 1.698$	20

plagioclase which has about the same range of composition as that of 4088, but the zoning goes deeper. The olivine is in somewhat larger crystals, often enclosing plagioclase or having its margin indented by plagioclase. In sections not cutting the large, ophitic, monoclinic pyroxenes, rhombic pyroxene is dominant over monoclinic. The rhombic pyroxene shows the lamella structure under crossed nicols (cf. p. 83) and contains the usual inclusions of monoclinic pyroxene; it probably finished crystallizing after the monoclinic. The monoclinic pyroxene, iron ore and associated biotite are the same as in 4088. As in the rocks of the outer marginal border group along the east and west margins, the large ophitic monoclinic pyroxene crystals which are as much as 2 centimetres in length, enclose abundant plagioclase but seldom olivine. Three other rocks belonging to the border group, 1709 from the northeast projection of Uttentals Plateau and 1678 and 1683 which are from the main part of the plateau, enclose gabbro-picrite masses and are similar to one or other of the rocks just described.

The main rock of the northern border group is thus similar to the outer part of the border group on the west and east margins (Table XXII). It is also similar to the lowest rock of the transitional layered series, 4087, except that the latter is coàrser in grain, has plagioclase which is more acid, and the rhombic pyroxene is less abundant relative to the monoclinic. Other points of distinction are that the transitional layered rocks do not contain the large, ophitic, monoclinic pyroxene crystals or so much biotite, and the lamella structure in the rhombic pyroxene is not found.

The gabbro-picrite, forming about a third of the border group on Uttentals Plateau, occurs in blocks from a small size up to 20 metres across which are embedded in the olivine-gabbro just described. The blocks, the larger of which are much veined by olivine-gabbro, are always rounded, and there is no sign of the olivine-gabbro being chilled against them. Similar gabbro-picrite masses occur at the foot of the point projecting north-east from Uttentals Plateau (1710) and in the border group to the north-north-east of Gabbrofjaeld (3030). Gabbro-picrite masses are occasionally found as rounded blocks in the eastern and western border groups near the outer contact (eg. 1820 and 2294). The gabbro-picrite masses in the border group of Uttentals Plateau are fairly constant in type and an example, 1682, collected a kilometre north of Uttentals Sund has been analysed. This rock (Pl. 19, fig. 2) contains 65 per cent. of olivine, the rest being plagioclase, rhombic and monoclinic pyroxene and iron ore. The olivine occurs in crystals which are usually about 4 millimetres across, and their distribution in space must be that of a pile of grains in contact, although in sections they are only occasionally seen to be touching. Looked at broadly the olivines show their characteristic crystal outline but in detail the margins have a cuspate form (better seen in 4090) which suggests re-solution. The composition of the olivine is about Fa_{20} ($\gamma = 1.709$, —ve $2V = 89°$). The crystals are intersected by cracks, the larger of which have been widened by decomposition, giving serpentine and iron ore; these are conspicuous in the microphotograph. The plagioclase and monoclinic pyroxene occur as small crystals, about 0.5 millimetres long, embedded in small poikilitic patches of rhombic pyroxene. The plagioclase is strongly zoned from An_{65} to An_{45}. The clinopyroxene, $\gamma' = 1.698$ and (+) $2V = 58°$, is definitely earlier than the rhombic and is in small rectangular crystals—an unusual habit for the pyroxenes of the Skaergaard rocks. The usual orientated schiller inclusions of iron ore are present. If it is assumed, as seems likely, that sesquioxides are unimportant, then by using the modified diagrams of Tomita (Deer and Wager 1938, figs. 2 and 3) the composition may be roughly estimated as Wo_{47}, En_{48}, Fs_5. The rhombic pyroxenes with —ve $2V$ about 82° have probably about 20 per cent. of the ferrosilite molecule. They do not contain blebby inclusions of monoclinic pyroxene as is usual in other Skaergaard rocks, but they show fine lamella structures under crossed nicols which is a particular feature of the olivine-gabbros and gabbro-picrites of the northern border. Iron ore, with which a little biotite is often associated, occurs as rare grains embedded in the olivine, or as small patches surrounding plagioclase and monoclinic pyroxene. It has also been formed along the cracks in the olivine.

Several similar rocks have been sectioned: 1676 is similar except that there is less ophitic rhombic pyroxene; 4094, from the largest continuous mass of gabbro-picrite about 20 metres across, is similar except that the proportion of olivine to the rest of the rock is only about 50 per cent. and the olivine has deep embayments suggesting solution along early cracks. Gabbro-picrite blocks are scarce along the eastern and western margins. The most southerly found along the eastern margin, 1820, one and a half kilometres south-south-east of Gabbro-fjaeld (Peak 1360 m.) is very similar to the analysed rock. The most southerly along the western margin, 2294, three-quarters of a kilometre north of the southern end of the outcrop on Kraemers Ö, also shows no significant difference from the analysed specimen.

All the gabbro-picrite has a mineral composition and microstructure which suggests that it is closely related to the border olivine-gabbro. In the field we tended to collect types clearly differentiated from the enclosing olivine-gabbro but intermediate rocks, less rich in olivine, could no doubt be found. Examination of thin sections suggests that the gabbro-picrites were formed as an accumulation of olivine crystals surrounded by average gabbro. The collecting together of the olivine crystals may have taken place approximately where the rock now occurs, but this cannot be stated for certain since the gabbro-picrite forms rounded inclusions which are often fractured and injected, indicating some degree of tectonic or magmatic movement which may have transported the material from the actual place where the rocks were formed. It is probably significant that gabbro-picrite masses occur most abundantly along the northern margin where the wall of the intrusion is dipping gently inwards, and where early cognate xenoliths might reasonably be expected to accumulate. They occur only rarely along the east and west margins where the dip is steep, and they have not been found along the southern margin where the dip of the contacts is outwards.

The analysis of the gabbro-picrite 1682 (Table XXIII) gives a norm differing considerably from the mode as determined on the integrating stage. Some of the differences may be ascribed to the partial decomposition of the olivine which may have resulted in the removal of some of the silica and oxidation of some of the iron, changes which might well cause nepheline to appear in the norm and would also cause more normative magnetite than is given in the mode, since magnetite from the decomposition of olivine is not included. If fifty per cent. of olivine containing twenty per cent. of fayalite—the actual composition of the olivine of the gabbro-picrites from optical determinations—is abstracted from the bulk composition of the rock, the remaining material has the composition given in column F, Table XXII. This shows an approach to the composition of the chilled marginal gabbro which is regarded as representative of the original Skaergaard magma. The low silica and high ferric iron is probably due to the partial serpentinisation of the olivine of the analysed rock. The percentages of lime and soda suggest a slight concentration of anorthite in the plagioclase and of diopside in the pyroxene. The approach to the composition of the initial magma gives support to the view, obtained from examination of thin sections, that the gabbro-picrite is essentially the olivine-gabbro of the chilled margin enriched with about 45 per cent. of extra olivine. Closely similar rocks have been described from Skye by Harker [1904, pp. 374—85] and Bowen [1928, pp. 145—59]. Comparison may also be made with harrisite from Rum described by Harker [1908, p. 80] and a picrite from the Shiant Isles described by Walker [1930, p. 364 et seq.]. Both Bowen and Walker ascribed the rocks, as we do, to enrichment of an original gabbro or alkaline gabbro magma by early formed olivine crystals.

The name gabbro-picrite is used for this Skaergaard rock in default of a better. The original picrite of Tschermak was an olivine-rich facies of a teschenite and some petrographers have considered that the name should be restricted to rocks related to alkaline basic rocks [see for

TABLE XXIII.

ANALYSIS OF GABBRO-PICRITE.

XIV (1682)		Norm			Mode (vol. %)	
SiO_2	41.27	Or.		0.56	Plag.	15.8
Al_2O_3	8.71	Ab.		3.14	Clino-pyrox.	13.8
Fe_2O_3	2.69	An.		20.29	Hyp.	4.7
FeO	10.52	Ne.		1.42	Oliv.	65.0
MgO	27.09	Di.	6.61	12.57	Ore	0.7[1]
CaO	6.59		4.90			
Na_2O	0.69		1.06			
K_2O	0.13	Ol.	43.96	54.57	Ratios	
H_2O+	0.87		10.61		$\dfrac{(FeO + Fe_2O_3) \times 100}{MgO + FeO + Fe_2O_3} \cdot\cdot$ 33	
H_2O-	0.07	Ilm.		2.89		
TiO_2	1.54	Mt.		3.94	$\dfrac{Fe_2O_3 \times 100}{FeO + Fe_2O_3}$ 20	
MnO	0.16	H_2O		0.94		
P_2O_5	0.02					
	100.35	Plag. $Ab_{13}An_{87}$			$\dfrac{K_2O \times 100}{Na_2O + K_2O}$ 16	
		Diop. $Wo_{52}En_{40}Fs_8$				
S. G.	3.18	Oliv. $Fo_{80}Fa_{20}$				

[1] Excluding ore formed by decomposition of olivine.

COMPARISONS.

	XIV	A	B	C	D	E	F
SiO_2	41.27	40.82	40.62	44.06	40.90	44.61	42.03
Al_2O_3	8.71	10.66	8.93	12.16	7.56	10.86	17.46
Fe_2O_3	2.69	1.80	0.57	4.85	3.01	2.31	5.38
FeO	10.52	8.92	12.61	5.48	7.31	7.46	6.81
MgO	27.09	28.08	26.31	18.21	29.63	21.06	8.07
CaO	6.59	6.11	5.64	9.50	5.40	9.01	13.20
Na_2O	0.69	0.58	1.32	} 0.98 {	0.98	1.15	1.37
K_2O	0.13	0.21	0.13		0.37	0.19	0.27
H_2O+	0.87	2.00	2.19	} 3.80 {	2.98	1.17	1.73
H_2O-	0.07	0.16	0.61		0.13	0.05	0.18
TiO_2	1.54	0.16	0.82	..	1.70	2.25	3.09
MnO	0.16	0.19	0.39	..	0.34	0.16	0.36
P_2O_5	0.02	..	0.15	..	0.10	0.10	0.05
S	..	0.02	..	..	..	..	..
Cr_2O_3	..	0.25	..	..	0.11	tr.	..
$(Ni,CO)O$	..	0.11	0.03	..	..	..	..
CuO	..	0.05	..	..	..	..	..
CO_2	..	..	0.03	..	..	..	..
Cl	..	..	0.01	..	..	..	..
	100.35	100.12	100.36	99.34	100.52	100.38	100.00

XIV. Gabbro-picrite, 1682. Uttentals Plateau, 1 kilometre N. of Uttentals Sund. Kangerdlugssuaq.

   A. Harrisite. Dornabac Bridge, N. of Harris, Rum Inverness-shire [Harker, 1908]. Anal. W. Pollard.

   B. Picrite. S. face of Garbh Eilean, Shiant Isles [Walker, 1930]. Anal. E. G. Radley.

   C. Olivine-dolerite, verging on picrite. Aodann Clach, Heast Road 2 miles S.E. of Broadford, Skye [Harker, 1904]. Anal. T. Baker.

   D. Peridotite dyke, Coir'a'Ghreadaidh, Skye [Bowen, 1928]. Anal. M. G. Keyes.

   E. Picrite-dolerite dyke, Coire Labain, Skye [Bowen, 1928]. Anal. M. G. Keyes.

   F. Analysis XIV after subtracting 50 % Fa_{20} and recalculating to 100 %.

example, Harker in "Petrology for Students", British Petrographical Committee 1921, and Trogger 1935]. Olivine-enriched olivine-gabbros like the rocks here described cannot fairly be named peridotite as their plagioclase, which is fairly abundant, is labradorite, and they cannot be named olivine-gabbros because they are too rich in olivine. Confusion in nomenclature has existed ever since Rosenbusch's extension of the meaning of the word picrite and various authors have used the latter term for rocks which have not been proved to have any genetic relation with alkaline gabbros [e.g. Harker 1904, pp. 374—85, Bowen 1928, pp. 145—59, and Holmes "Petrographic Nomenclature"]. Since the petrographic committee in 1921 recommended that the term picrite

be restricted to rocks related to alkaline gabbros it has not been used
by itself for the Skaergaard rocks but instead the term gabbro-picrite
has been adopted. This has been done because the Skaergaard rocks
bear the same relationship to gabbro as picrite sensu strictu bears
to teschenite.

3. South Eastern and Southern Margins.

In the southern third of the intrusion the marginal border rocks
are found either adjacent to the higher layered series or adjacent to the
upper border group. The latter are distinguished as the high marginal
border group in contradistinction to the low marginal border rocks
adjacent to the layered series (see Fig. 9). Only an approximate boundary
between the border group and the layered series can be drawn on the
Skaergaard Peninsula because the transition is so gradual. The same
is true of the junction on the ridge running north from Hammersfjaeldet
where the change from the marginal border rocks with steep outward
dips to the upper border group with outward dips of 20—30° takes
place over a distance of about 100 metres. On the west ridge of Tinden
we were not able to see the banding clearly enough to use this as a
criterion for distinguishing between the upper border and marginal
border group. The boundary which we have put on the map is roughly
the point where olivine-bearing rocks, regarded as belonging to the
marginal border group, give place to olivine-free rocks of upper border
type.

On the west wall of Udløberen the analysed, chilled marginal rock,
1825, gives place gradually to coarser types. A specimen, 1826, 20 metres
from the contact is still a chilled rock and is very like the other analysed
chilled rock, 1724, from the west face of Tinden. A similar rock, 1728,
was collected 40 metres from the contact at the head of Hammersdalen.
Since Udløberen covers the outer contact the distance from the margin
of other rocks of Kilen can only be roughly estimated. A specimen, 1827,
from about 100 metres from the margin, is a fairly coarse olivine-gabbro
containing occasional granophyre xenoliths, such as 1829, which have
little hybrid material round them. Two rocks, 1830 and 1832, from the
west wall of Udløberen at an estimated distance of 300 metres from the
margin, were collected 400 metres apart but are both very similar,
thoroughly coarse, olivine-gabbros (Table XXIV). The plagioclase is
strongly zoned, the inner part lying between An_{60} and An_{65}, while the
outer part is An_{40} or less. It is characteristic of many crystals that there
is a sudden change from the basic labradorite, forming three-quarters
of the crystals, to the more acid type, forming the outer quarter. The
plagioclase is dusty throughout with orientated, iron ore inclusions as
in the middle gabbros. The olivine crystals attain 8 millimetres in length

TABLE XXIV.

COMPARISON OF MINERALS FROM THE SOUTH AND SOUTH-EAST MARGINAL BORDER GROUP AND THE LAYERED SERIES.

		Plag. Mol. % An	Oliv. % Fa	Clino-pyrox. γ or γ'	Ortho-pyrox. % Fs	Quartz
Border group	Olivine gabbro, 2275. 200 m from S. margin.	55	51	$\gamma'=1.718$	? trace	present
	Olivine gabbro, 2274. 100 m from S. margin.	65	46	$\gamma'=1.714$	present	absent
	Olivine gabbro, 1830. About 300 m. from S.E. margin.	65—40	42	$\gamma'=1.713$	present	trace
Layered Series	Hortonolite-ferrogabbro, 1907.	40	61	$\gamma=1.749$	absent	trace?
	Middle gabbro, 3662.	45	abs.	$\gamma=1.735$	58	absent
	Hypersthene-olivine-gabbro, 4077.	56	37	$\gamma=1.722$	45	absent

and $\gamma = 1.756$ indicating Fa_{42}. The bulk of the pyroxene is a brown monoclinic type with schiller inclusions and $\gamma'= 1.713$. In both rocks a small amount of rhombic pyroxene occurs which has sheet inclusions of monoclinic pyroxene, reminiscent of the rhombic pyroxenes of the middle gabbros, and of 4298 and 1731 from the inner border group of the Skaergaarden. Iron ore is moderately abundant. Quartz and micropegmatite occur sparingly in the interstices between the felspars and are often associated with apatite. The olivine and rhombic pyroxene tend to be replaced by serpentine. Except that the quartz and micropegmatite are definitely late, no order of crystallisation can be decided upon. The next outcrop to the west, at the same level as these rocks, is a typical ferrohortonolite-ferrogabbro of the layered series with apatite as a primary phase, and micropegmatite in considerable abundance (eg. 1833). On the top and west side of Kilen the rocks belong to the upper border group, and are coarse olivine-free gabbros with much hybrid material.

A similar series of rocks is found at the foot of the West ridge of Tinden (Table XXIV). Chilled rocks such as 1724 collected 30 metres from the contact, give place gradually to coarse olivine-gabbros such as 2274, collected 100 metres from the contact. The olivine crystals of this rock, which are as much as 8 millimetres across and enclose much plagioclase, are Fa_{46}. Monoclinic pyroxene, in ophitic masses up to 15 millimetres across with, $\gamma' = 1.714$, is abundant, and rhombic pyroxene, with blebby inclusions of monoclinic pyroxene, forms perhaps 10 per cent. of the total pyroxene. The plagioclase, about An_{65}, occurs as small well-shaped crystals embedded in pyroxene and olivine, or as larger ill-shaped crystals. Iron ore is very scarce. Two hundred metres from the outer contact the rocks become thoroughly coarse and here the analysed rock, 2275,—an olivine-gabbro with a little interstitial quartz— was collected. The plagioclase, averaging An_{55}, is deeply zoned, but not strongly, except for a narrow outer margin adjacent to the interstitial areas of micropegmatite. It is crowded with orientated, iron ore inclusions. Here, where dykes are abundant, the inclusions might have been ascribed to reheating if it were not that similar inclusions are present in the middle gabbros and rocks from the south-east margin (1830 and 1832) where dykes are scarce. The olivine has $\gamma = 1.775$ indicating Fa_{51} and it is a good deal decomposed giving pseudomorphs consisting largely of iron ore. The pyroxene ($\gamma' = 1.718$) is similar in appearance and textural relations to that in the quartz-bearing olivine gabbros of the south-east margin. A little hypersthene, somewhat decomposed, is present. Iron ore is fairly abundant, and crystallised at about the same time as the olivine and pyroxene. Interstitial micropegmatite is present in considerable amounts. To the north the rocks are olivine free, so far as our observations go, and they are considered to belong to the upper border group.

The mineral associations found in the rocks of the main part of the border group along the south-east and south margins are not the same as in the layered series or in the marginal border group elsewhere (Table XXIV). The minerals present can be matched with examples from different parts of the layered series—this applies even to peculiar varietal features such as the orientated ore inclusions in the felspars and the blebby type of clinopyroxene inclusion in the orthopyroxene—yet the mineral associations in any particular rock are not the same as in any single rock of the layered series. In the eastern and south-eastern border gabbros the plagioclase and clinopyroxene have the refractive indices characteristic of these minerals in the lowest visible rocks of the layered series while the olivine is as iron rich as that of the middle gabbros and quartz is present interstitially as in the ferrogabbros. The one example of these rocks which has been analysed, 2275, shows marked divergence from the trend in the layered series. There can be no doubt that the rock has been derived in some way from the Skaergaard magma since it lies within the envelope of the chilled marginal gabbro. Evidence will be given below which indicates that certain features of this rock are due to hybridisation with acid gneiss from the metamorphic complex. The composition of 2275 (Table XXV) is somewhat similar to that

TABLE XXV.

ANALYSIS OF OLIVINE GABBRO FROM SOUTHERN MARGINAL BORDER GROUP.

XVIII (2275)		Norm			Mode (vol. %)	
SiO$_2$	50.41	Qu.		1.74	Qu.	6
Al$_2$O$_3$	18.30	Or.		2.78	Plag.	53
Fe$_2$O$_3$	1.96	Ab.		23.58	Clino-pyrox.	27[1]
FeO	8.13	An.		35.86	Oliv.	4
MgO	4.25	Di.	7.89	15.46	Ore	10
CaO	11.20			4.40		
Na$_2$O	2.74			3.17		
K$_2$O	0.46	Hy.	6.20	13.46	Ratios	
H$_2$O+	0.32			7.26		
H$_2$O—	0.19	Ilm.		3.34	$\dfrac{(FeO + Fe_2O_3) \times 100}{MgO + FeO + Fe_2O_3}$	70
TiO$_2$	1.75	Mt.		3.02		
MnO	0.16	Ap.		0.34	$\dfrac{Fe_2O_3 \times 100}{FeO + Fe_2O_3}$	21
P$_2$O$_5$	0.25	H$_2$O		0.51		
	100.12	Plag. Ab$_{40}$An$_{60}$			$\dfrac{K_2O \times 100}{Na_2O + K_2O}$	14
		Diop. Wo$_{51}$En$_{28}$Fs$_{21}$				
S. G.	2.98	Hyp. En$_{46}$Fs$_{54}$				

[1] Includes some chlorite.

XVIII. Olivine gabbro. 2275. 200 m from outer contact, foot of W. ridge of Tinden, Kangerdlugssuaq.

of the upper border gabbros and further discussion of this rock will be postponed until these have been described.

4. Summary and Comparison with the Layered Series.

The northern part of the border group is narrow and has a poorly defined junction with the layered rocks. The main unit is a basic olivine-gabbro, containing plagioclase which is a little more basic, and pyroxene and olivine a little more magnesium-rich, than these minerals in the lowest layered rocks. There is also more rhombic pyroxene relative to the monoclinic than in the lower layered rocks. The minerals which are solid solutions are higher temperature varieties than those of the adjacent layered rocks. The abundant gabbro-picrite inclusions in the northern border must also be considered an early differentiate since the olivine they contain is Fa$_{20}$.

The border group on the east and west margins has a width of 300 metres and a much more definite junction with the layered series. The gabbros composing it are more variable than those of the northern border. A little inwards from the chilled marginal gabbro, there is the perpendicular felspar rock which contains early crystallising minerals as judged by their compositions; thus the plagioclase is An_{65}, the olivine Fa_{30} and the monoclinic pyroxene has $\gamma' = 1.702$. The coarse olivine-gabbro lying further inwards has probably slightly more acid plagioclase (exact estimates are impossible because of zoning), the olivine is Fa_{37} and the pyroxene has γ' varying from 1.708 to 1.711. These minerals are on the whole solid solutions of lower crystallisation temperature than those of the perpendicular felspar rock. At increasing distances inwards the rock is composed of later and later crystallising minerals until at the inner contact the plagioclase is An_{48}, the pyroxene is iron-rich with $\gamma' = 1.735$, and no olivine is present. The outer, coarse olivine-gabbro is very similar to the lowest-exposed layered rocks and there is a gradual change to gabbros which are very like the uppermost middle gabbros of the layered series and, like them, free from olivine. In mineralogical details, such as the inclusions in the rhombic pyroxene and plagioclase, the innermost border gabbro of the Skaergaarden also resembles the middle gabbros. The layered gabbro, adjacent to the innermost border rock, is a slightly later crystallising ferrogabbro, having plagioclase which is An_{41}, pyroxene with $\gamma' = 1.739$, and olivine which is Fa_{60} ($\gamma = 1.797$).

A close relationship between the border rocks and the layered series is also apparent if the chemical compositions are considered. In order to make graphical comparison between border and layered rocks we have plotted them against the ratio of total iron oxides to total iron oxides plus magnesia—a ratio which for short will be called the iron-magnesium ratio. This values depends mainly on the composition of the pyroxenes and olivine, although the amount of iron ore also affects it. In the development of the layered series there is a steady increase in the iron component of the pyroxenes and olivine, so that the iron-magnesium ratio increases during the evolution of the series. In making comparisons we might have used the modal or normative composition of the plagioclase but this would be much affected by hybridisation with acid gneiss, whereas the ratio we have used is less affected. In the variation diagram, figure 31, the general trend of composition in the layered rocks is compared with the analysis of the gabbro-picrite, 1682, the perpendicular felspar rock, 1851, the outer olivine-gabbro, 1837, and the inner, olivine-free border rock, 4298. The outer olivine-gabbro and the inner olivine-free gabbro—a specialised rock type—give values which fall very close to the curves of variation of the layered rocks; the values for the perpendicular felspar rock, 1851, fall fairly close, while the values for the gabbro-picrite, as would be expected, show a considerable divergence since some special processes has produced a concentration of olivine in the rock. The two analyses, 1837 and 4298, of what may be described as normal marginal border rocks, fit closely to the curves for the layered series, and it is clear from microscopic examination that other marginal border rocks from the northern half of the intrusion would do the same. Thus, excluding the chilled olivine-gabbro at the margin, and excluding certain special types (the gabbro-picrite and perpendicular felspar rock), there is a gradual change from the outer towards the inner marginal rocks, which is essentially the same as that taking place from below upwards in the layered series. At the inner contact this sequence of changes stops just short of the layered rock which occurs adjacent to the border group.

Evidence for the variation in the high, marginal border group, that is where it is adjacent to the upper border group, is more fragmentary. The chilled marginal olivine-gabbro gives place to coarser olivine-gabbro and this to still coarser olivine-quartz-gabbros, which have some resemblance to the middle gabbros of the layered series. The parallelism between these border rocks and the layered series is, however, not close and the example analysed, 2275, when plotted on the variation diagram (Fig. 32) shows a marked difference from any layered rock. On the whole the high, marginal border rocks are more closely related to the upper border rocks than to the low, marginal border rocks from the northern two-thirds of the intrusion which are adjacent to the layered series. The composition of the high marginal border rocks will be more fully dealt with when the upper border rocks are considered (pp. 179—184).

(d) The Upper Border Group.

1. The Type Area of Brödretoppen.

The top 200 metres of both Brödretoppen and Osttoppen are formed from a horizon of the upper border group consisting of coarse quartz gabbros of moderately constant type with little or no flow structures. Reefs of more acid material, passing into augite granophyre, occur at intervals and give a poorly defined banding dipping south at 30°. Two examples of the dominant gabbro have been analysed, 3052 from the summit of Brödretoppen, and 3050 from 100 metres lower on the west face of this peak. These two rocks, which are very similar in hand-specimens show platy felspar, up to 0.75 square centimetres in area, embedded in a dark green diallagic pyroxene. A feeble parallelism of the felspars is seen in the lower rock, 3050, but none in the other, and in neither can the interstitial quartz and micropegmatite be detected without microscopic examination.

ANALYSES OF QUARTZ-GABBROS

	XIX (3052)	XX (3050)	XXI (4163)	Norms XIX		Norms XX	
SiO_2	48.02	49.16	57.00	Qu.	6.72		4.32
Al_2O_3	15.40	15.83	13.02	Or.	1.11		2.22
Fe_2O_3	4.35	4.13	2.15	Ab.	22.01		24.63
FeO	9.30	8.17	9.25	An.	29.75		28.36
MgO	4.06	4.36	2.64	Di. 5.80	11.38	9.51	18.48
CaO	10.05	10.81	5.75	3.20		5.80	
Na_2O	2.62	2.90	4.11	2.38		3.17	
K_2O	0.22	0.35	0.99	Hy. 6.90	11.78	5.10	8.00
H_2O+	0.94	0.62	0.97	4.88		2.90	
H_2O-	0.07	0.06	0.41	Ilm.	7.45		6.84
TiO_2	3.88	3.59	1.78	Mt.	6.26		6.03
MnO	0.25	0.21	0.12	Ap.	2.02		1.01
P_2O_5	0.92	0.39	2.11	Calc.	0.50		..
CO_2	0.19	..	..	Pyr.	0.24		..
S	0.14	..	..	H_2O	1.01		0.68
ZrO_2	0.01	..	..				
SrO	0.12	..	..	Plag. $Ab_{42}An_{58}$		$Ab_{46}An_{54}$	
BaO	0.04	..	..	Diop. $Wo_{51}En_{28}Fs_{21}$		$Wo_{51}En_{31}Fs_{18}$	
Cr_2O_3	nil.	..	..	Hyp. $En_{60}Fs_{40}$		$En_{64}Fs_{36}$	
CuO	nil.	..	..				
NiO	0.01	..	..				
	100.59	100.58	100.30				
S.G.	2.97	2.97	3.02				

	XIX	XX	A	B	C	D	E	XXI	F	G	H
SiO_2	48.02	49.16	50.12	49.60	47.24	51.92	52.89	57.00	54.00	56.22	59.82
Al_2O_3	15.40	15.83	15.98	15.06	18.55	14.12	14.07	13.02	13.09	12.45	12.15
Fe_2O_3	4.35	4.13	4.91	5.29	6.02	3.96	2.83	2.15	3.53	3.09	3.69
FeO	9.30	8.17	6.31	5.00	4.06	8.73	9.37	9.25	8.45	7.58	7.27
MgO	4.06	4.36	4.43	4.44	5.24	5.56	3.59	2.64	3.49	2.78	2.07
CaO	10.05	10.81	10.86	9.69	11.72	9.35	8.18	5.75	5.55	5.93	6.20
Na_2O	2.62	2.90	3.60	2.62	2.42	2.49	2.63	4.11	3.27	3.82	3.28
K_2O	0.22	0.35	0.70	0.70	0.15	0.98	1.67	0.99	1.80	2.67	1.44
H_2O+	0.94	0.62	0.53	1.29	2.24	..	1.44	0.97	1.71	1.35	0.88
K_2O-	0.07	0.06	0.46	2.65	0.21	..	0.20	0.41	1.26	0.44	0.24
TiO_2	3.88	3.59	1.76	2.38	1.46	2.46	2.20	1.78	2.83	2.74	1.93
MnO	0.25	0.21	0.18	0.19	0.31	0.16	0.15	0.12	0.37	0.43	..
P_2O_5	0.92	0.39	0.08	0.29	0.26	0.27	0.76	2.11	0.31	0.50	1.30
CO_2	0.19	..	0.21	0.44	0.19	..	..	..	0.25	0.05	tr.
S	0.14	..	..	0.40	..	..	..	..	..	..	..
FeS_2	..	..	0.05	..	..	..	..	..	0.14	..	..
ZrO_2	0.01	..	..	..	..	..	..	..	..	..	..
SrO	0.12	..	..	..	..	..	..	..	..	..	..
BaO	0.04	..	0.04	tr.	nil.	..	..	..	0.02	0.04	..
Cr_2O_3	nil.	..	0.04	0.02	..	..	..	..	..	..	..
CuO	nil.	..	..	..	..	..	..	..	..	..	..
NiO	0.01	..	..	..	0.05	..	..	..	..	..	..
	100.59	100.58	100.26	100.06	100.12	100.00	99.98	100.30	100.07	100.09	100.27

XXVI,
FROM THE UPPER BORDER GROUP.

XXI			Modes		
				XIX	XX	XXI
	13.14	Qu.	21.9[1])	3	21[3])
	6.12	Plag.	55.5	57	42
	34.58	Clino-pyrox.	20.6[2])	37	28[2])
	13.90	Apatite	..	..	..
0.93	1.89	Ore.	2.0	3	9
0.30					
0.66					
6.30	18.18	$\dfrac{(FeO + Fe_2O_3) \times 100}{MgO + FeO + Fe_2O_3}$	77	74	81
11.88					
	3.34				
	3.02	$\dfrac{Fe_2O_3 \times 100}{FeO + Fe_2O_3}$	32	33	19
	5.04				
	..				
	..				
	1.38	$\dfrac{K_2O \times 100}{Na_2O + K_2O}$	8	11	20

$An_{71}An_{29}$
$Wo_{49}En_{16}Fs_{35}$
$En_{35}Fe_{65}$

[1]) Micropegmatite with chlorite, apatite, and quartz.
[2]) About one tenth replaced by chlorite.
[3]) Mainly micropegmatite and includes dusty borders to plagioclase.

XIX. Quartz-Gabbro, 3052. Summit of Brödretoppen, Kangerdlugssuaq.

XX. Quartz-Gabbro, 3050. 200 m below summit of Brödretoppen, Kangerdlugssuaq.

   A. Quartz-gabbro, Ring-dyke, Centre 3, Ardnamurchan [Richey, Thomas etc., 1930]. Anal. B. E. Dixon.

   B. Porphyritic Dolerite. Major Intrusion in vent, Centre I, Ardnamurchan [Richey, Thomas etc., 1930]. Anal. B. E. Dixon.

   C. Basalt, Lava, Porphyritic Central Type, Cruach Choireadail, Mull. [Bailey, Thomas etc., 1924]. Anal. E. G. Radley.

   D. Average of 6 analyses, Whin Sill, North of England [quoted in Daly and Barth, 1930].

   E. Olivine free dolerite, W.N.W. of Högsater, Breven, Sweden [Krokström 1932]. Anal. N. Sahlbom.

XXI. Acid quartz-gabbro, 4163. North face of Osttoppen. 450 m below the summit.

   F. Quartz-dolerite, Garbad Sill. Arran, Scotland [Tyrrell, 1928]. Anal. E. G. Radley.

   G. Allied to Craignurite, Glen More Ring-dyke E.S.E. of cairn on Cruach Choireadail, Mull [Bailey, Thomas etc., 1924]. Anal. E. G. Radley.

   H. Average of three rocks of the intermediate zone between the acid and the basic members of the Sudbury complex, N. of Creighton, Sudbury [Phemister, 1937]. Anal. T. C. Phemister.

The summit rock, 3052, contains a plagioclase which is strongly zoned towards the margin, but the central three-quarters is fairly homogeneous and, from extinction angles, is estimated as An_{60}. The monoclinic pyroxene ($\gamma' = 1.716$) is present in light-brown, ophitic masses often becoming darkened towards the margin by schiller inclusions which, in this rock and similar upper quartz gabbros, are mainly parallel to (001), although examples parallel to (010) occur. These inclusions apparently consist of iron ore since, in some of the rocks, they become quite coarse and definitely opaque. They are thus different from the superficially similar inclusions of rhombic pyroxene in the monoclinic which is a feature of the pyroxenes in the lower layered series and outer marginal border group. Iron ore occurs in crystals of the same order of size as the plagioclase and pyroxene, and these three minerals, which together make up 80 per cent. of the rock, must have crystallised more or less simultaneously. The remaining 20 per cent., occurring as wedge-shaped masses between the plagioclase, consists of late-crystallising micropegmatite, quartz and chlorite. Adjacent to these areas the plagioclase, whose outer zone is oligoclase, passes into a cloudy felspar without lamellar twinning, and this is intergrown with quartz which is often in acicular units as in the basic hedenbergite-granophyre of the layered series. The chlorite and independent quartz crystals are usually found towards the centre of the interstitial areas. Stumpy apatite crystals embedded in the interstitial material are conspicuous and they may sometimes be seen in the pyroxene and felspar suggesting that some of the apatite crystallised fairly early. Late-stage effects are conspicuous. Thus the pyroxene, near microscopic belts of fracturing, is altered to chlorite dotted with ore and the fractured felspar crystals, sometimes sericitised, have chlorite in the cracks. In the upper border group late-stage effects are often more intense than in this particular rock.

The other analysed quartz gabbro, 3050, collected 100 metres below the summit of Brödretoppen, is not very different from the rock already described (Table XXVII and Pl. 22, fig. 1), and it has a closely similar chemical composition (Table XXVI). The plagioclase is less zoned, except adjacent to the late interstitial material, and from refractive index determinations the greater part is An_{56}. The brownish clinopyroxene ($\gamma' = 1.719$) occurs in more extensive ophitic masses and has less darkening due to schiller inclusions. A rather fresher specimen, 1222, from about the same horizon as 3050, provided the clinopyroxene which has been separated and analysed (Table XIX). The pyroxene proves to be intermediate in composition between that from the hypersthene-olivine-gabbro, 4077, and the middle gabbro, 3662. Like the other Skaergaard pyroxenes it is low in sesquioxides and can be calculated into the wollastonite, enstatite and ferrosilite molecules with the following result: Wo_{41}, En_{35}, Fs_{24}. Adjacent to the mesostasis, the brown pyroxene often gives place to a light green variety, without schiller inclusions, which is in optical continuity with the earlier brown pyroxene. These two types are generally to be found in the rocks of the upper border group, and the light green variety is always marginal and adjacent to the mesostasis. At the time of formation of the interstitial material the light green pyroxene was apparently the stable type. In some cases it appears to replace the brown pyroxene while in others it is probably the result of further deposition. It is noticeable that plagioclase when embedded in pyroxene is usually in smaller crystals than that outside, but there is nothing to prove that the pyroxene, plagioclase and also the iron ore, which occurs in crystals of comparable size, did not crystallise roughly simultaneously. The interstitial areas, which are much smaller than in 3052, are mainly composed of quartz and chlorite with only a little acid felspar. The higher silica and potash content of this rock compared with 3052 is not reflected in a greater amount of micropegmatite, at least in the sections which have been cut. However, there is less apatite than in 3052 and also less microfracturing and less late-stage alteration of pyroxene and plagioclase— observations which are in harmony with the lower P_2O_5 and water content shown by the analysis.

The rocks forming the upper 200 metres of Osttoppen are essentially similar to those just described (Table XXVII). A rock, 4149, collected 80 metres below the summit is a coarse quartz gabbro showing no parallelism of the felspars, and this is the dominant type. The plagioclase has the same average size, zoning and composition as that of 3052. The pyroxene is in relatively small ophitic masses with similar schiller inclusions. A green variety is developed near the interstitial areas as in 3050. The interstitial material is fairly abundant and consists of micropegmatite and chlorite. Late-stage sericitisation of the plagioclase and decomposition of pyroxene to give chlorite is similar in amount to that of 3052. A lower, but similar, rock, 4153, has abundant green pyroxene developed round the brown and $\gamma' = 1.721$ for the green pyroxene, while it is 1.711 for the brown. About 80 metres below the saddle between Osttoppen and Brödretoppen, that is about 160 metres below the summit of the eastern peak, the dominant rock, 4156, shows a slight parallelism of the platy felspars as in the lower analysed rock, 3050. Similarly the brown pyroxene, with patchy development of schiller inclusions, occurs in more extensive ophitic plates, which is a feature of the lower analysed rock of Brödretoppen, and the felspars are smaller when embedded in the pyroxene than outside. Micropegmatite, quartz and chlorite occur interstitially. Late-stage decomposition effects are slight. This rock thus shows detailed similarities with 3050. It would

TABLE XXVII.

COMPARISON OF MINERALS FROM THE UPPER BORDER GROUP AND LAYERED SERIES.

		Height[1]) in metres	Plag. Mol. % An.	Oliv. % Fa.	Clino- pyrox.	Ortho- pyrox. % Fs	Quartz
Brödretoppen	Quartz-gabbro, 3052, summit..........	3400	ca. 56	abs.	$\gamma' = 1.716$	abs.	present
	Quartz-gabbro, 3050, 200 m below summit	3250	ca. 54	abs.	$\gamma' = 1.719$	abs.	present
Osttoppen	Quartz-gabbro, 4153, 200 m below summit	3300	..	abs.	$\gamma' = 1.711$ brown $\gamma' = 1.721$ green	abs.	present
	Flow foliated quartz- gabbro, 4161, 325 m below summit.....	3200	ca. 65	abs.	$\gamma' = 1.720$	abs.	present
	Acid quartz-gabbro, 4163, 450 m below summit..........	3100	ca. 29	abs.	$\gamma' = 1.732$	abs.	present
Layered Series	Hortonolite-ferrogab- bro, 1907	1800	40	61	$\gamma = 1.749$	abs.	tr.?
	Middle gabbro, 3662..	1200	45	abs.	$\gamma = 1.735$	58	abs.

[1]) Approximate height on Layered Series scale.

seem that even in the heterogeneous upper border group, the rocks of some of the horizons are constant over considerable distances, a point of importance when we attempt to explain the origin of the rocks.

About the saddle between Osttoppen and Brödretoppen there is an admirable opportunity for examining the heterogeneity which is common to all the upper border rocks and which is believed to have been produced by hybridisation with acid gneiss inclusions. Lighter coloured bands, one or two metres thick and with indefinite boundaries, dip south at 30° (4151, 4152, and 4153); in other cases lighter material occurs as vague inclusions, and one of the more definite inclusions, 4154, which on a vertical face measured 1 × 0.5 metres, will be described. Where weathered it is light in colour, but the freshly broken surface is much darker and shows long-bladed pyroxenes up to 3 centimetres in length and a few large felspars. In thin section the rock is found to be made up of about 60 per cent. of micropegmatite with considerable areas of chlorite and quartz which are not intergrown with orthoclase. The orthoclase of the micropegmatite is cloudy and sometimes the quartz is acicular in habit. The plagioclase crystals are sporadic and much sericitised. They are strongly zoned and are basic andesine towards their centres. The bladed pyroxenes—so conspicuous in hand-specimens—form only 5 to 10 per cent. of the rock. The inner parts are brown with schiller inclusions; there is no ophitic habit, and the green outer parts are often euhedral towards the micropegmatite and quartz. Independent crystals of skeletal iron ore occur, and also patches of chlorite and iron ore, probably after pyroxene. Apatite is present but not particularly abundant. This rock which, from its form, was unmistakably an inclusion in the upper border gabbro and not a segregation from it, is believed to be either a metamorphosed fragment of rather basic grey gneiss from the metamorphic complex or normal grey gneiss made more basic by metasomatic processes on being immersed in the quartz gabbro. For this particular case we favour the former view and regard its persistence as a fairly definite inclusion as due to the original greater basicity of the rock which should have had the effect of making it melt at a rather higher temperature than more acid material.

The phases now present in this inclusion are the same as in the surrounding quartz gabbro but the proportions are very different. Among the lighter bands of the upper border group all gradations may be found between rocks of the composition of this inclusion and the dominant quartz-gabbro. For example, in the light band 4153, the quartz does not occur as separate masses as it does in the inclusion but is all intergrown to give a coarse micropegmatite which amounts to 30 per cent. of the rock. Strongly zoned plagioclase is more abundant and fresher, and pyroxene with similar characters is also more abundant. In this rock the green margin of the pyroxene may perhaps be due to recrystallisation of brown pyroxene containing schiller inclusions, as it contains a few larger ore inclusions which seem to have resulted from the aggregation of the finer schiller material. Apatite is particularly abundant and has hair-like inclusions parallel to the hexagonal axis. Another band, 4152, which is only slightly lighter than the average rock is essentially the same as the analysed, coarse quartz-gabbros 3050 and 3052, except that interstitial micropegmatite and chlorite amount to 30 per cent. An indefinite patch, 4155, towards the centre of the light band 4152, contains much quartz and epidote and a little highly alkaline hornblende and partly leucoxenised iron ore. Epidote is also found to be a fairly common accessory mineral in the altered inclusions of gneiss found in the marginal border group, The lighter bands of the upper border group, which form perhaps 20 per cent. of the total rock, are regarded as relics of acid gneiss which have become much hybridised and have been formed into bands or reef-like masses by slight flow. The

similarity of the acid material to the undoubted inclusions of granophyre, described below (pp. 185—99), provides further evidence for
this view.

Below the widespread quartz-gabbros of the type analysed the
upper border rocks become more variable. On Brödretoppen there is
a zone, extending downwards from nine hundred metres above sea-level
to the top of the Basistoppen raft at about seven hundred metres, which
consists of drusy granophyres and intermediate rocks. Some of these
have a mineral composition not unlike the small reefs of acid and intermediate rock which are considered to be derived from acid gneiss
inclusions. However, the acid material occurs, not as reefs, but as
extensive sheets, and basic granophyre resembling the basic hedenbergite-granophyre of the unlaminated layered rocks on Basistoppen
is associated with it. Although the relations of these rocks to the
surrounding quartz-gabbros were nowhere clear, we believe that they
are not to be correlated with the small acid inclusions and hybrids
but are late differentiates of the Skaergaard magma related to the
basic hedenbergite-granophyre of the unlaminated layered series (pp.
209—13).

On Osttoppen the quartz gabbros of the upper border group are
thick, and the two hundred metre horizon of granophyres and intermediate rocks is not found. Below 4156, already described, the foliation
on the average increases, although some bands are still massive. One
hundred metres below 4156 where the average foliation is at a maximum,
a series of specimens, collected in a vertical distance of 20 metres, may
be described as an example of the variation found. In the most foliated
band 4158 (Pl. 22, fig. 2) there are as many as 20 thin tabular plagioclase crystals within the thickness of 1 centimetre. These are flattened
parallel to (010); they range up to 1 square centimetre in area and are
only about 0.3 millimetres in thickness. Between the plagioclase in
some places there is pyroxene and iron ore in ophitic masses and in
other places there is micropegmatite. Every gradation exists between
this strongly foliated type and rocks of much the same composition
which have no parallel structures. Some of the latter, such as 4160,
are decidedly coarser than any found higher up, having tabular felspars,
at least 2 square centimetres in area, and diallagic pyroxenes of correspondingly large size. In a thin section of one of the moderately
foliated rocks, 4161, the plagioclase is seen to be but little zoned, and
from extinction angles the central half consists of An_{65}. The pyroxene,
$\gamma' = 1.720$, is similar to that in the analysed upper border gabbros;
it has ophitic habit and abundant schiller inclusions. There is no micropegmatite or quartz mesostasis. The slight zoning of the felspars and

the absence of micropegmatite are perhaps related facts. A coarse, unfoliated rock, 4160, from 10 metres higher, contains plagioclase which is a little more strongly zoned and which seems to have, on the average, a less anorthite content. The pyroxene is similar and is partly altered to chlorite adjacent to the interstitial, fine-grained micropegmatite and chlorite patches. Another eight metres above there is the extremely well foliated rock, 4158 already mentioned. This has similar plagioclase and pyroxene and in the one thin section examined there is at least 10 per cent. of quartz and micropegmatite. From the hand-specimen the distribution of the latter can be seen to be sporadic. Another moderately foliated rock, 4157, 3 metres higher is similar but has more micropegmatite and quartz. The section also shows a remarkable apatite crystal 1 centimetre long and slightly bent by pressure from adjacent plagioclase crystals. This crystal is the width of many others seen in these rocks, and its length is probably not greater than that of others which do not happen to be parallel to the slice. Much of the apatite in these rocks must be an early crystallised constituent as in the upper ferrogabbros.

A short distance below the zone of maximum foliation the rocks become massive once more. They are also thoroughly heterogeneous having so many blebby patches of granophyre and hybrid that a rock which could be regarded as free from contamination could not be collected. Two examples (4163 and 4164) of what is believed to be the less contaminated material have been sectioned, and one of them, 4163, has been analysed (Table XXVII). The plagioclase of these rocks is very strongly zoned, ranging from andesine to albite. In 4164 the pyroxene shows only slight ophitic tendencies, while in 4163 (Pl. 23, fig. 1) there is none. The disappearance of the ophitic texture is a feature of the most acid quartz-gabbros of the Skaergaard intrusion and of the hybrid rocks of intermediate composition. The pyroxene of 4163 has $\gamma' = 1.732$, indicating that it is an iron-rich type. Iron ore in both rocks is rather abundant. Very fine-grained micropegmatite with chlorite and quartz occurs interstitially. Both in mineral and chemical composition these rocks are intermediate between augite granophyres and quartz-gabbros, and we are aware of no satisfactory name which exactly fits them. They are too acid to be classed as quartz-gabbros and by some authors they might be named augite-diorites. Since, however, they are characteristically free from all traces of hornblende we shall distinguish them as acid quartz-gabbros. Below these rocks, where the north face of Osttoppen disappears under the glacier, a gabbro belonging to the Basistoppen Raft was found (1886).

2. Other Areas.

Rocks of the upper border group are also found on the west ridge of Tinden, on the south face of Sydtoppen, at Hammers Pas and on Kilen which extends north-north-west from Hammersfjaeldet. On the west ridge of Tinden they are not easy to study because the dyke swarm is dense. The rocks sectioned, 3053, 1717 and 2256, are all coarse quartz-gabbros but the plagioclase is a good deal replaced by chlorite. Some of the decomposition may be due to the dyke swarm. Vague patches of granophyre and hybrid rocks are abundant; one such rock, 3059, is similar to the inclusion, 4154, from Osttoppen; another, 1719 is a small granophyre mass with bladed augite crystals 5 centimetres long projecting into it and with epidote abundantly developed.

A coarse quartz-gabbro, 1765, collected from the south-east foot of Sydtoppen is fairly typical of the upper border gabbros. The clino-pyroxene is partly altered to hornblende—a rare change in these rocks. The felspars are much sericitised, and interstitially there is a good deal of orthoclase and chlorite with which is associated a little epidote and calcite but no quartz. The contact in this part of the intrusion was not examined and the presence of the marginal border group was not definitely proved; therefore, although probably present, it is not indicated on the map.

At Hammers Pas there is a coarse, quartz-gabbro, 1761, similar to those analysed from Brödretoppen. Associated with it are bands of a flow-foliated type with markedly tabular plagioclase crystals which are fresh and strongly zoned (e.g. 1758). Micropegmatite and chlorite is abundant interstitially. The horizon at Hammers Pas appears to be the same as the upper part of the foliated horizon below the saddle on Brödretoppen. Blebby inclusions of granophyre and hybrid are abundant (e.g. 1759), and also transgressive granophyre veins (e.g. 1757).

The top and west sides of Kilen consist of upper border rocks but the exposures are not so good as elsewhere owing to scree. Quartz-gabbros occur which roughly match the analysed rocks from Brödre-toppen. One example, 1876, consists of fairly fresh plagioclase, strongly zoned but only at the margins, ophitic pyroxene, with the usual patchy development of schiller structures, iron ore and interstitial micro-pegmatite and chlorite. The plagioclase has been shattered in places and veined by acid felspar. At the junction between brown pyroxene and the interstitial material, a green variety of pyroxene is sometimes developed but this is never found along the pyroxene plagioclase junctions. At other places, between the pyroxene and the interstitial chlorite and micropegmatite, a yellow-brown hornblende is developed. A little sphene is to be seen in the chlorite areas and there is abundant apatite with hair-like inclusions parallel to the hexagonal axis. Another rock, 1875, from this ridge resembles 1876 but, like 1765 from the S.E. ridge of Sydtoppen, orthoclase is present interstitially but not quartz. Epidote, with sphene embedded in it, is also an abundant interstitial mineral. Definite granophyre patches were not observed in the rocks of this ridge but coarse gabbros with bladed pyroxene 10 centimetre long are believed to be hybrids. An acid quartz-gabbro, 1874, is an abundant type low on the west side of the ridge and this resembles the analysed rock, 4163, from low on Osttoppen. The pyroxene is in independent, brown to green crystals without ophitic habit. Iron ore is rather abundant, and coarse micropegmatite, with apatite but not much chlorite, forms probably 40 per cent. of the rock. At the north-west point of this ridge there is an indefinite junction between the upper border rocks and the layered series; here occurs the coarse fayalite bearing quartz-gabbro, 1513 which has already been mentioned (pp. 112—13).

3. Summary and Preliminary Conclusions on the Origin of the Upper Border Rocks.

The upper border rocks frequently show flow structures, and parallel to these are sheet, or reef-like masses of more acid material. The direction of these features is roughly constant over considerable areas and the upper border rocks are, therefore, considered to be a succession of sheets whose relative thickness may be estimated from the observed dips. From the thickness and dips the relative heights of various specimens have been roughly calculated with the same datum line as used for the layered rocks (Table XXVII). Excluding the granophyre reefs and obvious hybrid rocks which probably amount to 10 per cent. of the total, there is apparently a fairly constant type of gabbro developed at any one horizon. Thus about the summits of Osttoppen and Brödre-toppen, the dominant quartz-gabbros are coarse rocks with ophitic pyroxene, zoned plagioclase of which the more basic, inner half is labra-dorite, iron ore, and interstitial micropegmatite and quartz. Neither olivine nor orthopyroxene, or their alteration products, have been identified with certainty[1]. A characteristic of the rocks is that they have suffered from considerable late-stage reactions. The brown pyroxene has been altered to, or surrounded by, more ferriferous green pyroxene and is scarcely ever altered to hornblende, nor is the iron ore altered to sphene. These two minerals are only found in obvious hybrids and are there associated with epidote. The dominant rock at lower horizons is acid quartz-gabbro. The pyroxene is not ophitic and, from refractive

---

[1] Olivine or othopyroxene, probably the former, was present in 1222 but is now completely decomposed.

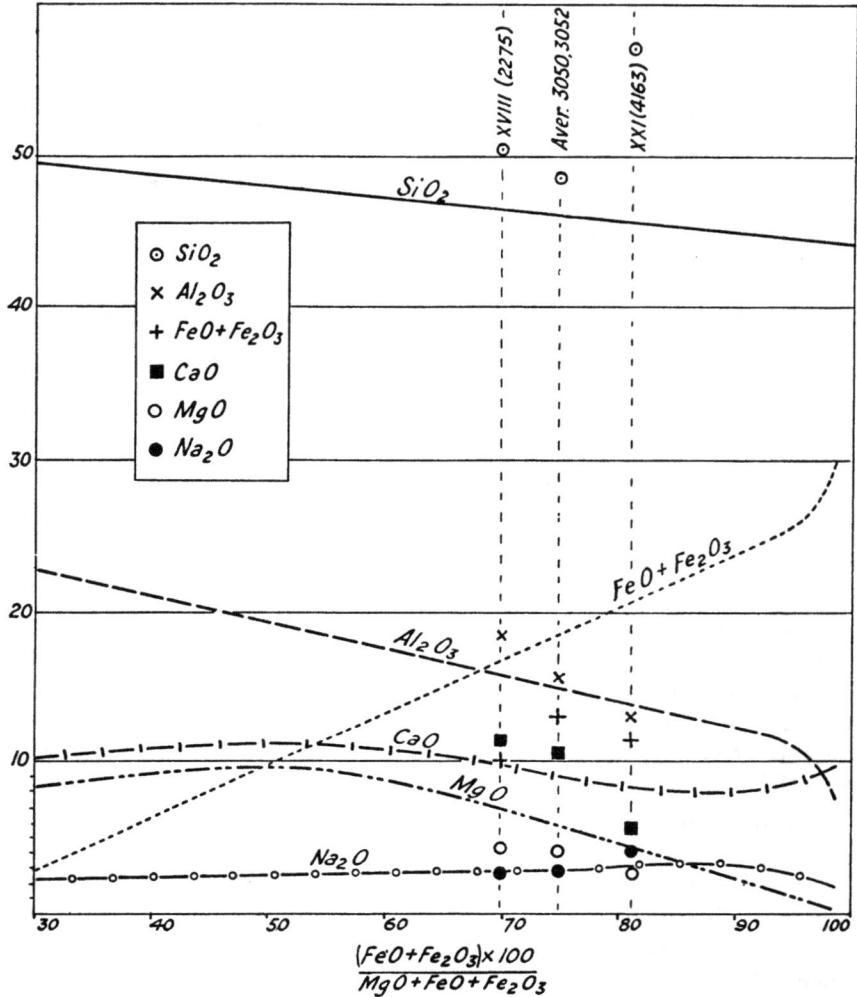

Fig. 32. Variation diagram of the layered series (taken from fig. 30) with plots of the analysed upper border rocks, and of the high marginal border rock, 2275.

index tests, is more iron-rich than in the higher rocks. The plagioclase is more acid and micropegmatite more abundant. The composition of the minerals in some of the dominant rocks at various horizons is given in Table XXVII. Changes in the composition of the minerals are by no means as regular as in the layered rocks but, on the whole, the plagioclase becomes more acid downwards and the pyroxene more ferriferous, changes which are the reverse of those taking place in the layered series.

There is also a fairly steady variation in bulk chemical composition with height. The amount of silica, soda and potash increases in passing from 3052 to 4163, while the amount of iron, magnesia and lime decreases. The percentages of the important oxides of the upper border rocks and of the high marginal border rock, 2275, are plotted in figure 32 on the variation diagram of the layered rocks which has already been used. It will be seen that both magnesia and total iron are low compared with layered rocks of similar iron-magnesium ratios, and it is also

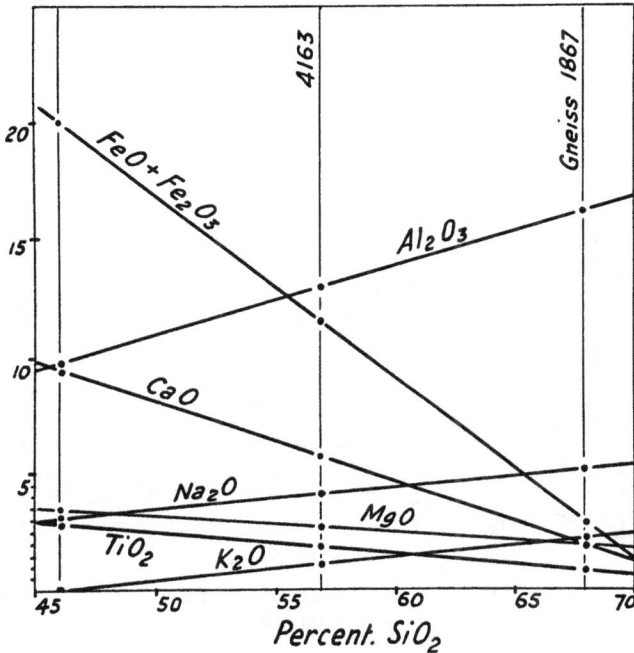

Fig. 33. Graph showing the composition of the material which, if added to acid gneiss, would give the quartz-rich upper gabbro, 4163.

clear from the graph that the amount of these oxides cannot simultaneously fit both the curves at whatever iron-magnesia ratio they may be plotted. The field evidence suggests the possibility that these rocks have been contaminated with acid gneiss. This hypothesis has been tested by drawing straight line graphs showing the material which, added to acid gneiss, would yield the various upper border rocks (Figs. 33, 34 and 35). In all cases the analysis of the gneiss which has been used is an adjusted analysis, since the particular specimen analysed seems to have been richer in K_2O than the average grey gneiss surrounding the intrusion. The value of K_2O has been taken as 2.0 which is close to the average of three determinations for the grey gneiss of Mellemö, and of several determinations for the granophyre inclusions which are later shown to have been originally grey gneiss (Table XXX and pp. 194—96). The amount of Na_2O has been adjusted so that the molecular proportion of Na_2O plus K_2O in the adjusted analysis is the same as in the original. The composition of the material to be added to grey gneiss to give the various upper border rocks and the high marginal border rock is given in Table XXVIII. In the case of 4163 the material is related in composition and magnesium-iron ratio to high hortonolite-ferrogabbros; in the case of 3050 and 3052 to high middle gabbros, and in the case of 2275 to low middle gabbros. These

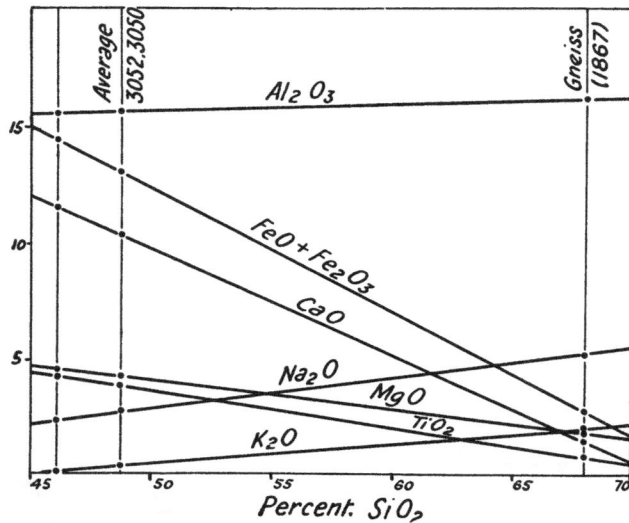

Fig. 34. Graph showing the composition of the material which, if added to acid gneiss, would give the upper quartz-gabbros 3050 and 3052.

results suggest that the upper border rocks and high marginal border rocks have been produced by hybridization between grey gneiss and the Skaergaard magma as developed at different stages of the differentiation. The abundant xenoliths of acid material and bands of intermediate rock which are found intimately associated with the upper border rocks in the field are in harmony with this hypothesis. The material which became mixed with the acid gneiss to give the upper border rocks is likely to have been a residual liquid of the Skaergaard magma and not the crystal fraction represented by the layered rocks, and comparison of Table XXVIII should therefore be made with the calculated composition of successive residual magmas made on later pages (pp. 217—24).

The analysed quartz-gabbros from Brödretoppen show two differences from the composition of the contemporaneous layered rocks which may not be the effects of hybridization; thus they are richer in water and richer in ferric oxide than the hypersthene-olivine-gabbros and middle gabbros of the layered series. The higher water content may be due to the fact that any water escaping from the Skaergaard magma would have to pass through the upper border group and there, during cooling, an opportunity would occur for some of it to be fixed

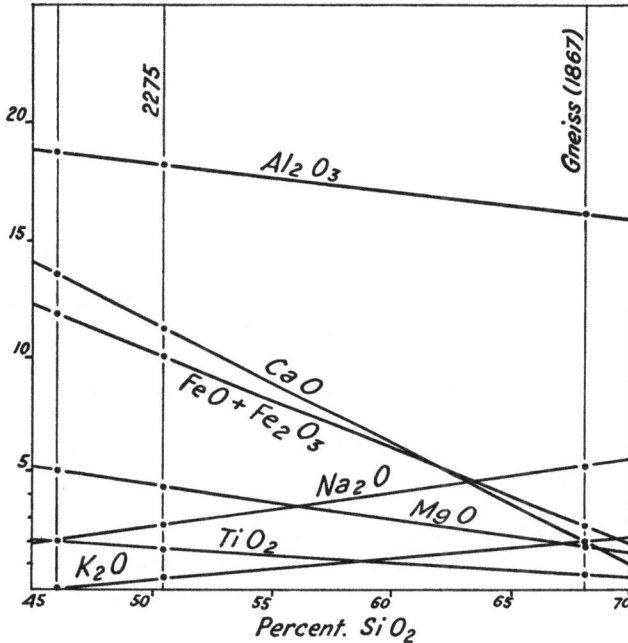

Fig. 35. Graph showing the composition of the material which, if added to acid gneiss, would give the quartz-gabbro, 2275, from the high marginal border group.

in hydrated, lower-temperature minerals. The equilibrium between ferrous and ferric iron in igneous rocks must be largly controlled by the reversible reaction:—

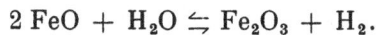

$$2\,FeO + H_2O \leftrightarrows Fe_2O_3 + H_2.$$

During the crystallization of the primary minerals of the layered series it is most unlikely that there would be separation of a gas but during late-stages of the solidification of the upper border group a gaseous phase may well have formed. When this occurred the above equilibrium would be controlled by the partial pressure of the H_2O and H_2. In places the gases would no doubt be able to escape by mass flow through fairly large openings in the rock but in other places the gases would only be able to diffuse through pores. Since the hydrogen molecule is much smaller than the water molecule, hydrogen would more easily diffuse

TABLE XXVIII.

	A	B	C
SiO$_2$	46	46	46
Al$_2$O$_3$	10	15.5	19
Fe$_2$O$_3$.........................	} 20	14.5	12
FeO			
MgO.........................	3.5	4.5	5
CaO	9.5	12.5	13.5
Na$_2$O	3	2.5	2
K$_2$O	0	0	0
TiO$_2$.........................	3	4.5	2
$\dfrac{(\text{FeO} + \text{Fe}_2\text{O}_3) \times 100}{\text{MgO} + \text{FeO} + \text{Fe}_2\text{O}_3}$	85	76	71

A. Possible composition of liquid (see Fig. 33) giving with acid gneiss the acid quartz-gabbro, 4163 of upper border group. (The total is low as P$_2$O$_5$ and other minor constituents have not been considered).

B. Possible composition giving average of the quartz-gabbros 3050 and 3052 of the upper border group (see Fig. 34).

C. Possible composition giving the quartz-gabbro, 2275, of the high marginal border group (see Fig. 35).

through the pores of the rock and escape. Thus, in so far as the escape of the volatile constituents was a matter of diffusion, there would be an increase in partial pressure of the water and the above reaction would tend towards the continuous production of ferric oxide. It seems possible that this process has contributed to the higher degree of oxidation of the iron in the upper ferrogabbros. The ratio of ferric to ferrous iron in basic rocks has been discussed by Phemister [1934, pp. 40—43] and the above point noted as a possibility. We put forward this explanation of the high $\dfrac{\text{Fe}_2\text{O}_3}{\text{FeO} + \text{Fe}_2\text{O}_3}$ ratio of some of the upper border rocks as a tentative hypothesis only, since the high ratio is not shown by the analysed upper border rock, 4163 as would have been expected. Some of the changes in the ratio of ferric to ferrous iron which are shown by the Skaergaard rocks are to be ascribed to crystallisation differentiation (see pp. 229—30) and this factor also seems quite as likely to have controlled the case discussed by Phemister [1934, p. 41] as the variation in the partial pressure of water and hydrogen.

VIII. INCLUSIONS AND HYBRID ROCKS

(a) Granophyre Inclusions and Hybrids.

Patches of granophyre occur sporadically throughout the border group but are most conspicuous in the outer part along the east, west and south contacts. In the inner part of the border group there are abundant patches of coarse hybrid material apparently formed from granophyre inclusions. In the upper border group granophyre is abundant, sometimes as definite inclusions but usually as small reef-like masses. In the layered series granophyre patches also occur and they are surrounded, and sometimes almost completely replaced, by hybrid material. They are frequent in the hypersthene olivine gabbros of the southern slopes of Uttentals Plateau; they get rarer as the series is ascended and are not found above the hortonolite ferrogabbro horizon.

1. In the Layered Series.

The granophyre and hybrid material which has been sectioned and partially analysed from the lower layered series was collected on the south-east slopes of Uttentals Plateau and five hundred metres from the shore of Uttentals Sund. The patches are here a few metres across, the outer part consisting of a thoroughly coarse plagioclase-pyroxene rock, of which 4072 is an example. Some of the plagioclase crystals are intergrown with quartz, and quartz, felspar and some decomposed ferromagnesian mineral occur interstitially. In this rock there are drusy cavities into which project well-shaped quartz, plagioclase and orthoclase crystals. The quartz has the form of the β-polymorph, the face (1121) being occasionally developed. A golden-brown, micaceous mineral is sprinkled over the orthoclase in the drusies and this is apparently the same as a mineral, found in a drusy granophyre inclusion from Mellemö, which has been analysed and shown to be stilpnomelane (p. 188). A granophyre, 4074, from the centre of a neighbouring mass is largely micropegmatite, single felspars of which extend over several square centimetres. In sections the felspars are seen to be

Fig. 36. Vesicular and porphyritic granophyre, 1841—part of an inclusion in the outer border group, Mellemö. Width of specimen 9 centimetres. The analysed stilpnomelane was extracted from the vesicles of this rock.

orthoclase and oligoclase, both of which are cloudy with decomposition products; dark green hornblende, iron ore and brown chlorite are also present in considerable abundance while zircon and apatite are accessory minerals. The proportion of orthoclase to plagioclase is roughly indicated by the alkali percentages which are: Na_2O, 4.35; K_2O, 2.83.

Granophyre and hybrid patches become less abundant as the layered series is ascended and the highest examples were found in the hortonolite-ferrogabbro near the Basishusene. At this horizon the amount of coarse hybrid material is often much greater than the granophyre, and in some cases all the granophyre has been assimilated in the production of hybrid. Drusy cavities have not been found in these granophyres and hybrids. When making the detailed map of the trough banding on the scale of 12 inches to a mile a coarse hybrid patch was found every hundred metres or so. A typical example, one and a half metres across and six metres high, occurs two hundred and fifty metres westsouth-west of the main house and about five metres from the coast. The outer thirty centimetres of this mass (2585) consists of a coarse plagioclase-hortonolite-pyroxene rock in which cross sections of the plagioclase crystals attain four square centimetres in area, and the olivine and pyroxene are on a similarly large scale. Extreme coarseness is a characteristic of the outer, hybrid part of these inclusions and makes them conspicuous in the field. The minerals of the outer hybrid part appear to be closely similar to those of the surrounding ferrogabbro but detailed work on them has not yet been done. The central part of this mass, 2584, is finer-grained and consists of about fifty per cent. of felspar, in crystals up to two centimetres in length, thirty per cent. of ferromagnesian minerals and about twenty per cent. of quartz. The total percentage of alkalies in this rock, 5.97, is lower than in most of the granophyre inclusions and this is reflected in the greater proportion of ferromagnesian minerals seen in the slice. The amounts of soda and potash are: Na_2O, 4.01; K_2O, 1.96. Plagioclase occurs in fresh strongly-zoned crystals ranging in composition from acid-andesine to albite. Outside this the felspar appears to be a perthite and then follows cloudy orthoclase and quartz. Pyroxene and amphibole form the bulk of the ferromagnesian constituents. The pyroxene is brownish centrally and becomes pale green towards the margin, thus resembling the pyroxene in the hybrids of the upper border group. The amphibole is also variable, the bulk being a greenish-brown variety but some is colourless and fibrous, and some dark green. Skeletal crystals of iron ore, and brown and green chlorite also occur. Apatite and zircon are accessory minerals.

2. In the Marginal Border Group.

Granophyre patches are particularly abundant in the marginal border group on Mellemö and Ivnarmiut. In the outer part of the border group they are rounded in form and have only a narrow development of coarse hybrid material (Pl. 13, fig. 1), while in the inner part, granophyre is only present as a head to an extensive streak of coarse, hybrid material. Drusies are always found in the granophyre masses; in some cases, 318 and 1840, the cavities are highly irregular; in others, 1836 and 1841, they are approximately spherical. The almost spherical form of the cavities in 1841 (Fig. 36) affords a proof that at one time the material of the rock was in a state of complete fusion. The fact that the cavities are still scattered irregularly throughout the rock indicates the high viscosity of the regenerated magma; with low viscosity the bubbles would have succeeded in amalgamating into one large cavity. The rock with spherical drusies, 1841, and also 1836, with moderately rounded drusies, have conspicuous porphyritic felspars while the two rocks, 318 and 1840, with irregular drusies have no porphyritic elements. The groundmass of those granophyre inclusions with porphyritic crystals is finer in grain than those without porphyritic crystals.

The granophyre inclusion, 1841, is seen under the microscope to consist of decomposed phenocrysts of plagioclase in a base of separate, granular crystals of quartz and felspar usually about 0.3 millimetres across (Pl. 24, fig. 2). A brown micaceous mineral, in minute ragged crystals often grouped together in bunches, is abundant along the junctions between the quartz and felspar, and is apparently similar to the stilpnomelane from the drusies which is described below. In this particular rock there is no tendency to micropegmatitic or granophyric textures. The groundmass felspar is so cloudy with decomposition products that it is difficult to determine, but it is probably microperthite. The inner nine-tenths of the phenocrysts, which are very definitely euhedral, is only weakly zoned but there is strong zoning outside this, and the outer felspar of the phenocrysts has the same relationships to the quartz as have the grains of alkali felspar in the base. Occasional sagenetic webs of ilmenite occur but no sphene. A little apatite and chlorite, perhaps after hornblende, is also present. The alkali percentages for this rock are: $Na_2O = 5.45$; $K_2O = 2.40$ (see Table XXX).

The conspicuous minerals of the drusy cavities are quartz, felspar and the golden-brown, micaceous mineral. The latter was teased out of the cavities in 1841 with a needle, and small felspar and quartz impurities were removed by hand picking. In this way 0.3 grams were obtained on which a partial analysis was made (Table XXIX). There was not sufficient material to allow the alkalies to be determined but small quantities are probably present and the rather high total, which their presence would produce, must be accounted for by the difficulties of making the analysis on such a small amount of material. The analysis lies within the range shown by three recent analyses of stilpnomelane by Hutton [1938]. The formula worked out on the basis of 14(O,OH) is: $(OH)_{3.4}(Al,Fe,Mg,Mn)_{3.4}[(Si,Al)_4O_{10}]$ which compares satisfactorily

TABLE XXIX.
ANALYSIS OF STILPNOMELANE.

	Wt. %	Mol. %	Metals to 14 (O,OH)	
SiO_2	45.29	.755	3.824	} 4.0
Al_2O_3	5.57	.054	.547 { .176	
TiO_2	0.02	..	.. .371	
Fe_2O_3	23.95	.151	1.538	
FeO	8.99	.125	.633	
MnO	1.14	.016	.081	} 3.4
MgO	3.30	.082	.416	
CaO	4.28	.076	.388	
Na_2O	n.d.	..	..	
K_2O	n.d.	..	..	
H_2O+	6.12	.339	3.436	3.4
H_2O-	1.79	.100	..	
	100.45			

COMPARISONS.

	1	A	B	C
SiO$_2$........................	45.29	45.24	44.67	44.99
Al$_2$O$_3$........................	5.57	6.73	6.83	6.99
TiO$_2$........................	0.02	0.33	tr.	st.tr.
Fe$_2$O$_3$........................	23.95	25.34	22.04	31.67
FeO........................	8.99	3.45	13.73	1.29
(Ni, Co)O	n.d.	tr.	tr.	n.d.
MnO	1.14	0.60	0.06	2.57
MgO	3.30	7.67	2.33	2.89
CaO........................	4.28	1.91	0.83	0.36
Na$_2$O........................	n.d.	0.03	nil	0.39
K$_2$O........................	n.d.	1.67	1.36	0.87
H$_2$O+........................	6.12	6.72	6.74	7.94
H$_2$O—........................	1.79	0.76	1.57	0.44
	100.45	100.45	100.16	100.40
α........................	1.626	1.599	1.595	1.625
$\beta = \gamma$........................	1.745	1.680	1.685	1.735
$\gamma - \alpha$........................	0.119	0.081	0.090	0.110
2V........................	0°	0°	0°	0°
α........................	golden-yellow	bright golden-yellow	pale yellow	bright golden-yellow
$\beta = \gamma$........................	dark reddish-brown	deep reddish-brown	deep olive-brown	deep reddish-brown
sp.gr........................	n.d.	2.78	2.84	2.83

1. Stilpnomelane from spherical vesicles in the granophyre inclusion, 1841, which was enclosed in marginal olivine-gabbro, Mellemö, Kangerdlugssuaq, E. Greenland.
A. Stilpnomelane from albite-stilpnomelane-actinolite-schist, western Otago, New Zealand [Hutton, 1938].
B. Stilpnomelane from vein, Zuckmantel, Silesia [Hutton, 1938].
C. Stilpnomelane from quartz-garnet-schist, western Otago, New Zealand [Hutton 1938].

with Hutton's ideal formula $(OH)_4K_{0.33}(Al, Fe, Mg, Mn)_{3.33}[Si_4O_{10}]$. The low value of the OH group may be connected with the high content of Fe''', due to replacement of Fe'' by Fe'''; this will occur if some of the OH is replaced by O.

The optical properties of the mineral from the Skaergaard granophyre inclusion are also given in Table XXIX. It is difficult to estimate the birefringence in thin sections owing to the intense absorption and the value given is obtained from the direct determination of α and γ refractive indices. The values for γ and for the birefringence, when plotted against the ratio of (Fe, Mg, Mn)O to (Al, Fe)$_2$O$_3$, fall close to the extension of Hutton's curves [1938, fig. 5, p. 187] which were largely plotted from the comparative data given in Table XXIX. The mineral from the cavities in the granophyre, 1841, is clearly stilpnomelane, and it is apparent, from observation of optical properties that stilpnomelane is widespread in the granophyre inclusions of the Skaergaard intrusion. So far as we are aware this type of occurrence of stilpnomelane has not been definitely shown before.

The central granophyre, 1836, of the inclusion of Pl. 13, fig. 1, is similar to 1841 in that the quartz and felspar have neither micropegmatitic nor granophyric relationships but occur as independent crystals. A rough estimate of mineral composition is: felspar, sixty per cent.; quartz, thirty per cent.; stilpnomelane, hornblende, sphene and apatite, ten per cent. Analysis gave Na$_2$O = 6.14 per cent. and K$_2$O = 1.50 per cent. Most of the felspar is microperthite but a little plagioclase is present which, from extinction angles, is An$_{35}$. Probably the rare plagioclase phenocrysts of this rock have about the same composition but none are present in the section. Stilpnomelane is present and has a similar form and distribution to that in 1841. There is also pale greenish-brown, weakly pleochroic, hornblende in scattered crystals which are comparable in size with the quartz and felspar crystals and are thus much larger than the stilpnomelane. Granular sphene and a little apatite also occur but no iron ore has been noted.

Another central granophyre mass, 318, surrounded by coarse, hybrid material (see Figs. 37 and 38) is different in texture from the above, though probably not very different in composition, since the alkali percentages are Na$_2$O = 5.35, K$_2$O = 2.62. Excluding the accessory minerals, the rock consists of about eighty per cent. of fine-grained micropegmatite, ten per cent. of quartz (not intergrown with felspar) and ten per cent. of idiomorphic, acid plagioclase. The most abundant ferromagnesian constituent is stilpnomelane which forms wispy crystals up to 0.1 millimetres in length which are aggregated into small groups. A few acicular crystals of pale, greenish-brown hornblende occur which resemble the hornblendes of 1836 and have $\gamma \wedge c = 17°$. Grains of sphene and of epidote, which in some of the larger crystals pass into a darker brown pleochroic type, presumably allanite, are common; iron ore occurs rarely. This rock does not seem to differ essentially in composition from the two previously described granophyres which formed under similar conditions. That one consists mainly of a fine-scale intergrowth of quartz and felspar and the other of separate crystals may be connected with a difference in the amount of volatile constituents but we have no direct evidence to offer for this hypothesis.

The hybrid material surrounding these granophyre patches is thoroughly coarse, like the similar hybrids in the layered series. In the case of the granophyre patch, 1836, (Pl. 13, fig. 1) which is only forty metres from the outer contact, the hybrid zone is not extensive, but well-shaped plagioclase up to ten centimetres in length and well-shaped stumpy, pyroxene about half as long have grown inwards from the margin of the inclusion, and are surrounded by rather coarse granophyre. The inner three-quarters of the large plagioclase crystals is almost free from zoning and is about An_{45} while the outer part is strongly zoned and gives place ultimately to what appears to be microperthite, dusty with decomposition products. The pyroxene, which shows slight colour zoning, has $\gamma' = 1.718$ while the pyroxene of the surrounding olivine gabbro has $\gamma' = 1.708$. A small amount of a pale pink, isotropic mineral with low refractive index is associated with the quartz and may be fluorite. The material to produce these large felspar and pyroxene crystals must have been partly derived from the olivine gabbro and partly from the granophyre, and the crystals must have grown slowly as material of the right composition diffused towards them.

Another example of coarse hybrid material between a granophyre inclusion and the surrounding olivine-gabbro is shown in figures 37 and 38. In a zone adjacent to the olivine gabbro there is a coarse quartz-gabbro, with small drusies. This gives place to a zone with large plagioclase crystals projecting into granophyre (317), and here are found also conspicuous apatite crystals two centimetres long and 0.2 centimetres broad. A neighbouring coarse gabbro, 319, forming a patch 20 by 5 metres, shows bladed pyroxene fifty centimetres long. The interstitial granophyre of this hybrid rock has small drusies into which quartz projects. Acidification of the olivine-gabbro first caused the gradual disappearance of the olivine without any increase in the coarseness of the rock (e.g. 1249 and 1856, described on p. 156). At a later stage, when quartz became sufficiently abundant to have produced pyroxene from all the olivine and some also occurs interstitially, the hybrid rock becomes particularly coarse in texture. Since the zones and reefs of hybrid quartz-gabbro may be as much as two metres across it is likely that mechanical mixing as well as diffusion has contributed to their formation.

Somewhat different granophyric patches and hybrids which are small and not drusy, are characteristic of the high marginal border group. In some cases, such as 1719 from near the southern margin, long-bladed augites project into the granophyre material, and large plagioclase crystals as in the Mellemö examples are not developed. A small patch, 1829, eight centimetres across, collected about a hundred

Fig. 37. Granophyre and hybrid—part of an inclusion in the outer border group, Mellemö. Upper specimen, 318, is non-porphyritic granophyre forming centre of the mass. Lower specimen, 317, is a hybrid rock containing large plagioclase, augite and apatite crystals; this rock immediately surrounds the granophyre (Figure, two-thirds natural size).

Fig. 38. The upper specimen, 316, is coarse quartz-gabbro, due to hybridisation and occurs in a zone surrounding the granophyre and hybrid shown in the previous figure. The lower specimen, 315, collected 10 centimetres from the hybrid is typical olivine-gabbro of the outer border group on Mellemö. (Figure, two-thirds natural size).

metres from the south-eastern margin, also has long bladed augite crystals. This inclusion is surrounded by coarse olivine-gabbro, having a trace of quartz like the two rocks 1830 and 1832 already described, and the transition zone between them has been examined in detail. In passing from the olivine-gabbro towards the inclusion there is first a zone of olivine-free gabbro consisting of plagioclase and ophitic pyroxene with schiller inclusions, then a zone consisting of partly chloritised pyroxene and large perthite crystals with some interstitial chlorite and coarse micropegmatite. Ilmenite in small sagenetic webs, and small granules of sphene, also occur here. This gives place to the granophyre material which contains 4.65 per cent. Na_2O and 0.44 per cent. K_2O. The granophyre consists mainly of a fine-grained micropegmatite with here and there patches of chlorite and quartz, free from felspar. Scattered throughout are numerous small acicular crystals of pale-green hornblende and a few grains of sphene, apatite and zircon.

3. In the Upper Border Group.

A basic granophyre, 4154, occurring as an indefinite block in the upper gabbros of Osttoppen, a smaller patch, 1759, from Hammers Pas, and certain quartz-gabbros regarded as hybrid in origin from the same horizon have been described with the upper border gabbros in which they are found (pp. 174—178). Similar granophyre masses and hybrids occur everywhere in the upper border group and have been collected on the west side of Kilen, on the northern side of the west ridge of Tinden and in the low, upper border gabbros below the saddle between Brödretoppen and Osttoppen. These other examples need not be separately described.

Besides granophyres which have the field relations of inclusions there is a considerable range of variable, basic granophyres on the west face of Brödretoppen which are quite different in habit (p. 176). The amount of these basic granophyres seems to preclude their formation by hybridisation in situ and, although the field relations are obscure, they are regarded as transgressive sheets related, on the one hand, to certain late transgressive granophyres of acid composition and, on the other, to the basic hedenbergite-granophyre of the unlaminated layered series. These rocks are described in a later section (pp. 203—13).

4. Origin of the Granophyre Inclusions and Hybrids.

The leucocratic minerals characteristic of the granophyre patches, wherever they may occur in the intrusion, are acid plagioclase, orthoclase and quartz. Stilpnomelane is the commonest ferromagnesian mineral and pale hornblende, sometimes replaced by green chlorite, is also a fairly frequent accessory. Pyroxene is found in the rather basic granophyre inclusion in the hortonolite-ferrogabbro near the Basishusene. Iron ore, some of which, from the sagenetic web habit appears to be ilmenite, is usually present in small amounts. Sphene, zircon, apatite and sometimes zoisite and epidote are accessories. In one rock the zoisite is seen to pass into allanite. This assemblage of minerals is similar to that of many of the British Tertiary granophyres.

The hybrid material, when still in its position of formation between the granophyre and enclosing gabbro, is usually a much coarser textured rock than either of its parents. It varies in composition, as would be expected, according to the nature of the surrounding gabbro. Thus the hybrid round a granophyre immersed in the hypersthene-olivine-gabbro of the lower layered series is a coarse but otherwise not exceptional, olivine-free gabbro, while a coarse quartz-gabbro of hybrid origin, surrounding granophyre in a hortonolite-ferrogabbro, contains iron-rich olivine which was apparently stable in the presence of quartz. The hybrids of Mellemö show, first, the gradual disappearance of olivine without other noticeable changes, and then the development of an extremely coarse textured, and often drusy, quartz-gabbro, which may have bladed pyroxene up to fifty centimetres long. In the narrow hybrid zone round the chlorite- and calcite-bearing granophyre from the southeast margin, uralitisation of the pyroxene, and the partial replacement of the hornblende by chlorite, are usual features. While the granophyre patches are fairly constant in mineralogical composition, no matter in what rock they may be immersed, the hybrids are varied, depending on the nature of the enclosing gabbro.

The granophyre masses occur embedded in very different rocks ranging, in the layered series, from olivine-hypersthene-gabbros to ferrogabbros and, in the border group, from olivine-gabbros to quartz-gabbros. It is unreasonable to postulate that such similar granophyres would be differentiated from such diverse parent rocks. This suggests that the granophyres are inclusions, a view supported by the form of certain masses in the outer border group. These are not streaked out but have a rounded form, and this is no doubt because the surrounding gabbro solidified before the form of the inclusion was completely altered by magmatic flow. Elsewhere the granophyre patches have usually the form of reefs or schlieren which are considered to be due to the flow of once, block-like masses. The percentage of alkalies in the granophyre inclusions is variable (Table XXX) but on the average there is a decided similarity with the alkali percentages of the three analysed examples of grey gneiss from the metamorphic complex. There must be a considerable variation in the alkali percentages of a heterogeneous group such as the metamorphic complex, and the range shown by the granophyre

13*

TABLE XXX.

ALKALIES IN GREY GNEISS FROM THE METAMORPHIC COMPLEX AND IN GRANOPHYRE INCLUSIONS FROM THE SKAERGAARD INTRUSION.

		Na$_2$O	K$_2$O	Total
Grey gneiss from the metamorphic complex. Mellemö.	1867 (large specimen)	4.40	3.32	7.62
	1867 (small specimen)	4.81	1.90	6.71
	1865	3.98	2.93	6.91
	Average...	4.40	2.72	7.08
Granophyre inclusions in layered series.	4074 in hyp.-ol.-gabbro	4.35	2.83	7.18
	2584 in hort.-ferrogabbro	4.01	1.96	5.97
Granophyre inclusions in border group.	318 Mellemö	5.35	2.62	7.97
	1836 —	6.14	1.50	7.64
	1840 —	5.08	1.02	6.10
	1841 —	5.45	2.40	7.85
	1829 S.E. margin	4.65	0.44	5.09
	1759 Upper border group	5.14	1.13	6.27
	Average...	5.02	1.74	6.76
Average for the gneiss and the granophyre inclusions..		4.85	2.00	6.84

masses could undoubtedly be matched from among the gneisses. Combining the results of alkali determinations with the observed amounts of ferromagnesian and accessory constituents which are found in the grey gneiss, we are led to the conclusion that the granophyres have the same general bulk composition and range as the grey gneisses of the metamorphic complex. These lines of evidence indicate that the granophyre patches were originally blocks of the grey gneiss which became incorporated in the Skaergaard magma during the act of intrusion.

On Tinden, as already recorded (p. 57), there occurs a very large inclusion of acid gneiss several hundred metres high. This has not been re-fused, doubtless owing to its size. The presence of this undoubted grey gneiss inclusion within the Skaergaard intrusion supports our interpretation of the small granophyre patches.

Experimental evidence, and also certain field observations such as those quoted by Daly [1933, pp. 426—36], suggest that relatively acid material like the grey gneiss should be molten at the temperature of a basic olivine-gabbro magma such as gave rise to the Skaergaard intrusion. However, there is always the possibility that volatile constituents might lower the temperature at which a basic magma could exist so that enclosed acid material would remain unfused. It is, therefore, of value to have definite evidence from the spherical vesicles in the granophyre, 1841, that some of the gneiss inclusions became completely liquid. At the same time the fact that the vesicles have not coagulated into a single, large cavity at the top of the local patch of granophyre magma proves the high viscosity of the re-formed acid magma. Experimental work on melts approaching the composition of igneous rocks seems effectively to preclude the possibility of molten granophyre being immiscible with molten gabbro [Greig 1927] and we suggest that the reason why the granophyre tended to remain as distinct masses, instead of being evenly distributed through the basic magma, was the high viscosity of the newly formed acid magma which only allowed mixing to take place with difficulty.

From the evidence of the drusy cavities, volatile constituents were clearly present in many of the granophyre inclusions. Only in the cases where the drusies are spherical can we be certain that the vapour phase separated while the rock was molten, and even then a few early, porphyritic crystals may have been present. In most cases the drusies are of irregular shape, and the separation of the vapour phase probably took place at a fairly late stage of the crystallisation. The gneiss blocks, from which the granophyres formed, must have contained a small amount of volatile constituents, present in certain of the minerals, and they may also have contained meteoric water in cracks and small open spaces. Some of the volatile constituents which later produced the drusies in the granophyres may be accounted for in this way, but it does not seem likely that this will account for all, since the minerals of the granophyre are almost as rich in volatile constituents as those of the gneiss, and the granophyre is drusy while the gneiss is not. It may be that a late-crystallising liquid, embedded in an earlier crystallising basic rock, behaves as a sink towards which water and other volatile constituents from the crystallising rock would diffuse. If this be the case the volatile constituents at one time must have been completely dissolved in the acid magma from which they separated only as crystallisation took place. This would account for the usual, late, drusy cavities of irregular shape and indeed it might also account for the rarer spherical type, since separation of phenocrysts, a few of which are not incompatible with spherical cavities, may have been sufficient to cause a vapour phase to form.

In connection with the volatile constituents, it is interesting to compare the metamorphism of the grey gneiss, adjacent to the intrusion, with the material, presumably similar in composition, which became completely immersed in the magma. The highly metamorphosed grey gneiss, a metre from the margin (p. 12) preserves some of the texture of the original gneiss, and contains hypersthene which may be assumed to have formed at higher temperatures than the common constituents of the. granophyres, hornblende and chlorite. These differences may be partly ascribed to the fact that the granophyre masses retained, ,and possibly also absorbed, volatile constituents and formed a liquid which only solidified at relatively low temperatures. The volatile constituents of the gneiss bordering the intrusion, would probably be expelled into the surrounding rocks before the temperature was raised to the point at which a hornfels was formed and free passage of the gases was prevented. The recrystallisation of the gneiss adjacent to the intrusion may therefore have taken place at a higher temperature than that of the blocks immersed in the intrusion. That the inclusions re-crystallised at a lower temperature than the surrounding rocks may, on first consideration, seem paradoxical but the evidence of the horn-blende and chlorite which are present in the granophyre inclusions and of the hypersthene in the contact-metamorphosed grey gneiss seems to be definite. .It would be of great interest if we could decide whether the newly-developed granophyre magma had absorbed significant quantities of water vapour from the crystallising gabbro surrounding it, but there seems no satisfactory way of estimating the volatile con-stituents present in the granophyre and its drusies. In this connection it should be noted that the content of water in the analysed gneiss, 1867, cannot be taken as an indication of that in the average metamor-phic complex, since it may have had some of its water expelled during slight metamorphosism by the Skaergaard intrusion from which it was only a hundred and fifty metres distant.

The hybrid material round the granophyre patches is always coarse and sometimes extremely coarse. This is probably due to the acid material acting as a flux. During the long period of time between the solidification of the olivine-gabbro surrounding·the granophyre patches and the solidification of the granophyre itself, there would be ample opportunity for large crystals to grow. Basic inclusions in the Skaergaard magma behave in a different way from the acid inclusions and tend to have fine-grained margins (p. 200). Hybridisation between solidifying, basic rock and enclosed patches of molten, acid material is a very different matter from the better-known case in which acid magma has digested basic xenoliths without re-fusion.

During the production of hybrid material at the junction between gabbro and granophyre, bodily movement of liquid, as well as diffusion, has taken place. Evidence for this is afforded by the arrangement of the granophyre and hybrid material in the larger patches of the inner border group on Mellemö, Ivnarmiut and elsewhere, where the granophyre tends to form a cumulose head to an extensive streak of coarse hybrid. It is clear that the light granophyre has floated upwards and the hybrid material, formed from the granophyre and olivine-gabbro, has slid round the head of molten granophyre and accumulated as a trailing tail to the inclusion. The hybrid material has certainly been aided in its formation by mechanical mixing between a partly solidified gabbro magma and a viscous, but fluid, acid magma. Since mechanical processes as well as diffusion have contributed to the formation of the hybrid material it is not surprising to find it occurring in considerable quantities, as in the upper border group.

The mineralogical and textural characteristics of the granophyre inclusions which have been produced by re-fusion of acid gneiss immersed in the Skaergaard magma are remarkably similar to those of many of the granophyres of the British Tertiary province. Despite the fact that there is no comparison in relative amounts between the large, independent granophyre intrusions of Skye, for example, and the granophyre inclusions in the Skaergaard intrusion, we feel that the close similarities in texture and composition add some support to the hypothesis which ascribes the Tertiary granophyre intrusions to re-fusion of acid gneisses by basic magma; further evidence for this view is given in the concluding section of this paper.

(b) Inclusions of Basalt, Early Sill Gabbro, Sediments, etc.

In the border group of Kraemers Ö about seven hundred metres from the south coast and a hundred metres from the contact there are several small blocks of a fine-grained, basic rock some having light patches with the form of amygdalés (2296, 2297). On sectioning, these prove to be olivine-pyroxene-plagioclase-granulites with no trace of original basalt structures except that some of the felspar is segregated in amygdale-shaped patches. Rocks of similar grain-size and composition are found as abundant inclusions in the gabbro of the Eskimo ruins macrodyke of Mikis Fjord, a kilometre northeast of the limits of the map. In the inclusions the amygdular form of the light patches, now mainly felspar, is well preserved and examples of pipe amygdales were found. Although strongly metamorphosed, there is no doubt that the blocks in the border group of Kraemers Ö, like those of the Eskimo ruins macrodyke, were originally basalt.

The huge inclusion of the Basistoppen raft, composed of part of an early, differentiated sill, has already been described and compared with the similar sills outside the intrusion (pp. 57 and 58). Away from the contact with the Skaergaard magma, the gabbros of the Basistoppen raft appear to be unaltered, but within fifty metres of the contact there is a breakdown of the ophitic texture of the gabbro into granulitic texture which becomes steadily finer-grained towards the margin. Thus a medium-grained gabbro, 4123, from near the top of the large inclusion on the first nunatak east of Basistoppen, has strongly zoned plagioclase up to two millimetres in length and ophitic augite altering to hornblende. The outer several centimetres of this inclusion (e.g. 4125) is a fine-grained granulite consisting of small, zoned plagioclase crystals, 0.3 millimetres in length, with granular pyroxene and magnetite of about the same size. General reduction in grain size is a constant feature of the gabbros of the Basistoppen raft where in contact with the layered series; other examples sectioned are 4126 and 1900.

Two large inclusions of moderately coarse gabbro were found embedded in the layered rocks near the neck of the Skaergaard Peninsula Away from the margin these are of types that can be matched among the rocks of the Basistoppen raft. One of these, 4276, is an olivine-gabbro with zoned plagioclase crystals two millimetres long; the pyroxene is ophitic and the olivine is partly altered to talc. The margin in contact with the hortonolite-ferrogabbro is a very fine-grained granulite with plagioclase less than 0.1 millimetres long. A fine-grained granulite, 2268, was found as a small inclusion in the layered rocks on the island north-east of Ivnarmiut and it was at first believed to be a metamorphosed basalt but we now consider that it is just as likely to be a completely granulitised fragment of coarse sill gabbro. The only criterion which we have accepted for the basalt origin of basic granulite inclusions is the existence of palimpsest amygdales.

Certain large masses which behave as inclusions have been mentioned as affecting the disposition of the layering (p. 42). We have not been able to decide on the origin of these but a thin section of the block at the foot of Pukugaqryggen does not resemble any of the layered rocks, and suggests that this is an accidental xenolith perhaps from the sill gabbros. The mass on Kraemers Ö seems related to the middle gabbros and may be a cognate xenolith; the plagioclase in a specimen of it proves to be An_{52}. Other examples were seen from a distance but were not reached and the available material is not sufficient to come to any definite conclusions about the origin of these masses.

Xenoliths of sedimentary origin have only been found in the outer part of the border group and even here they are rare. They usually occur as small, but conspicuous, rounded, or sub-angular masses of light colour which are resistant to weathering. Examples collected on Ivnarmiut about three metres from the contact (4291, 4292) are similar in appearance both in hand specimen and under the microscope to the thin metamorphosed sediment, 312, overlying the thin conglomerate on Mellemö. In all of these there is a fine-scale mesh of sillimanite while in 4292 corundum and spinel are present as well. Several sedimentary inclusions were found in the border group of Kraemers Ö near the basalt fragments and one of these, 2299, which has been sectioned, consists of needles of sillimanite in felspar with a few grains of green spinel. Somewhat similar xenoliths (e. g. 1726) were found in a cluster in the chilled, marginal rock at the west foot of Tinden. These contain, among other minerals, sillimanite, green spinel and corundum, and a similar assemblage is found in 4091 from the northern border group. A rock, 4092, from the considerable raft of metamorphosed sediments in the border group at the top of Uttentals Plateau contains, quartz, cordierite and sillimanite with spinel in places; another specimen, 1679, is essentially a plagioclase-spinel hornfels.

All the sedimentary xenoliths which we have found seem to have been derived from the Kangerdlugssuaq Sediments with the possible exception of 4091 which might be a mass of the porphyroblast schist from the metamorphic complex of Uttentals Plateau. The considerable reef-like mass from the border group at the top of Uttentals Plateau has still the appearance of a variable series of bedded rocks such as the sandy, ferruginous shales of Vandfaldsdalen. This mass must have sunk through the Skaergaard magma from its original position above the metamorphic complex. The small, sedimentary xenoliths of Kraemers Ö, Ivnarmiut and the west ridge of Tinden are so similar to the metamorphosed sediment still in its original position on Mellemö that they may be assumed, with little doubt, to be derived from the late Cretaceous Kangerdlugssuaq Sedimentary Series. The specimens on Kraemers Ö, like the neighbouring basalt xenoliths, must have sunk slightly from their original position while those of Ivnarmiut must have risen slightly; those of the west ridge of Tinden must have ascended about two kilometres from their presumed source.

There remain for brief description certain, melanocratic inclusions which were particularly noticable in the outer border group on Mellemö (e. g. 1838, 1839) and on Kraemers Ö (345). These are rather basic rocks with a patchy variation in composition and they form definite inclusions up to two or three metres across. The dominant constituent is augite which amounts to fifty or even seventy per cent. Olivine and hypersthene are also present in considerable quantities, and iron ore and basic plagioclase in small quantities. The ferromagnesian minerals do not seem to have any close resemblance to those occurring in the olivine-hypersthene-gabbro by which the inclusions are surrounded. Alterations of the pyroxene to hornblende, the olivine to talc, and the felspar and ferromagnesian minerals together, to chlorite and carbonates has taken place to a considerable extent in 345 and to a less extent in 1838. Another small basic xenolith deserving mention, 1844, consists of a fine-grained mat of hornblende. Unlike the gabbro-picrite of the northern border group, these xenoliths show no internal evidence that they have been derived from the Skaergaard magma; they are probably from basic bands of the metamorphic complex which must have formed inclusions in a similar way to the acid material which is now granophyre. The three examples, 345, 1838 and 1839 may well be derived from the fairly abundant amphibolite bands of the metamorphic complex while the unique type, 1844, may perhaps represent one of the ultrabasic patches which are found occasionally in the gneisses of the Kangerdlugssuaq region.

The granophyre inclusions described in the preceding section are abundant and widely distributed throughout all the Skaergaard rocks except the upper layered series. The xenoliths described in the present section are far less abundant and their known distribution may be connected with the detail with which the various parts of the intrusions have been mapped. The rareness of any xenoliths derived from the basalts is surprising and suggests that the agglomerate, which we have postulated to have been produced from the material formerly occupying the space of the Skaergaard intrusion, was roughly stratified, the upper layers being mainly basalt and the lower mainly gneiss fragments. A distribution of this kind would allow abundant gneiss blocks to become involved in the Skaergaard magma while basalt would be expected to be rare. It would imply that during the explosion which produced the intrusion the materials forming the agglomerate kept roughly the same relative position as they had before the disturbance.

IX. TRANSGRESSIVE GRANOPHYRES AND BRÖDRETOPPEN INTERMEDIATE ROCKS

(a) Small Veins near Granophyre Inclusions.

Irregular veins of granophyre, extending for a few metres only and then dying out, are found sporadically in the upper border group and the adjacent marginal border rocks. These veins are not chilled against the gabbros they penetrate but tend to merge with them; they have the same appearance in hand specimen as the granophyre from neighbouring inclusions. One such narrow vein, 1828, collected on Kilen near the granophyre inclusion, 1829, (see p. 191) is closely similar to the inclusion in appearance and composition. It consists of fine-grained micropegmatite with granophyric structure in which are embedded cloudy acid plagioclase crystals, showing rectangular form. Areas of clear quartz, calcite, and green chlorite of late solidification occur. Green chlorite is also found in patches replacing some other mineral, probably hornblende. A little apatite and iron ore occur, the latter sometimes in sagenetic webs, partly altered to sphene. The percentage of alkalies in this rock are: Na_2O 5.40, K_2O 1.62.

Similar narrow and discontinuous veins of granophyre were found in the upper border group at Hammers Pas. One of these, 1757, collected near the granophyre patch 1759 (see p. 194) has a slightly coarser general texture than the example just described though the micropegmatite part is on a finer scale. Green chlorite occurs in patches replacing some ferromagnesian mineral and also in isolated flakes embedded in interstitial calcite. A little iron ore, some in the form of sagenetic webs, is widely scattered but no apatite is to be seen. The percentages of alkalies present are: Na_2O 5.92, K_2O 0.41.

These two granophyre veins are very similar to the granophyre inclusions which were found near them. The differences between the alkalies of the vein material and that of the neighbouring inclusions (Table XXXI) are slight, being probably no greater than the variation in different parts of a single granophyre inclusion or vein. The mineralogical differences are also slight. The close similarity between these small

TABLE XXXI.

ALKALIES IN SMALL GRANOPHYRE VEINS AND NEIGHBOURING INCLUSIONS.

	Kilen		Hammers Pas	
	Granophyre inclusion, 1829	Granophyre vein, 1828	Granophyre inclusion, 1759	Granophyre vein, 1757
Na₂O.........	4.65	5.40	5.14	5.92
K₂O..........	0.44	1.62	1.13	0.41
Total...	5.09	7.02	6.27	6.33

acid veins and the neighbouring granophyre inclusions suggest that the veins are granophyric material from the inclusions which has been injected into cracks in the surrounding, solid, but still heated, gabbro. They resemble contemporaneous veins except that the source of the material, instead of being the late crystallising interstitial magma of the surrounding rock, was acid magma formed from inclusions.

(b) Larger Veins and Tinden Granophyre Sill.

The thick sill on Tinden and Brödretoppen which cuts the upper border rocks but not the marginal border group is a minutely-drusy granophyre which is unchilled against the enclosing rocks. The upper part, represented by 360 and 3058, contains few ferromagnesian constituents. It consists of small rectangular crystals of acid plagioclase which pass into a fine-grained micropegmatite, while interstitially a little quartz not intergrown with felspar is developed. The rock, 3058, has been analysed (Table XXXIII) and the percentages of alkalies—Na₂O = 4.24; K₂O = 3.85—give an indication of the nature of the felspars. A dirty brown mineral, in small flakes and associated with a little iron ore, is probably stilpnomelane; it is widely scattered but is not so abundant as the almost colourless chlorite which occurs interstitially with calcite and quartz. The lower part of this sill, 359, contains rather more melanocratic constituents but is otherwise similar. The melanocratic material is largely stilpnomelane with green chlorite in interstitial patches. Here and there, well-shaped albite and quartz crystals, not intergrown, project into the interstitial material which consists of calcite and chlorite. A little iron ore and zircon also occur.

Drusy acid granophyres are found above the sill on Brödretoppen where they are intimately associated with the more basic granophyres

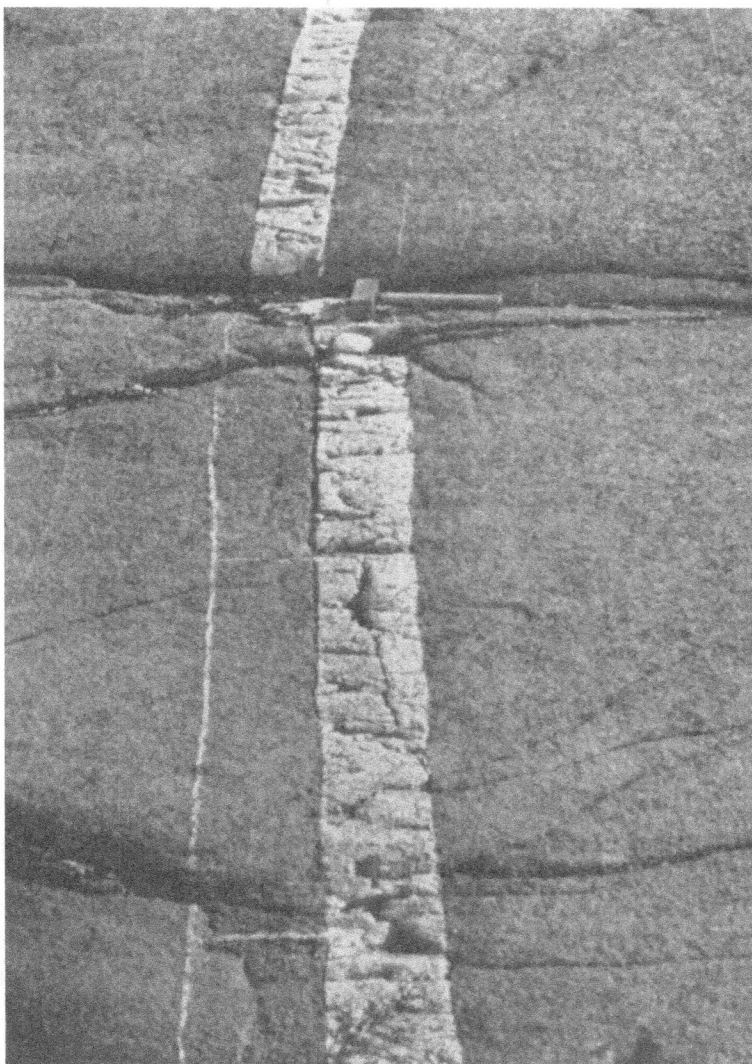

Fig. 39. Transgressive granophyre vein cutting ferrogabbro, near coast, north of
Basishusene. Vein 25 centimetres across.

described in the next section. Two of the acid granophyres, 1226 and
3045, consist mainly of a fine-grained micropegmatite, acid felspar and
quartz. Embedded in this are euhedral acid plagioclase crystals and a
little green hornblende altering to chlorite, a little acicular iron ore
altering to sphene, grains of zircon of considerable size, epidote, and
iron pyrites. Both brown and green chlorite are developed in interstitial
patches. The other drusy granophyre, 1224, has abundant, small, acicular
hornblendes. The finer-grained micropegmatite part of these rocks is

like the sill granophyre, while in their ferromagnesian constituents they resemble certain transgressive granophyres, such as 2562 now to be described.

Acid granophyre also occurs as transgressive veins in the layered rocks along the coast of Uttentals Sund north of the houses, at the foot of Pukugaqryggen, on the opposite side of Uttentals Sund and else-where. These granophyre veins are not found in the rocks surrounding the intrusion, and the magma which gave rise to them does not appear to have had an outside origin. They vary from a few centimetres to a few metres in thickness and they cut sharply through the layered rocks (Fig. 39). Since the vein material is not chilled it must be concluded that the layered rocks were solid, but at a high temperatures, when the veins were injected.

Three examples of the sharply-transgressive, granophyre veins have been sectioned and analysed for alkalies. Two of these, 4064 from the west side of Uttentals Sund and 2562 from just north of the houses, are very similar, being greenish-grey, medium-grained rocks without granophyric texture. The rock, 4064 (Pl. 24, fig. 1) contains about sixty per cent. of felspar, some of which is acid plagioclase and the rest apparently perthite, although decomposition makes its exact nature uncertain. The alkali percentages are Na_2O 5.14; K_2O 3.51. Quartz, present to the extent of about thirty per cent., shows no tendency to be intergrown with the felspars. The remaining ten per cent. consist of dark green and pleochroic chlorite (in small flakes having the habit of biotite and usually enclosing abundant sphene), a little iron ore in small crystals, and calcite which occurs interstitially. Allanite is relatively abundant and a few crystals of apatite and zircon were noted. The rock, 2562, which is very similar in hand specimens has also a similar alkali percentage: $Na_2O = 4.58$; $K_2O = 3.44$. Under the microscope this granophyre is seen to be somewhat finer-grained and the ferromagnesian constituents consist of about equal amounts of altered biotite and well-shaped, acicular, brown hornblende.

A rock, 3659, from near 2562 has similar relations to the layered series and is a rather coarser grained, pinkish rock without granophyric texture. The alkali percentages are: Na_2O 4.25, K_2O 4.18 and it is thus the richest in potash of any of the granophyre veins which has been analysed. In thin section some unaltered acid plagioclase is seen to be present though the bulk of the felspar is microperthite. The dominant dark mineral is biotite some of which is still fresh but other crystals have altered to a green chlorite like that in 4064 and 2562. Some iron ore, apatite, zircon and allanite are also present.

The larger transgressive veins and the granophyre sill of Tinden and Brödretoppen represent magma available when all the gabbroic rocks of the Skaergaard intrusion were solid and capable of being cleanly fractured, but they were injected while these rocks were still hot for their contacts are unchilled. The magma which gave rise to the acid granophyres does not appear to have been abundant but it is unlikely that the source was re-fused acid gneiss inclusions, as seems to be the case with the smaller transgressive veins which have been described.

TABLE XXXII.

ALKALIES IN THE LARGER GRANOPHYRE SILLS AND VEINS AND COMPARISONS WITH ALKALIES IN THE GREY GNEISS AND THE GRANOPHYRE INCLUSIONS.

	Sill Tinden (360)	Vein Kraemers Ö (4064)	Vein Basishousene (2562)	Vein Pukugaqryggen (3659)	Aver. for the sills and veins	Aver. for Grey Gneiss (see Table XXX)	Aver. for Grey Gneiss and granophyre inclusions (see Table XXX)
Na_2O	4.33	5.14	4.58	4.25	4.58	4.40	4.85
K_2O	3.77	3.51	3.44	4.18	3.73	2.72	2.00
Total...	8.10	8.65	8.02	8.43	8.31	7.08	6.84

All the larger transgressive granophyres whether from the Tinden sill or the veins on the shores of Uttentals Sund have apparently a very similar composition. Thus the alkali percentages of four granophyre veins set out in Table XXXII show a much less range than that shown by the acid gneiss and the granophyre inclusions, and the potash content is greater than any of the eight, partially analysed, granophyre inclusions.

Since the larger, transgressive acid granophyre, veins are not found outside the Skaergaard intrusion, the source of the magma giving rise to them seems to have been within the intrusion. The highest unlaminated layered rocks are basic hedenbergite-granophyres (e. g. 4137) and these have a much higher potash content than any of the earlier rocks. The trend of differentiation of the Skaergaard magma in the later stages of the formation of the layered series gave rise to basic granophyres and we consider that the acid granophyres have been developed by further differentiation in the same direction. On this hypothesis, the basic granophyre from Brödretoppen shortly to be described represents a stage between the basic hedenbergite-granophyre of the layered series and the acid granophyre of the transgressive veins.

The analysis of the acid granophyre, 3058, from Tinden (Table XXXIII) is probably typical of the late transgressive granophyres of

TABLE XXXIII.
ANALYSIS OF ACID GRANOPHYRE.

XXIII (3058)		Norm		Mode not determined
SiO_2	75.03	Qu.	33.96	
Al_2O_3	13.17	Or.	22.80	Ratios
Fe_2O_3	1.56	Ab.	35.63	
FeO	0.58	An.	3.61	$\dfrac{(FeO + Fe_2O_3) \times 100}{MgO + FeO + Fe_2O_3} \cdot\cdot\ 93$
MgO	0.15	Cor.	0.92	
CaO	0.69	En.	0.40	
Na_2O	4.24	Ilm.	0.61	
K_2O	3.85	Mt.	0.93	$\dfrac{Fe_2O_3 \times 100}{FeO + Fe_2O_3}\ \cdots\cdots\ 73$
H_2O+	0.28	Hae.	0.96	
H_2O-	0.13	H_2O	0.41	
TiO_2	0.31			$\dfrac{K_2O \times 100}{Na_2O + K_2O}\ \cdots\cdots\ 48$
MnO	0.01	Plag. $Ab_{91}An_9$		
P_2O_5	0.02			
	100.02			
S. G.	2.54			

	XXIII	A	B	C	D	E	F	G	H	I
SiO_2	75.03	73.36	70.34	73.32	71.53	71.60	75.20	71.51	76.50	73.70
Al_2O_3	13.17	11.99	13.18	12.25	12.00	13.60	12.65	12.82	12.10	12.87
Fe_2O_3	1.56	2.76	2.65	2.77	2.90	} 2.40 {	1.53	2.09	1.43	3.76
FeO	0.58	0.83	2.24	2.20	2.02		0.28	1.40	..	0.31
MgO	0.15	0.24	0.40	0.11	0.62	0.21	0.26	0.17	0.23	0.11
CaO	0.69	1.78	1.24	1.65	2.33	2.30	0.60	1.09	1.31	0.14
Na_2O	4.24	3.99	3.61	3.92	4.27	5.55	5.67	4.24	4.06	3.63
K_2O	3.85	2.99	4.90	2.34	3.06	3.53	4.14	4.52	3.40	4.56
H_2O+	0.28	2.08	0.76	0.35	0.36	} 0.70	0.12	} 1.23 {	0.83	} 0.57
H_2O-	0.13	0.15	0.46	0.35	0.13				0.26	
TiO_2	0.31	0.15	0.46	0.51	0.64	..	0.12	0.10	..	0.12
MnO	0.01	0.02	0.19	0.12	0.36	..	0.10	..	..	0.07
P_2O_5	0.02	0.15	0.10	0.10	0.17	..	..	tr.	..	tr.
BaO	..	..	tr.	0.09	0.08	..	..	..	..	..
Cl	..	..	0.02	..	..	..	..	..	..	..
Li_2O	..	..	tr.	..	..	..	..	..	..	..
CO_2	..	..	..	0.06	nil.	..	..	..	..	..
(Ni, Co)O	..	..	..	..	0.02	..	..	..	..	..
	100.02	100.49	100.55	100.14	100.49	99.89	100.67	99.17	100.12	99.84

XXIII. Granophyre, 3058, from large sill. W. ridge of Tinden, Kangerdlugssuaq.

    A. Rhyolite, Cap Franklin, E. Greenland : [Backlund and Malmquist 1935].Anal. N. Sahlbom

    B. Hornblende-granophyre, Druim Eadar da Choire, Skye [Harker, 1904]. Anal. W. Pollard.

C. Augite-granophyre, Knock Ring-dyke, Beinn Bheag, Mull [Bailey, Thomas etc., 1924]. Anal. E. G. Radley.

D. Quartz-porphyry. W. slope of Ashval, Rum [Harker, 1908]. Anal. W. Pollard.

E. Augite-granophyre, Carrock Fell, Cumberland [Harker, 1895]. Anal. G. Barrow.

F. Aegerine-felsite, Cnoc an Droighinn, Assynt, Sutherland [Quoted Mem. Geol. Surv. Chem. Anal. Igneous Rocks, 1931, p. 14, no. 49].

G. Granophyre, S. of Säterstugan, Breven, Sweden [Krokström, 1932]. Anal. K. Winge.

H. Micropegmatitic mesostasis from mottled diabase south of Cobalt, Ontario [Phemister, 1937]. Anal. T. C. Phemister.

I. Red Rock (porphyritic), Little Brick Island, Pigeon Point. Anal. L. G. Eakins [Quoted from Grout, 1918c p. 653].

the Skaergaard intrusion. Examination of the analysis shows that the rock is very rich in silica and that soda is dominant over potash. A Tertiary rhyolite from East Greenland (Anal. A) is similar. The larger acid masses of the British Tertiary province differ from this Skaergaard granophyre in having potash dominant over soda but the closest example is quoted (Anal. B). Two smaller Tertiary masses from Mull and Rum (Anals. C and D) are nearer to the Skaergaard rock in this respect as also is the granophyre from Carrock Fell (Anal. E) and an aegerine-felsite from Assynt (Anal. F). It is interesting that among the rocks of the Breven dyke a rather similar type exists (Anal. G). Segregation veins from the Sudbury intrusion and an example of "Red Rock" from Duluth also approach the Skaergaard granophyre in composition and these may have had a somewhat similar origin (Anals. H and I).

(c) Hedenbergite-Granophyres of Brödretoppen.

The rocks now to be described are less basic than the unlaminated layered rocks which have been called basic hedenbergite-granophyre and they are decidedly less acid than the granophyres described in the previous section. They are therefore distinguished from both these groups by naming them hedenbergite-granophyre. They form a complex mass about 200 metres thick, and probably lenticular in shape, lying between the analysed quartz-gabbros of Brödretoppen and the top of the Basistoppen Raft; they are well seen in the tongue of the rock which stretches almost across the upper part of Brödregletschen. At the foot of this tongue is the prominent acid granophyre sill which extends to the west ridge of Tinden. Above the acid granophyre of the sill is a basic, fine-grained granophyre, then for 100 metres the rocks consist of a succession of indefinite bands, varying from granophyres, with small drusy cavities and little melanocratic material (e.g. 1224, 1226 and

3045 already described) to rocks of intermediate composition, with
bladed augite crystals up to three centimetres in length; the latter are
more basic granophyres and contain hedenbergite and sometimes fayalite
(1223, 3046, and 3347). Examples were found of acid granophyre veining
hedenbergite-granophyre—the contacts being rather indefinite (3047 and
3048). From field observations it was not possible to prove whether the
hedenbergite-granophyres were transgressive towards the surrounding
border group, or whether they were sheets, of approximately contemporaneous formation, like the Tinden granophyre sill. Microscopic and
chemical examination suggests that they are related, on the one hand,
to the acid granophyres just described, and on the other hand, to the
basic hedenbergite-granophyres of the layered series.

　　The analysed example of the hedenbergite-granophyres, 3047,
(Pl. 23, fig. 2) contains about twenty-five percent. of ferromagnesian
minerals, a greater proportion than in the acid granophyres, and dark-
green, feathery crystals of pyroxene as much as two centimetres long
are conspicuous in hand specimen. The leucocratic seventy-five per cent.
of the rock consists of euhedral, strongly-zoned, plagioclase crystals
which pass into cloudy felspar intergrown with quartz as a coarse micropegmatite. The dominant ferromagnesian mineral is a green pyroxene

TABLE XXXIV.
ANALYSIS OF HEDENBERGITE-GRANOPHYRE

XXII (3047)		Norm.		Mode (vol. %)	
SiO_2	58.81	Qu	14.58	Qu.	36[1]
Al_2O_3	12.02	Or	14.46	Plag.	40
Fe_2O_3	5.77	Ab.	33.01	Clino-pyrox	13
FeO	9.38	An.	8.06	Ore	11[2]
MgO	0.72	Di 5.22	10.90		
CaO	5.03	0.80			
Na_2O	3.91	4.88		$\dfrac{(FeO + Fe_2O_3) \times 100}{MgO + FeO + Fe_2O_3}$	95
K_2O	2.39	Hy. 1.00	6.94		
H_2O+	0.21	5.94			
H_2O-	0.19	Ilm.	2.43	$\dfrac{Fe_2O_3 \times 100}{FeO + Fe_2O_3}$	38
TiO_2	1.26	Mt.	8.35		
MnO	0.21	Ap.	1.68		
P_2O_5	0.71	H_2O	0.40	$\dfrac{K_2O \times 100}{Na_2O + K_2O}$	38
	100.61				
		Plag. $Ab_{80}An_{20}$			
S.G.	2.82	Diop. $Wo_{48}En_7Fs_{45}$			
		Hyp. $En_{14}Fs_{86}$			

[1]) Includes micropegmatite and dusty borders to plagioclase.
[2]) Includes some chlorite.

COMPARISONS.

	XXII	A	B	C
SiO$_2$	58.81	62.37	57.18	60.16
Al$_2$O$_3$	12.02	12.04	10.75	13.18
Fe$_2$O$_3$	5.77	1.87	4.96	8.88
FeO	9.38	5.81	6.24	3.15
MgO	0.72	0.97	2.15	1.03
CaO	5.03	3.51	5.73	3.89
Na$_2$O	3.91	3.47	4.62	3.42
K$_2$O	2.39	2.34	2.67	3.53
H$_2$O+	0.21	5.54	1.31	} 1.90
H$_2$O—	0.19	0.44	0.33	
TiO$_2$	1.26	1.06	3.25	0.20
MnO	0.21	0.24	0.32	0.22
P$_2$O$_5$	0.71	0.30	0.46	tr.
CO$_2$	..	..	0.08	..
BaO	..	0.07	0.06	..
FeS$_2$	..	nil.	..	..
(Co,Ni)O	..	nil.	..	..
	100.61	100.03	100.11	99.56

XXII. Hedenbergite-granophyre, 3047. About 800 m on west face of Brödretoppen, Kangerdlugssuaq.

  A. Innimorite-Pitchstone, half a mile S.W. of Pennygheal, Mull [Bailey, Thomas etc., 1924]. Anal. E. G. Radley.

  B. Allied to Craignurite, Glen More Ring-Dyke, quarter of a mile W.S.W. of Corra-bheinn, Mull [Bailey, Thomas etc., 1924]. Anal. E. G. Radley.

  C. "Intermediate" rock from S. of Svärdfallet, Breven, Sweden [Krokström, 1932]. Anal. K. Winge.

occurring in spongy masses and having a patchy development of coarse orientated inclusions of iron ore; it has a high refractive index and is clearly iron-rich. A little iron-rich olivine is also present and has the same habit as the pyroxene. Iron ore is abundant in compact masses or sagenetic webs of early crystallisation, and as late crystallising material intergrown with quartz. Dark brown, strongly-pleochroic hornblende is found associated with the pyroxene, and greenish-brown chlorite or stilpnomelane, associated with ore, is a late crystallising constituent. Apatite is an abundant accessory. The neighbouring rock, 3046, is similar except that the ferromagnesian minerals are rather altered, and another neighbouring rock, 1223, with the feather-like pyroxenes at least three centimetres in length is also similar except that no olivine is to be seen in the section. A rock from Basistoppen, 3021, found between the layered series and Basistoppen raft, is similar to the analysed hedenbergite-granophyre and, like it, contains an iron olivine.

14*

The analysis of the hedenbergite-granophyre (Table XXXIV) compares moderately closely (except that the Skaergaard rock contains more total iron) with two rocks from Mull: an inninmorite-pitchstone (Anal. A) and a rock allied to Craignurite (Anal. B). An intermediate rock from Breven (Anal. C) is also fairly close, and even approaches the

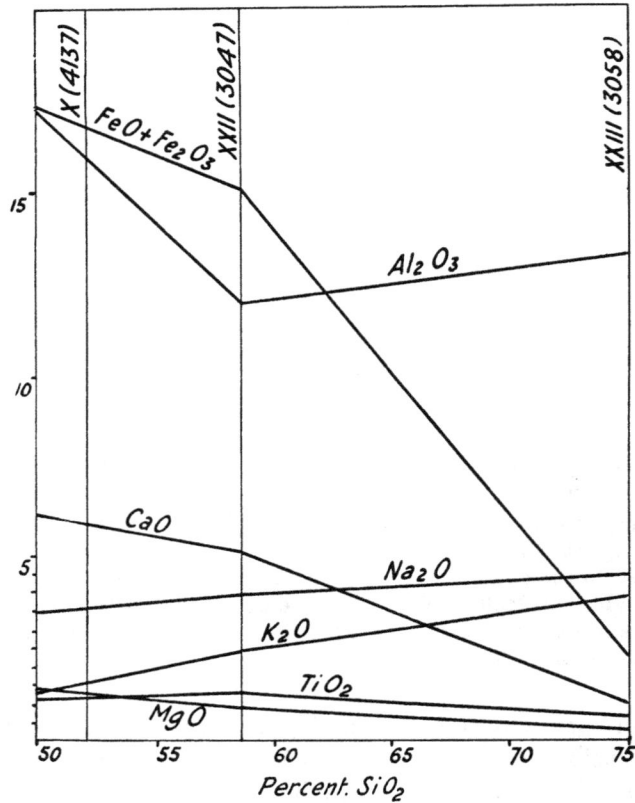

Fig. 40. Variation diagram of the basic hedenbergite-granophyre, 4137, from the unlaminated layered series, the hedenbergite-granophyre, 3047, from Brödretoppen, and the acid granophyre, 3058, from the Tinden sill.

analysed hedenbergite-granophyre in total iron content, although the bulk is ferric iron and not ferrous as in the Greenland rock.

Among these intermediate rocks of Brödretoppen are types which link up the basic hedenbergite-granophyre, formed as the latest layered rock, and the more normal acid granophyres of Tinden sill and the transgressive veins. The variation diagram (Fig. 40) of three late rocks of the Skaergaard complex—the basic hedenbergite-granophyre of the layered series (4137), the hedenbergite-granophyre (3047), and the transgressive granophyre of Tinden sill (3058)—is reasonably smooth, and the trend is roughly the familiar one shown by normal basic to acid rock series, except that at the basic end the rock is rich in iron instead of magnesium. The hedenbergite-granophyres of Brödretoppen are regarded as an intermediate stage in the differentiation which led from the basic hedenbergite-granophyres to the acid granophyres. The quantity of the acid granophyres and intermediate hedenbergite-granophyres is small relative to the rest of the Skaergaard intrusion—probably less than one per cent.

X. THE TIME SEQUENCE OF THE ROCKS, AND THEIR COMPOSITION COMPARED WITH THAT OF THE VARYING MAGMA FROM WHICH THEY WERE DERIVED

(a) Sequence of Solidification.

Injection of one rock type by another, or the inclusion of one as blocks in the other, often provides satisfactory evidence for the order of solidification of rocks in plutonic complexes. This kind of evidence is for the most part lacking in the Skaergaard intrusion and the relative order of solidification has to be deduced by other means. The order of solidification of plagioclase, olivine and certain pyroxene, solid solution series is definitely known from experimental investigations and it has been shown (pp. 125—27) from the variation in composition of these minerals in the layered series that the order of solidification of the layered rocks was from the bottom upwards.

The same method, combined with certain textural and structural features, can be used to decide the order of solidification of the border group. The chilled marginal gabbro which was obviously the first part of the Skaergaard magma to solidify, is succeeded along the northern margin by a gabbro with more anorthite-rich plagioclase and more forsterite-rich olivine than the lowest visible layered rocks (Table XXII), and it must be concluded that this part of the marginal border group was formed before any layered rocks now visible[1]). The gabbro-picrite, since it occurs as blocks in the northern border group, must have solidified before, or contemporaneously with, the northern border group. The transitional layered rocks are seen to be banked up against the border group and were formed later than the northern border group.

---

[1]) By extrapolation it is reasonable to conclude that layered rocks of an earlier period of solidification exist below those now visible. These are called the hidden layered series; they are considered to be more basic than the visible layered rock, and to be roughly comparable with outer border rocks (see pages 220—24).

Along the east and west borders a similar sequence of solidification may be deduced. Thus the perpendicular felspar rock and adjacent gabbros contain plagioclase which lies between An_{62} and An_{66}, olivine between Fa_{30} and Fa_{37} and clinopyroxene having γ' between 1.702 and 1.711 (Table XX). These minerals are higher-temperature solid solutions than those in the lowest visible rocks of the main layered series. The time of formation of this part of the western border group must, therefore, be correlated with that of the hidden layered rocks. Within the western border group later solidifying types are found in proceeding inwards. The innermost border rock on Ivnarmiut contains plagioclase and pyroxene crystals which have the same composition as those of the upper middle gabbros and from this evidence it may be concluded that this rock solidified at approximately the same time as the upper middle gabbros. On Ivnarmiut layered rocks are banked up against the border group and from their position these must have been formed later than the adjacent border group. How much later may be deduced from the mineral compositions. Thus the layered rocks which are banked against the border group on Ivnarmiut belong to the lower ferrogabbros. In the layered series these immediately succeed the upper middle gabbros with which the inner border rock was contemporaneous, and therefore the time interval between the layered rocks and the adjacent border group on Ivnarmiut must have been short.

Along the south and south-east margins, the marginal border group is adjacent to the upper border group, and it shows the following rock types which grade into each other: first, chilled olivine-gabbro; secondly, olivine-gabbros with rather basic plagioclase and olivine-rich forsterite, and finally, quartz-bearing olivine gabbros in which the olivine is decidedly rich in fayalite (see Table XXIV). Again it may be concluded that the rocks solidified successively from the margin inwards.

Variation in the upper border group is much complicated by hybridization with acid gneiss inclusions. The uppermost gabbros contain zoned plagioclase which is on the average labradorite, and the pyroxene is only moderately ferriferous, while the lower rocks have more acid plagioclase and more ferriferous pyroxene. From data of this sort, summarised in Table XXVII, it may be concluded that the upper border rocks solidified from above downwards. On Kilen the marginal and upper border rocks are in contact and the latter are similar to these occurring low on Osttoppen. They are presumably later solidifying than the adjacent olivine-gabbros of the marginal border group which contain only a little quartz. Similar evidence comes from the west ridge of Tinden, and it seems safe to conclude that, at any one point, the upper border rocks had a later period of solidification than the adjacent marginal border group.

The time of crystallisation of the upper border rocks relative to the layered series can only be roughly estimated, since there is no exact parallelism in the composition of either the rocks or minerals. In attempting a correlation it is safest to use features which would be little affected by hybridisation with the acid gneiss inclusions from the metamorphic complex. Thus it is better to use the iron-magnesium ratio rather than the composition of the plagioclase. The reason for this is that the total amount of iron oxides present in the gneiss is small and the proportion of iron to magnesium is about equal; this ratio will, therefore, not be much affected by hybridisation. It would not be safe to use the composition of the plagioclase as this is abundant in the gneiss and might have a considerable effect on the plagioclase composition of the hybrid. For the two highest, analysed, upper border gabbros, 3052 and 3050, the iron-magnesium ratios are 76 and 74 respectively and these correspond with the values for the middle gabbros. Another chemical feature linking these upper gabbros with the middle gabbros is high titanium content. Normative plagioclase in these two rocks is Ab_{58} and Ab_{54} respectively, which also corresponds fairly well with that of the middle gabbros although we consider that the significance of the plagioclase composition is less than that of the iron-magnesium ratio. The plagioclase crystals of the upper border gabbros are zoned—the inner half being between An_{55} to An_{60}; this is the same as the modal plagioclase in the lowest layered rocks. Hybridisation with acid gneiss would decrease the average anorthite content of the rock but it would probably not affect the composition of the inner zones. This line of evidence therefore suggests similarity in time of solidification with the lower layered rocks. In the later, quartz rich, upper border gabbros such as 4163, the magnesium-iron ratio is 81, which corresponds with that for the hortonolite-ferrogabbros. In this case normative plagioclase shows no correspondence, a discrepancy which we ascribe to hybridisation. The available evidence thus indicates that the highest upper border gabbros probably formed at about the same time as the upper hypersthene-olivine-gabbros of the layered series, while the later quartz-rich upper border gabbros such as 4163, probably formed contemporaneously with the hortonolite-ferrogabbros.

The latest rocks to solidify, excluding the transgressive granophyres, were the upper ferrogabbros and the unlaminated layered rocks. There were probably small amounts of marginal and upper border rocks contemporaneous with these but we have not identified them. The last rocks to solidify were the markedly transgressive granophyres such as the sill on Tinden and Brödretoppen and the veins found widely throughout the intrusion. From chemical and mineralogical evidence we should expect the hedenbergite-granophyres such as 3047 to have formed later than any of the layered rocks and earlier than the acid granophyre; the available field evidence is not incompatible with this view.

The sequence of solidification which has been deduced for the rocks of the Skaergaard intrusion is summarised in figure 41. The solidification is from the margin inwards, as would naturally be expected. There is, however, the unusual feature of a large amount of material—that form-

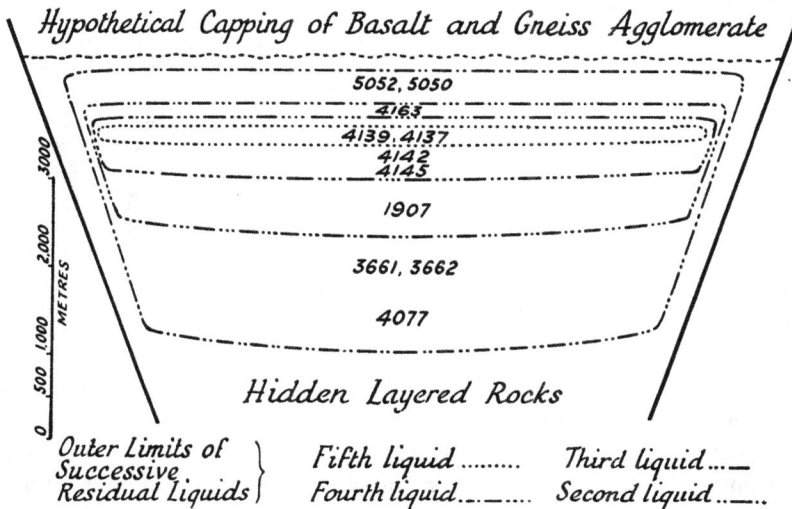

Fig. 41. East-west section across the Skaergaard intrusion showing successive stages of solidification and the approximate position of the analysed rocks used in calculating the composition of the successive residual magmas.

ing the layered series—which solidified from the bottom upwards. This material is considerably more abundant than that forming the upper border group which solidified from above downwards.

(b) Composition of Successive Residual Magmas and of the Hidden Layered Rocks.

The volumes of the various rock types produced during the differentiation of the Skaergaard magma may be approximately estimated from the form and dimensions of the intrusion and the observed thicknesses of the various types. For some of the rock groups, such as the hypersthene-olivine-gabbros, the middle gabbros and the hortonolite-ferrogabbros, the estimates of volume can be made with considerable accuracy. For the upper layered rocks the estimate depends on whether the sheets are considered as extending to the marginal border group or whether they are regarded as wedging out, and for the upper border group the estimates vary according to the assumed bulk of the inclusions

THE COMPOSITION OF SUCCESSIVE RESIDUAL

	1 % 4137	1/2 % 4139	1¹/₂ % Fifth Liquid	Comp. Fifth Liquid	1 % 4142	2¹/₂ % 4145	5 % Fourth Liquid	Comp. Fourth Liquid	6 % 1907
SiO₂	.521	.226	.747	49.8	.441	1.115	2.303	46.1	2.689
Al₂O₃	.159	.047	.205	13.7	.079	.292	.576	11.5	.828
Fe₂O₃.......	.056	.029	.080	5.7	.041	.051	.172	3.4	.225
FeO	.112	.119	.230	15.4	.266	.567	1.063	21.3	1.000
MgO........	.011	.002	.013	0.9	.003	.043	.059	1.2	.332
CaO	.058	.045	.098	6.6	.100	.218	.416	8.3	.512
Na₂O	.0363	.0121	.0484	3.23	.0215	.0737	.1436	2.87	.2010
K₂O	.0138	.0024	.0162	1.08	.0047	.0087	.0296	0.59	.0198
TiO₂........	.0114	.0083	.0197	1.32	.0248	.0607	.1052	2.10	.1530
MnO........	.0030	.0016	.0046	0.31	.0048	.0052	.0146	.29	.0102
P₂O₅........	.0070	.0045	.0115	0.77	.0161	.0462	.0738	1.47	.0048

such as the Basistoppen raft and the Tinden gneiss block. The bulk of the marginal border group is relatively small and has been ignored. Estimates of the volume of the layered rocks which must be presumed lying below those visible, depend on whether the walls maintain the dips observed at the present surface, or whether inward bending, as in cone sheets, or outward bending is assumed. From estimates of the volume of the various rock types, based on the intrusion form given in figure 15, and from the sequence of solidification deduced in the previous section, the composition of the magma existing at various stages has been calculated. The composition of the hidden layered rocks has also been calculated by subtraction of the visible material from that originally present, assuming the composition of the latter to have been that of the chilled marginal gabbro. All reasonable assumptions about the relative volumes of the visible rocks have little effect on the estimated composition of the hidden layered rocks, but different plausible assumptions about the relative volumes of visible to hidden rocks make big differences. If the boundaries are imagined extrapolated as in figure 15, the volume of visible to hidden layered rocks is approximately 3:2, and with this assumption certain oxides in the estimated composition of the hidden rock, such as FeO, Fe₂O₃, K₂O, TiO₂ and P₂O₅, are reduced to vanishing point or beyond; for this reason the form assumed in these figures must be wrong in certain essentials. If the intrusion is assumed to have somewhat outward-flaring boundaries in depth, then the volume of visible to hidden material becomes less, and a more reasonable composition of the hidden layered series is

XXXV.

MAGMAS AND OF THE INVISIBLE LAYERED SERIES.

1 % 4163 (modified)	12 % Third Liquid	Comp. Third Liquid	9 % 3661 & 3662	12 % 4077	7 % 3050 & 3052	40 % Second Liquid	Comp. Second Liquid	First Liquid Av. 1724 & 1825	60 % Invisible Rock	Comp. Invisible Rock
.490	5.482	45.7	4.221	5.564	3.401	18.668	46.7	47.92	29.25	48.7
.107	1.521	12.7	1.490	2.018	1.093	6.122	15.3	18.86	12.74	21.2
.030	.427	3.6	.266	.182	.300	1.175	2.9	1.18	0.00	0.0
.140	2.203	18.2	1.096	1.253	.611	5.163	12.9	8.66	3.50	5.8
.030	.421	3.5	.522	1.153	.295	2.391	5.9	7.82	5.43	9.0
.072	1.000	8.3	.870	1.355	.730	3.955	9.9	10.46	6.50	10.8
.0350	.3796	3.16	.2673	.2940	.1932	1.1341	2.83	2.44	1.31	2.2
.0030	.0524	0.44	.0189	.0240	.0196	.1149	0.29	0.18	0.06	0.10
.0250	.2832	2.36	.2349	.0948	.2611	.8730	2.18	1.35	0.48	0.8
.0016	.0264	0.22	.0126	.0108	.0161	.0659	0.16	0.10	0.03	0.05
.0320	.1106	0.92	.0054	.0072	.0455	.1687	0.42	0.07	..	..

obtained if the ratio is assumed to be 2:3. The relative volumes which we have finally assumed are given in Table XXXV, where is also set out the calculation of the composition of successive residual magmas and the composition of the hidden layered material.

In constructing this table, the unlaminated layered rocks, forming $1^1/_2$ per cent. of the total intrusion, are taken as giving the composition of a late residual magma (the fifth of our classification). The laminated fayalite-ferrogabbros and the ferrohortonolite-ferrogabbros, together with the fifth residual magma, give the composition of fourth residual magma. Included with this there should probably be a little upper border material but we have no definite evidence of its nature, and it has been ignored. The third residual liquid is estimated to consist of the previous residual magma plus six per cent. of 1907 and one per cent. of upper border material, which is taken as the composition of 4163, less forty per cent. of average grey gneiss, 1867. The second residual liquid amounting to forty per cent. of the whole is taken as the previous residual magma, together with nine per cent. of the average middle gabbro, twelve per cent. of 4077 and seven per cent. of upper border gabbro taken as the average of 3052 and 3050. The first liquid, that which originally filled the Skaergaard intrusion had the composition of the chilled marginal olivine gabbro, and this is taken as the average of the analyses of 1825 and 1724. At the third liquid stage the amount of P_2O_5 in the calculated liquid exceeds the total amount in the original magma and yet analyses show that there is also a small amount of P_2O_5 in the early rocks. As the total P_2O_5 in the estimated residual magmas exceeds the total amount in the original magma, either the assumed relative volumes or the average compositions must be wrong, or there has been addition of P_2O_5 to the magma. It is likely that a small amount of acid gneiss has been completely incorporated in the magma (see below p. 292), but this can scarcely be sufficient to account for the whole discrepancy in the calculation. Slight errors in the estimated volume of one of the late apatite-rich rocks would readily produce this difficulty with the P_2O_5 content of the residual magmas, and we do not consider that the abnormal behaviour of the P_2O_5 content of the calculated liquids is sufficient to vitiate the the general deductions to be drawn from these estimates of the composition of successive residual magmas.

By subtracting the total composition of the forty per cent. of visible rocks from the composition of the material originally occupying the Skaergaard intrusion (average of 1825 and 1724) the total composition of the sixty per cent. of hidden layered rock is obtained, and this may be recalculated in percentages. The figures so obtained (Table XXXV) are of interest in throwing light on the probable composition of the hidden layered rocks and also as a check on the reasonableness of the relative volumes of the various rocks which have been assumed. The estimated composition of the hidden layered rock is that of a felspar-rich olivine-eucrite having a norm as set out in Table XXXVI. The hidden layered rocks are the earlier fractions separating from an olivine-gabbro whose composition is that of the chilled marginal gabbro. From general considerations and also from the sequence of crystallisation determined in thin sections of the marginal olivine-gabbro, we should expect the earliest crystal fraction to be felspar-rich, like the estimated rock, and we should expect that the felspar would be basic—yet not anorthite—corresponding to the composition of the centres of the zoned felspars in the chilled marginal gabbro. It might perhaps have been expected that the earliest fraction of the Skaergaard magma would have been a troctolite rather than eucrite. However, no reasonable estimate of relative volumes will reduce the silica percentage of the hidden rocks so that it becomes troctolitic in composition, and we there-fore believe that the early fractions, separating from the Skaergaard magma, consisted of plagioclase, olivine, and pyroxene, in such pro-portions as to give a eucrite. Microscopic examination of the chilled marginal gabbro (see p. 138) suggested the possibility that pyroxene was an early mineral to separate.

It has been suggested that the outer border rocks, excluding the chilled marginal gabbro, should correspond to the hidden layered rocks, just as the inner marginal border rocks correspond to parts of the visible layered series. In the variation diagram, figure 30, the analysed layered

TABLE XXXVI.

COMPOSITION AND NORMS OF SUCCESSIVE RESIDUAL MAGMAS, AND OF THE INVISIBLE LAYERED ROCK.

	First Liquid Av.1825 and 1825	Second Liquid	Third Liquid	Fourth Liquid	Fifth Liquid	Invisible Rock
SiO_2	47.92	46.7	45.7	46.1	49.8	48.7
Al_2O_3	18.86	15.3	12.7	11.5	13.7	21.2
Fe_2O_3	1.18	2.9	3.6	3.4	5.67	0.0
FeO	8.66	12.9	18.2	21.3	15.37	5.8
MgO	7.82	5.9	3.5	1.2	0.87	9.0
CaO	10.46	9.9	8.3	8.3	6.57	10.8
Na_2O	2.44	2.83	3.16	2.87	3.23	2.2
K_2O	0.18	0.29	0.44	0.51	1.08	0.10
TiO_2	1.35	2.18	2.36	2.10	1.32	0.80
MnO	0.10	0.16	0.22	0.29	0.31	0.05
P_2O_5	0.07	0.42	0.92	1.47	0.77	..

Norms

	First Liquid Av.1825 and 1825	Second Liquid	Third Liquid	Fourth Liquid	Fifth Liquid	Invisible Rock
Qu.	..	..	..	..	5.28	..
Or.	1.11	1.67	2.8	3.34	6.67	0.56
Ab.	20.69	23.58	26.7	24.10	27.25	18.34
An.	39.75	28.40	19.2	16.96	19.46	47.82
Ne.	..	..	..			
Di.	4.99 9.74 / 2.90 / 1.85	7.66 15.28 / 3.40 / 4.22	7.1 14.6 / 1.8 / 5.7	6.26 13.20 / 0.60 / 6.34	3.48 7.34 / 0.30 / 3.56	2.44 4.70 / 1.60 / 0.66
Hy.	7.50 11.92 / 4.42	4.7 10.40 / 5.7	3.1 12.6 / 9.5	2.40 29.46 / 27.06	1.90 20.25 / 18.35	11.00 15.22 / 4.22
Ol.	6.37 10.96 / 4.59	4.6 10.9 / 6.3	2.7 11.7 / 9.0	.. / ..	.. / ..	7.00 9.86 / 2.86
Ilm.	2.66	4.0	4.6	3.95	2.43	1.52
Mg.	1.86	4.2	5.3	4.87	8.35	..
Ap.	0.20	1.0	2.0	3.70	2.02	..

rocks and the estimated composition of the average hidden layered rock are plotted against iron-magnesium ratios. On a similar graph (Fig. 31) is plotted the composition of the analysed outer border rock, 1837. As anticipated, the points for this rock fall on, or close to, the curves for the layered series and its position on the diagram corresponds to that of a layered rock a little below any at present visible. The perpendicular felspar rock, 1851, and the gabbro-picrite, 1682, of the northern border group are two abnormal facies of the outer border group, and these are also plotted on figure 31. The perpendicular felspar rock falls less close to the curves for the layered series than the normal outer border rock but it roughly corresponds to the estimated composition of a still lower hidden layered rock. The analysis of the gabbro-picrite does not even approximate to the curves; in this rock there has been a strong concentration of olivine and it does not approach the composition of any complete crystal fraction. In iron-magnesium ratio it corresponds, as would be expected, to the composition of a very low hidden layered rock.

The estimated average composition of the hidden layered rocks gives a normative mineral composition resembling the early border rocks (Table XXXVI). Thus the normative plagioclase of the outer marginal border rock, 1837, is An_{67}; for the perpendicular felspar rock it is An_{78}, while the value for the hidden layered rock is An_{73}. This is an average—actually there will be a range from somewhat below, to somewhat above this value. Thus the estimated composition of the plagioclase of the average hidden layered rock is in harmony with the normative plagioclase composition of these two border rocks. The normative composition of the plagioclase for the perpendicular felspar rock should correspond to that of the lowest hidden layered rocks. The probable range in composition of the hidden layered rocks may be judged from the extrapolated variation diagram (Pl. 27). The composition of the first fraction obtained by means of this graph has normative plagioclase which is An_{78}; as anticipated this is the same as for the perpendicular felspar rock. The iron-magnesium ratio of the gabbro-picrite is 33, while that for the average hidden layered rock is 39. Again, there would be a range in this ratio for the hidden layered rocks and the value might reasonably be expected to fall to about that found in the gabbro-picrite: indeed, by extrapolation (see Pl. 27) the iron-magnesium ratio of the earliest hidden rock is 29. The iron-magnesium ratio for the perpendicular felspar rock does not correspond with that of the estimated lowest layered rock, probably because the perpendicular felspar rock consists of early plagioclase surrounded by the products of crystallisation of the original magma whose composition is that of the chilled marginal gabbro. The perpendicular felspar rock is the result of selective fractionation (for the definition of this term see p. 282), whereas the layered rocks consist of all the minerals crystallising at the particular time (non-selective fractionation, see same page). The normative plagioclase of the gabbro-picrite is not a significant figure because nepheline is present in the norm; also it seems likely that the gabbro-picrite is composed of early olivine surrounded by the products of crystallisation of the original magma and is thus another case of selective fractionation.

The variation diagram for the rocks of the intrusion, plotted against their estimated relative quantities, gives a graphical check on the

calculation of the composition of the hidden rocks; it also enables the
effects of different assumptions about the relative quantities of the
different rock types to be visually perceived. The area between the
curve for any oxide and the horizontal axis of reference is proportional
to the total amount of that oxide in the whole intrusion. This should
equal the original amount present in the intrusion which is indicated
by the area of the rectangle whose sides are the horizontal axis of reference
and the amount of the oxide present in the original magma. The case
of iron is shown separately in figure 42. The total amount of iron present

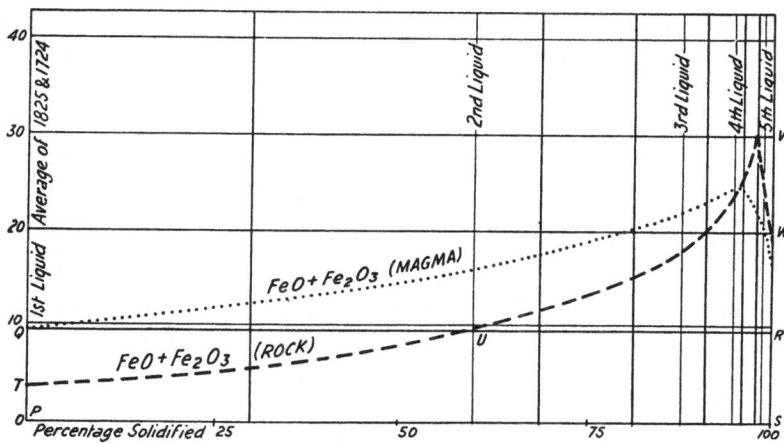

Fig. 42. Variation diagram for total iron in the magma (dotted) and in the layered
rocks (broken line), plotted against percentage solidified.

in the intrusion is proportional to the area of the rectangle PQRS and
this must equal the area PTUVWS which represents the total iron in
the rocks. In other words the area TQU must equal UVWR whatever
may be the assumptions made about relative volumes. If the proportion
of hidden layered rocks to the visible layered rocks is greater than
has been assumed then there will be a relative extension of the left-
hand part of the graph and to maintain the equality of the areas the
point T will approach Q. If this is the case then the hidden layered
rocks will be on the average richer in iron. On the other hand, if the
proportion of hidden rocks to the visible is less than has been assumed
then the point T must approach P and the rocks of the hidden layered
series will be, on the average, less rich in iron. Judging by the known
basic igneous rock it would be reasonable that the point T should vary
in position to some extent from that actually shown, but it is unlikely
that the amount of iron would fall below one or two per cent. and it
certainly should not vanish. Consideration of other oxides gives a similar
result and we are led to the conclusion that the amount of reasonable

variation in the relative volumes of the different rock types which have
been postulated cannot be great. The assumptions which have actually
been made with regard to the volumes of the different rocks may
therefore be taken as leading to an essentially true picture of the variation in the composition of the rocks of the whole intrusion.

(c) Variation in Composition of the Residual Magma and its Relation to the Composition of the Layered Series.

The calculated composition of successive residual magmas is plotted
against the percentage solidified in figure 43. As we have just shown
reasonable changes in the assumed quantities of the different rock types
will not affect the curves significantly. The composition of the first liquid
is, of course, not an estimate but the determined composition of the
chilled marginal gabbro. These curves, which show the variation in
composition of the magma during the differentiation of the intrusion,
may be taken as correct within narrow limits.

While 95 per cent. of the magma was solidifying the curves for
the major oxides show only slight changes in direction. While the
remaining 5 per cent. solidified there was a strong change in the
direction of the curves. Apparently the general trend of the differentiation was constant while the greater part of the magma solidified, and
this we shall call the early and middle stages of the differentiation.
While the last 5 per cent. of the magma solidified the curves for the
major oxides take on new directions and this we shall call the late stage
of the differentiation. The variation diagram given in figure 42 combined
with that for the latest differentiates given in figure 40 shows the trend
of variation for the different oxides and there is no need to describe
these trends in words. In the present section an attempt will be made
to interpret the trends in terms of the composition of the layered series
whose formation controlled the variation in composition of the successive residual magmas. This can best be done with the help of the variation diagram given in plate 27 which shows both the composition of
the magma and the composition of the layered rocks plotted against
the percentage solidified.

The magma variation diagram for any oxide depends mainly on
the relative amounts of the primary phases separating and on their
composition. These in turn depend on the composition of the magma
undergoing fractional crystallisa..on. In the present section we shall
consider these two mutually related series of changes with a view to
elucidating the change in composition of the magma, while in the next
section we shall consider the same matter from the opposite aspect,
namely the trend in detailed composition of the minerals in relation to

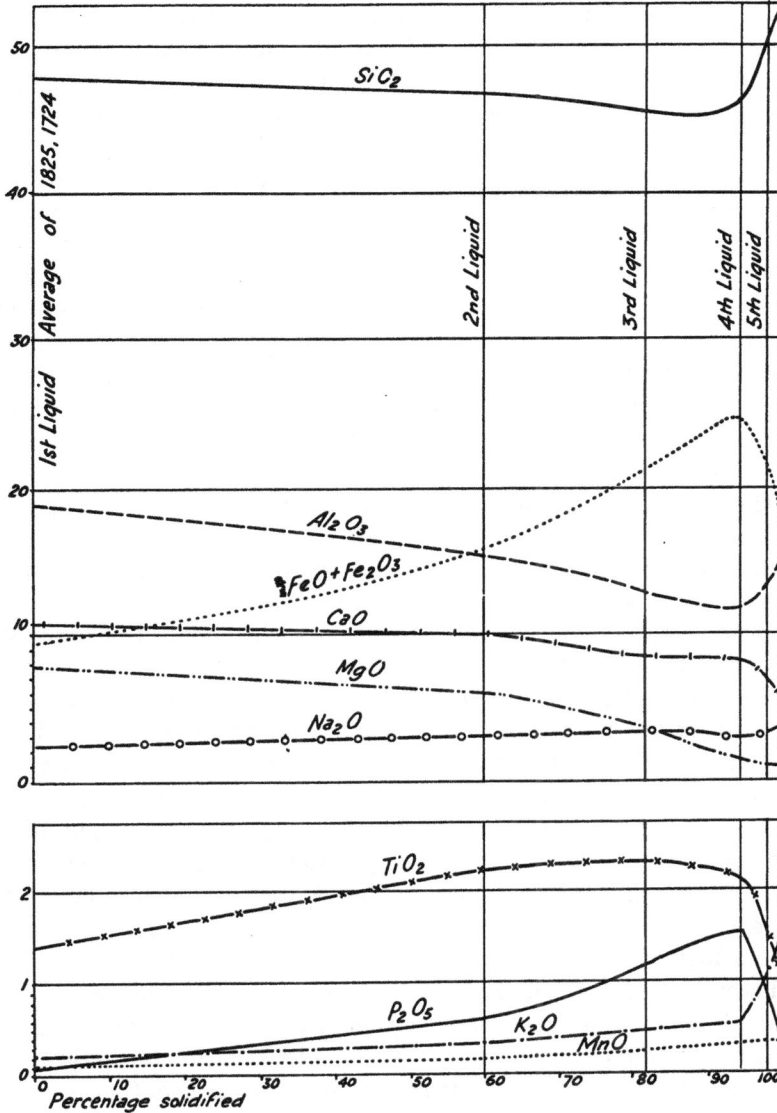

Fig. 43. Variation diagram of the successive residual Skaergaard magmas (see Table XXXV) plotted against the percentage solidified.

the varying composition of the magma. Throughout this discussion the liquid is assumed to be effectively stirred so that it is homogeneous, and the layered series is assumed to be the crystal fraction separating at any moment together with about 20 per cent. of the contemporaneous liquid; the evidence for the latter assumption has been given already (pp. 125—27).

Since silica occurs in most of the phases separating from the magma, its variation is the result of a complex balance in the relative proportions, and composition, of the minerals separating. The most significant feature shown by the curve for silica is the small amount of variation, while ninety-five per cent. of the magma solidified; what slight variation there is during the early and middle stages, leads to successively lower values. The silica content only begins to rise while the last five per cent. of the magma solidified. To explain the silica curve in terms of the minerals separating and their relative amounts seems to be too complex a matter to be profitable at the moment. It is, however, important to note that for basic magmas differentiating as has the Skaergaard magma, a variation diagram plotted against silica percentage would be unsatisfactory, and it is possible that the significance of variation among basic rocks may have been partly missed by considering the variation in terms of silica percentage.

The curves for total iron and magnesium show an antipathetic relationship. The amount of iron separating in the early fractions is less than was present in the original magma; hence the iron content of the residual magma rises, and this goes on until the concentration in the magma is twenty-four per cent. At this point the amount of iron being precipitated in the solid phases exceeds that in the magma, and the curve changes direction so that the final liquid, represented by the transgressive granophyre, contains only 2 per cent. of total iron oxides. The curve for MgO in the magma falls during differentiation because the amount separating in the early fractions was greater than that in the original magma. The curves for iron and magnesium in the magma and in the rocks are fairly smooth, and, since these oxides occur in several minerals, the cause of the variation can only be assessed in a general way as the result of the iron-rich pyroxenes and olivines having a lower melting point than the magnesium-rich. This results in iron becoming concentrated in the residual magmas during the early and middle stages of the differentiation. In the latest stage iron compounds also become relatively insoluble, and the ultimate liquid is poor in both magnesium and iron. It is remarkable that the content of magnesium is lower in the fayalite-ferrogabbros, 4142 and 4139, than in the basic hedenbergite-granophyre 4137. This is not due to a lower content of ferromagnesian minerals, the reverse being the case. It may be due to the fact that the mineral separating from the magma was β-wollastonite which contained less magnesium than the pyroxenes separating from the magma both before and after this stage.

The curve for Al_2O_3 in the magma falls owing to the separation of abundant plagioclase in the early stages. In this respect the Skaergaard magma resembles the Porphyritic Central magma of Mull, which shows early separation of plagioclase. The upward trend of the Al_2O_3 curve after the early and middle stage of the differentiation is due to the concentration of alkali felspar in the final residual magma. The curve for Na_2O shows only a very gradual upward tendency, and at about 90 per cent. solidified there is even a temporary downward tendency. This illustrates the way in which the relative quantities of the minerals separating controls the direction of the curves as much as the composition of the minerals. As would be expected from the experimental evidence, the later residual liquid contains a felspar which is more albite-rich than the earlier liquid but there is a stage when this felspar is present in the liquid in relatively small amounts and the total Na_2O in the liquid is less than that in earlier liquids. At the late stage of the differentiation the curve for soda and also that for potash rises. This is due to the concentration of the soda and potash felspar in the latest stage due to the relatively low temperature at which these minerals solidify. The curve for lime shows little variation until the very latest stages (Fig. 40) and this is perhaps surprising, since the lime enters into the plagioclase and pyroxene which are both varying in composition and amount throughout the differentiation.

The curves for certain minor constituents MnO, P_2O_5, TiO_2 and Fe_2O_3 are particularly interesting. For MnO the curve follows that for FeO. There is no separate manganese minerals but only partial replacement of iron and magnesium by manganese, and it is interesting that the proportion of manganese to iron remains almost constant throughout the series. The curve for P_2O_5 is very different in character from that for manganese, due no doubt to it forming a separate solid phase. P_2O_5 is present only in small amount in the original magma and no separate primary mineral containing P_2O_5 is formed in the early layered rocks; nor does P_2O_5 enter appreciably into the composition of any of the primary minerals. The P_2O_5 of the early layered rocks is believed to have existed only in the 20 per cent. of interprecipitate magma and to have crystallised from it as rare, small, apatite crystals. The amount of P_2O_5 occurring in the lower layered rocks is less than existed in the magma, and hence the amount in the successive residual magmas increased. The amounts of P_2O_5 deposited with the successive layered rocks only increases very slightly up to the hortonolite-ferrogabbro horizon and the slope of the curve for magmatic P_2O_5 becomes gradually steeper. When the amount in the magma reached about 1.5 per cent., apatite began to precipitate as a primary phase, and after this the curve for magmatic P_2O_5 falls gradually because the amount separating was a little more than the amount present in the magma. The falling curve for magmatic P_2O_5 after apatite begins to separate as a primary mineral, indicates that the solubility of apatite in the changing residual

15*

magma becomes progressively less; by the time the acid granophyre is reached it is very small indeed. Vogt [1921, pp. 324 and 645] has expressed rather similar views on the way in which apatite separates from magmas but he considered that the solubility of apatite in magmas was of the order of .01 per cent. This is probably correct for very acid magmas but our evidence for basic magmas indicates that the solubility is 500 times as much. The statistical study of the amount of P_2O_5 in British Tertiary igneous rocks made by Richey [1937] may also be interpreted as indicating greater solubility of apatite in intermediate than in either basic or acid rocks. During the middle stage of the differentiation of the Skaergaard magma there is a marked difference between the amount of P_2O_5 in the magma and that separating with the layered rocks. Richey [1937, pp. 49 and 50] has stressed the fact that basalts are, on the average, richer in P_2O_5 than gabbros and he suggested "that gravitative settling of early formed crystals is not a factor which leads to enrichment in apatite". The Skaergaard intrusions support this deduction over the range of normal basic rocks, while it shows that, in the unusual ferrogabbros, there is a stage when gravitative settling produces a sudden increase is the P_2O_5 content of the rocks. A comparison of the P_2O_5 content of the layered rocks with certain of the border rocks can be made to throw light on the relative proportions of the primary crystals and of the interprecipitate material, a point to which we shall return in attempting to explain the composition of border rocks (p. 279).

The curves showing the amount of TiO_2 in the magma and in the layered rocks have a similarity with those for P_2O_5. The amount separating in the early fractions was less than in the original magma, hence the curve for magmatic TiO_2 rises. When it reached about 2.5 per cent. titanium began to precipitate from the magma in much greater amounts than previously. The amount precipitating was then a little more than that present in the magma, and thus the curve for magmatic TiO_2 thereafter falls. The small amount of TiO_2 separating with the early fractions must have been partly in solid solution in the primary phases, such as the olivine and pyroxene. However the greater part was probably present as ilmenite in the interprecipitate material as in the case with the apatite of the lower layered rocks. At the stage where the TiO_2 content of the layered rocks suddenly rises, ilmenite becomes a primary phase (see p. 128). The increase in the TiO_2 content of the pyroxene comes at the later, hortonolite-ferrogabbro stage and cannot be connected with the sudden change in TiO_2 content of the rocks shown at the middle gabbro horizon, at which stage the pyroxene is still relatively poor in titanium (cf. Table VII). It appears that the conditions for the separation of ilmenite from magmas are rather similar to those for apatite. For crystals to precipitate from a basic magma the concentration of TiO_2 must apparently reach about 2.5 per cent.; it also appears from the decline in the amount of TiO_2 in successive residual magmas that the solubility of ilmenite falls off as the magma changes in composition due to fractional crystallisation.

The variation of Fe_2O_3 in the magma and in the rocks of the layered series is rather like that of TiO_2. In the original magma the amount of Fe_2O_3 was small, and little Fe_2O_3 is precipitated in the early layered rocks. The Fe_2O_3 in the layered rocks up to the middle gabbros is present partly in the pyroxene and partly in iron ore which occurs as an interprecipitate mineral only. During this period, that is, while about 75 per cent. of the magma was solidifying, the amount of Fe_2O_3 precipitated was a little less than in the magma and the amount in the magma rose slowly until it attained about three and a half per cent. At the stage, when the middle gabbros were forming, magnetite appears as a primary precipitate. As in the case of TiO_2 we interpret this to mean that the solubility of magnetite in basic magma of this kind is such that the concentration of Fe_2O_3 must exceed three and a half per cent. before magnetite is precipitated. This is only a rough figure but it corresponds well with Vogt's estimate [1921, p. 133] that "In gabbroidic rocks with much hypersthene, diallage or olivine, the magnetite seems already to have commenced its crystallisation at about four per cent. Fe_3O_4." With the changing composition of the magma the amount of Fe_2O_3 in the magma rises until in the fifth liquid it is five-and-a-half per cent. As magnetite remains a primary phase this must be interpreted to mean that the solubility of magnetite in the changing magma increases. The same value of about five per cent. Fe_2O_3 is found in the hedenbergite-granophyre, 3047, which is believed to represent a still later liquid but the amount falls after this as Vogt maintained, until it is only one-and-a-half per cent. in the acid granophyre, 3058. This explanation of the variation in the amount of Fe_2O_3 in the magma is only a first approximation, since Fe_2O_3, like TiO_2 is present in small amounts in the other minerals, especially pyroxene, and since there must be complications due to the mutual solubility of ilmenite and magnetite.

The relative amount of ilmenite and magnetite in certain rocks of the layered series has been estimated by a method already explained (p. 83); the results are given in Table XXXVII. In the hypersthene-olivine-gabbro there is 0.9 per cent. of ilmenite and 0.6 per cent. of magnetite. When, in the middle gabbros, iron ore becomes a primary phase the amount both of ilmenite and magnetite increases abruptly to 5 per cent. in each case. Thereafter the amount of ilmenite in the crystal fractions separating as the layered series, falls off steadily, while the amount of magnetite increases slightly throughout the period of formation

TABLE XXXVII.

AMOUNTS OF ILMENITE AND MAGNETITE IN ROCKS OF THE LAYERED SERIES.

Rocks in order of Height	4077	3661	1907	4145	1881	4139
$\dfrac{\text{Il.} \times 100^{1)}}{\text{Il.} + \text{Mt.}}$	60	50	35	30	30	5
Total Iron ore in mode (= Il. + Mt.)	1.5	10[2]	8	6.5	8	12
Il. in mode	.9	5	2.8	2.0	2.4	.6
Mt. in mode	.6	5	5.2	4.5	5.6	11.4

[1]) ± 5.
[2]) Average of 5 middle gabbros.

of the layered series but falls suddenly in the acid granophyre. The behaviour of the iron ore will be considered again when the trend in composition of the Skaergaard magma is compared with other examples.

The compounds in the norm of the layered series and of the successive residual magmas may be divided roughly into those separating from the magma at relatively high temperatures and those separating at relatively low temperatures. In making this division we have been guided partly by the results of experimental work and partly by the sequence of the primary phases of the layered series. The group which for short will be termed the high temperature compounds, consists of anorthite, wollastonite, enstatite and forsterite, while the group of low temperature compounds consists of quartz, orthoclase, albite, ferrosilite, fayalite, ilmenite and magnetite. Since these two groups contain all the important compounds of the norm, the total approximates to one hundred per cent. and in plotting it does not matter which group is chosen. If the percentage of the low temperature compounds in the layered series and successive magmas is plotted against percentage solidified two smooth curves are obtained (Fig. 44) both showing, as would be expected, increase in the amount of low temperature minerals as differentiation proceeds. While ninety-eight per cent. of the magma solidified there are two curves, one showing the composition of the magma and the other an approach to the composition of the solid fraction separating. Again as would be expected the solid fraction is poorer in the low temperature minerals than the contemporaneous liquid. The curves resemble roughly the liquidus and solidus curves for a series of solid solutions such as the plagioclases.

Since the iron-magnesium ratio is a fairly regular function of the percentage solidified, it is possible to put down, along side the percentage solidified, a scale showing the corresponding iron-magnesium ratios. Using this scale, points for three lower marginal border rocks (1851, 1837 and 4298) have been plotted and as expected they fall near the curve for the layered series showing that these rocks roughly correspond to crystal fractions separating from the magma.

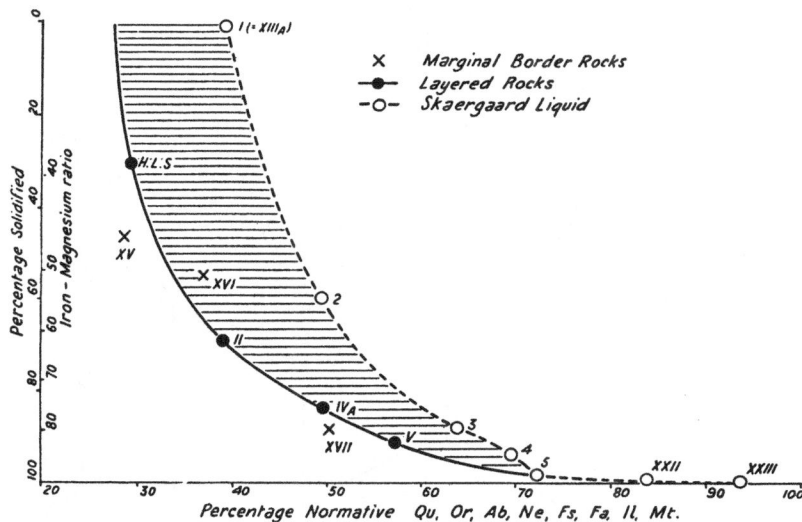

Fig. 44. Percentage of normative low temperature compounds, plotted against estimated percentage solidified.

The normative values for the Skaergaard rocks may also be divided into three groups, those tending to separate at high, intermediate and low temperatures. The group of the more refractory compounds is taken as consisting of anorthite, enstatite, forsterite and wollastonite; the group of compounds, intermediate in their refractory nature, is taken as consisting of fayalite, ferrosilite and iron ores, and the third group of less refractory substances is taken as consisting of albite and orthoclase[1]). This division of the normative compounds according to the range of temperature at which they separate is, of course, only rough and moreover may not be applicable to other rock series. The justification for this grouping in the case of the Skaergaard intrusion is that it shows, better than others, the trend during differentiation, and moreover the significance of the trend may be partly understood. The percentage of

---

[1]) Though we have used this type of diagram it would probably have been better to have put normative quartz with the acid felspars and half the wollastonite with the iron compounds.

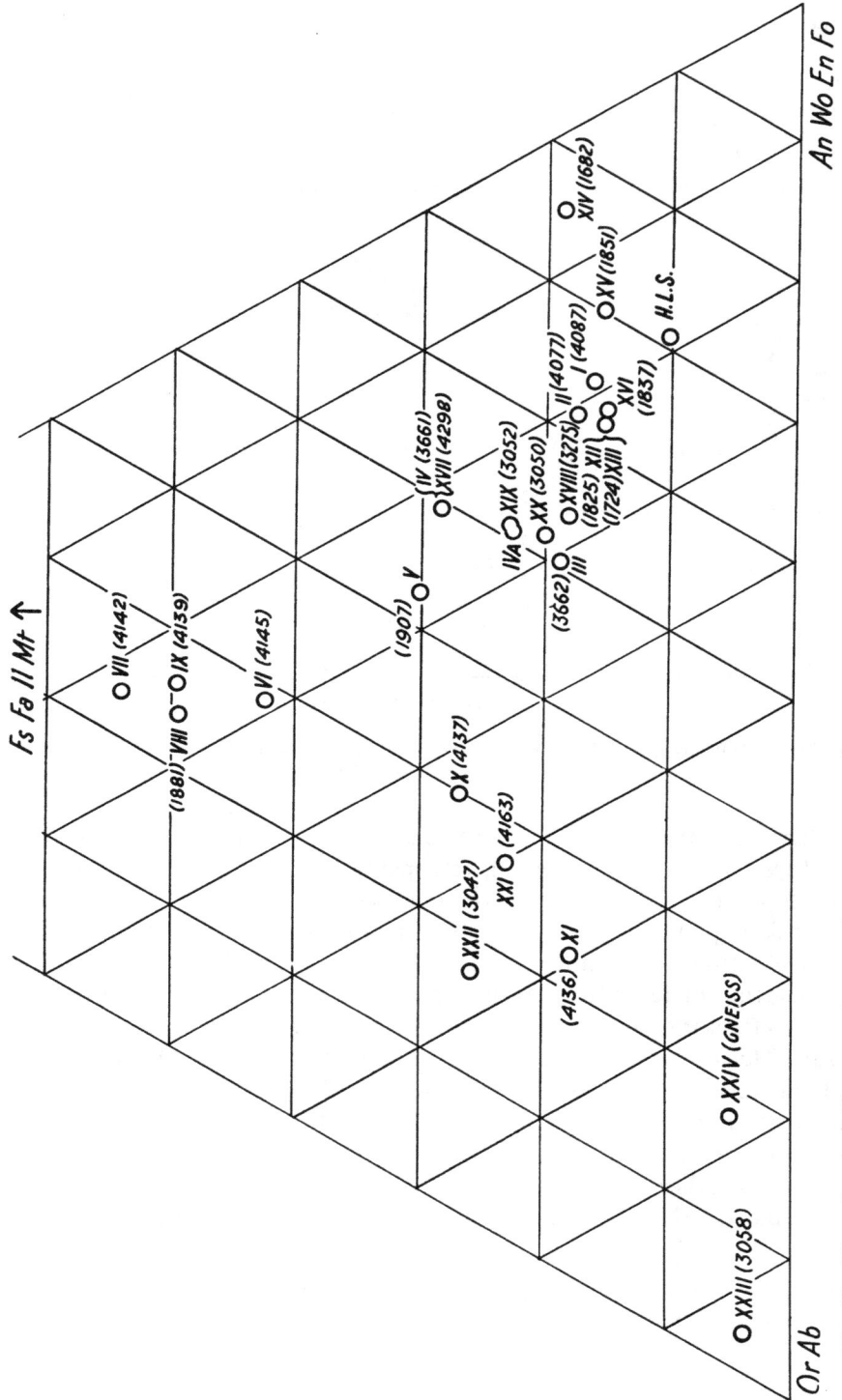

Fig. 45. All analyses of Skaergaard rocks plotted on a triangular diagram showing the more refractory, medium refractory and less refractory normative compounds.

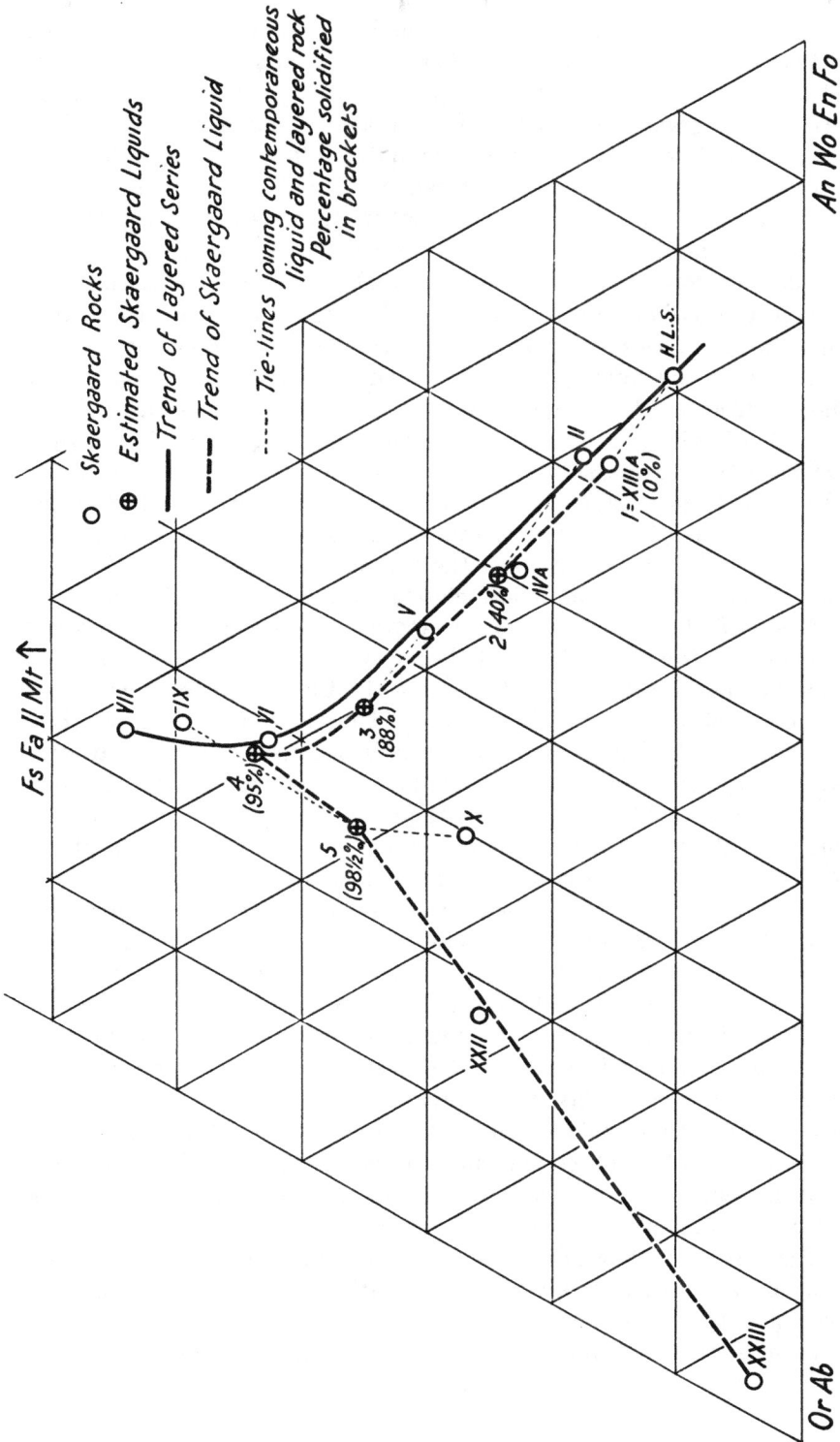

Fig. 46. Analyses of Skaergaard layered rocks and estimated compositions of residual liquids plotted as in figure 45.

each of these three groups for all the Skaergaard analyses has been plotted on a triangular diagram (Fig. 45). The analyses of the layered rocks and of the corresponding liquids, and also the analyses of the hedenbergite granophyre, 3407, and the acid ·granophyre, 3058, which are considered to show the general trend of differentiation in the later stages (p. 212) have also been plotted in a similar way (Fig. 46). The later stages of the differentiation are further considered below (pp. 292-93) where it is suggested that some of the constituents of the rocks have been derived from acid gneiss incorporated in the magma. If this were the case the assimilation resulted in a homogeneous magma and it is proper for our present purposes to plot the late stage rocks as late differentiates of the Skaergaard magma. On the triangular diagram the points indicating the composition of the layered rocks (about eighty per cent. of which is the solid fraction separating from the magma) have been tied to the points representing the magmas from which the rocks were formed, and the approximate proportion solidified, when each of the calculated magmas was in existence, is given in brackets.

The trend in composition of the layered rocks and in the deduced composition of the liquid follow closely similar courses. For both the trend is at first away from the refractory components mainly towards the medium refractory but also slightly towards the least refractory, the orthoclase and albite. During this stage the tie lines show clearly that the solid fraction is richer than the liquid in the more refractory components. After the fourth liquid the trend in the magma curve is directly towards the least refractory constituents, while for a short time the solid fraction, represented by analyses of the upper ferrogabbros VII and IX, approaches the medium refractory components. The composition of the liquid when the rock giving analysis VII was forming has not been calculated but would lie between liquids 4 and 5; again the liquid was richer in the less refractory substances than the contemporaneous layered rock. After the fifth liquid two curves, one showing the composition of the solid fraction and the other the composition of the liquid fraction, cannot be given, but only one, showing the general trend of the differentiation, and this is towards the least refractory constituents, the orthoclase and albite. This type of diagram is not only of value for the Skaergaard rocks but will be used to compare the trend of the Skaergaard magma with that of other magma and rock series since it produces a spread of points, the significance of which can be interpreted in terms of crystal fractionation of normal basic magma.

The variation diagrams, figures 40, 43 and plate 27, and the triangular diagram, figure 46, show that, whether the rocks or the magma from which they were separating is considered, there was an abrupt change in the trend of compositions when about ninety-five per cent. of the magma had solidified. The same abrupt change is shown by other methods of plotting. Thus in figure 47 percentages of CaO, MgO and FeO for all the Skaergaard rocks and the calculated liquids are plotted, and the trend of the differentiation is indicated. Again there is a change in trend after the fourth liquid. Rather similar results are obtained by other

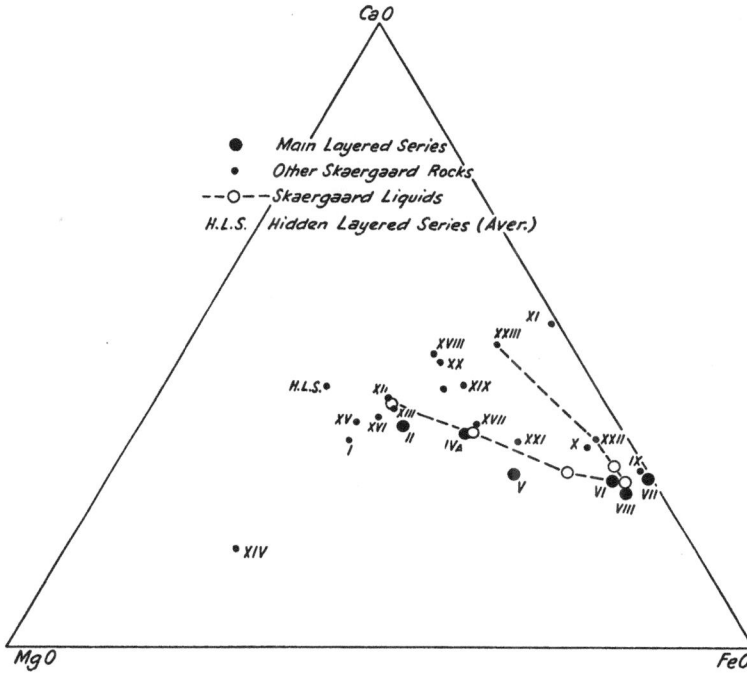

Fig. 47. Skaergaard rocks and liquids plotted on diagram showing weight percentages of CaO, MgO, FeO.

common methods of plotting normative compositions (Figs. 64 and 65). For convenience of description we shall divide the differentiation trend of the Skaergaard magma into three stages: the early stage during which the first sixty per cent. of the magma solidified; the intermediate stage during which from sixty to ninety-five per cent. of the magma solidified, and the late stage during which the final five per cent. solidified. During the early stage the magma varied little and the rocks were, eucrites and olivine-gabbros, the lower of which belong to the hidden layered series and are only inferred. During the intermediate stage the change in composition of the magma was considerable and that of the rocks was still more so, although the direction of change was essentially the same as in the early stage. The rocks formed vary from olivine-gabbros through olivine-free rocks to ferrogabbros, the latest of which are very rich in iron. During the late stage there is an abrupt change in the course of the differentiation, however we may look at it, and this results in a succession of rocks passing from basic hedenbergite-granophyre through hedenbergite-granophyre to acid granophyre. The quantity of the late stage intermediate and acid rocks, is small, probably amounting to no more than two or three per cent. of the total original magma.

XI. THE TREND IN COMPOSITION OF SOME OF THE MINERALS OF THE LAYERED SERIES COMPARED WITH THAT OF THE MAGMA

In the preceding section we have shown how the trend in composition of the Skaergaard magma is related to the composition of the layered series which approximates in composition to the crystal fractions successively separating from the magma. The variation in composition of the magma is a function of the minerals forming as primary phases in the layered series. Equally the minerals precipitating as primary phases are related to the course taken by the composition of the magma. In the previous section the way in which the particular minerals separating from the magma influenced the changes in its composition has been considered. In the present section the same matter will be considered from the reverse aspect, namely, the way in which the changing composition of the magma affected the composition of the minerals separating. In most cases the relations are simple and as anticipated but in the case of the clinopyroxenes the results are more complex and are only partly elucidated by experimental investigations of melts. As the data is available the clinopyroxenes will be dealt with in some detail.

(a) Plagioclase, Olivine and Orthopyroxene.

In this natural example of fractional crystallisation the variation in the plagioclase, olivine and orthopyroxene, which are relatively simple solid solution series, follows the course which is expected on the basis of other petrological and experimental investigations. Except in the hedenbergite-andesinite, which has been formed differently from the rest of the layered series, the albite content of the plagioclase increases steadily in the successive fractions; this is the result anticipated from the experimental investigations if the presence of other phases has little effect on the plagioclase equilibrium. The composition of the normative plagioclase of the original Skaergaard magma was An_{65}. The earliest

plagioclase to form from a magma of this composition may be estimated
from the thermal diagram for the plagioclase series [Bowen 1928, p. 34]
as An$_{85}$ providing that the other constituents of the magma have little
effect. This estimated value is fairly close to that found in the perpendicular felspar rock An$_{78}$ which is regarded as representing one of
the earliest fractions. It is also close to the original composition of the
plagioclase which we have deduced from extrapolation of the variation
diagram of the layered rocks. As a result of strong fractionation it is
theoretically possible to develop albite and this has formed in the latest
Skaergaard rocks, the acid granophyre.

No analyses have been made to determine definitely the potash
content of the plagioclase, and optical work is not yet sufficient to give
reliable evidence. The initial magma was exceptionally low in potash
and an early rock such as the perpendicular felspar rock contains even
less. Some of the potash in the later magmas and rocks may be derived
from solution of acid gneiss as discussed below (pp. 292—93). However,
that may be it is clear that the plagioclase of the early fractions contains
very small amounts of potash felspar, for occasional flakes of brown
mica are present in the rocks, and therefore not all the small amount
of K$_2$O of the analyses is present in potash felspar. In the later layered
rocks there is a micropegmatite containing untwinned felspar—no doubt
perthite or orthoclase—and here also the amount of potash felspar in
solid solution in the andesine must be small. Although there was so
little potash in the initial magma yet the quantity precipitated in the
early rocks under the conditions of formation of the layered series was
still smaller so that there was concentration of potash in the later
residual magmas. However, the concentration from this cause, together
with any absorption from the acid gneiss xenoliths, was not sufficient
to produce dominance of potash over soda even in the extreme differentiates.

The olivines of the layered rocks also remain a fairly pure series of
two components: Mg$_2$SiO$_4$ and Fe$_2$SiO$_4$. With increasing richness in
iron there is a steady increase in the amount of replacement by Mn
as shown by the analyses (Table VI). There is also an increase in the
amount of replacement by Ca, and in one of the last formed olivines,
that from 4139, the percentage of CaO is 2.18. There seems to be no
reason why the analysed sample of olivine from 4139 should be more
contaminated with lime minerals than the earlier olivine samples and
it must be concluded that this olivine has a relatively high content
of lime.

The recent experimental work has shown that the Mg$_2$SiO$_4$—
Fe$_2$SiO$_4$ series of olivines form a series of solid solutions of Roozebooms
type IV. If the other constituents of the magma had little effect on the

olivine equilibrium, then, according to Bowen and Schairer's diagram [1935, p. 163], the composition of the first olivine to crystallise from the initial Skaergaard magma should be Fa_{15}. The actual olivine in the gabbro-picrite is Fa_{20}. From the experimental evidence and the observed nature of the zoning the later fractions should contain more iron-rich olivine and this is shown by the successive layered rocks. Experimental work on the melt system CaO—FeO—SiO_2 indicates that hedenbergite-ferrosilite solid solutions are in equilibrium with La_5Fa_{95}. There is an increase in the larnite molecule in late fractions of the Skaergaard magma but it does not attain so high a value as is anticipated from the experimental work.

It is a remarkable fact, already commented upon (p. 131), that in respect of the olivines the natural Skaergaard magma behaves during fractionation in a way strictly analogous to the simple system MgO—FeO—SiO_2 investigated by Bowen and Schairer [1935]. In this system, under conditions of strong fractionation, there is a range of liquids which first precipitate olivine, of changing composition and these pass into the pyroxene field when clinopyroxene is alone precipitated. They then return to the boundary curve between the olivine and pyroxene fields and both olivine and pyroxene are precipitated simultaneously, the olivine being more iron-rich than before its temporary disappearance. It seems that the origin of the olivine-free middle gabbros is due to analogous behaviour in the natural magma.

Orthopyroxene is not usually so abundant in the Skaergaard rocks as clinopyroxene and no example has been separated and analysed. The composition of the orthopyroxene has only been estimated from optical properties. This shows the interesting fact that the percentage of ferrosilite in the orthopyroxene is greater than that in the clino-pyroxene of the same rock (Table III). This is also true if the $CaSiO_3$ component is ignored and the amount of $MgSiO_3$ and $FeSiO_3$ in the clinopyroxene is recalculated to one hundred per cent. (Table XXXVIII). The trend in the composition of the orthopyroxene with fractionation is towards more iron-rich types as would be expected. After the ortho-pyroxene has reached about Fs_{70} it ceases to form. From other rocks, such as eulysites, where mineralising solutions have been active, ortho-pyroxenes as rich in iron as Fs_{80} have been recorded [Henry 1936], and an orthopyroxene close to ferrosilite has been found in lava cavities by Bowen [1935]. In the earliest, outer border rocks the orthopyroxene is often abundant and it is a fairly early constituent, essentially con-temporaneous with the clinopyroxene. In the lowest layered rocks exposed it is present only in small amounts and is an interprecipitate mineral. In the middle gabbros the textural relationships existing between the ortho- and clino-pyroxenes are difficult to interpret. It seems possible

TABLE XXXVIII.

	4087	4077	3662	1907
Clino-pyroxene En and Fs re-calculated to 100 per cent. .	$En_{76}Fs_{24}$	$En_{64}Fs_{36}$	$En_{47}Fs_{53}$	$En_{31}Fs_{69}$
Ortho-pyroxene.............	$En_{59}Fs_{41}$	$En_{55}Fs_{45}$	$En_{42}Fs_{58}$	Not present

that clinopyroxene was alone precipitated from the magma and that parts inverted to orthopyroxene with abundant clinopyroxene inclusions (see p. 96).

The composition of the olivine and orthopyroxene precipitated simultaneously with clinopyroxenes of varying composition is discussed in the following section.

(b) Clinopyroxene.

The Skaergaard intrusion is well suited for a study of the course of pyroxene crystallisation resulting from the strong fractionation of a magma approximating closely to normal basalt composition. The range shown by the clinopyroxenes is long and those from the ferrogabbros show an extension of the range previously known from natural magmas. The primary minerals of the layered series have apparently crystallised under such conditions that there was true equilibrium between them and the liquid and it is possible to make a fairly close correlation between the compositions of the clinopyroxenes and the other solid solution series with which they are in equilibrium.

It has long been recognised that the earlier crystallising clino-pyroxenes of basalts are richer in the enstatite and diopsidic components than the clinopyroxenes of later crystallisation which are enriched in the ferrosilite molecule, and reference to Table III shows that such a trend has been followed by the clinopyroxenes of the layered series. A number of diagrams (Figs. 48—52) have been constructed to show graphically the detailed trend of the clinopyroxene crystallisation and its relationship to that of the parent rocks, the residual liquids and the compositions of the olivine and orthopyroxene in equilibrium with the clinopyroxenes at the various stages of the differentiation.

The analysed clinopyroxenes together with their parent rocks, have been plotted in figure 48 on the basis of the Wo, En, and Fs weight per cent. Pyroxene I from the lower hypersthene-olivine-gabbro 4077, represents an early stage in the pyroxene crystallisation of the lower layered series. It falls in the well known field of clinopyroxenes from olivine basalts and related rocks, and analyses of similar clinopyroxenes can be duplicated from many localities. It is a moderately iron poor type of diopsidic-augite somewhat richer in the wollastonite than the enstatite molecule, and it is associated with olivine (Fa_{37}) and ortho-pyroxene (Fs_{45}).

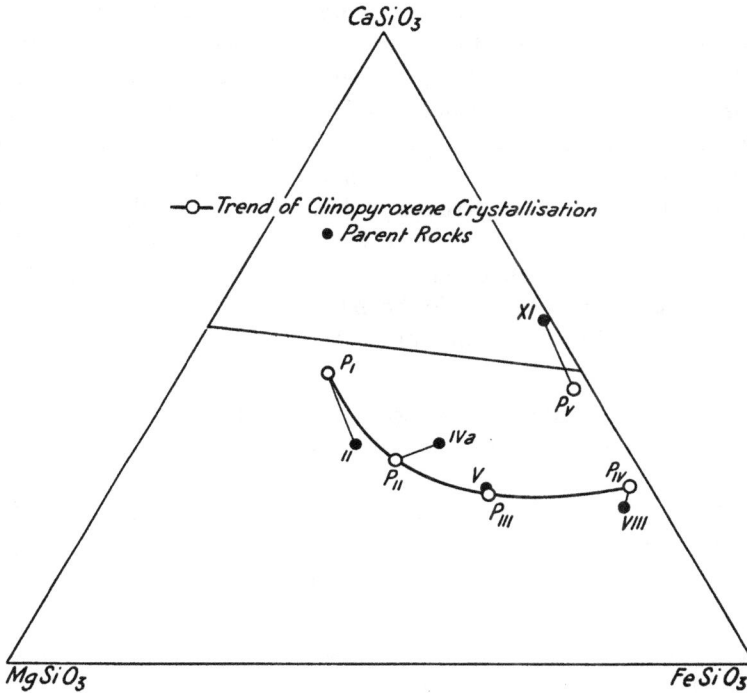

Fig. 48. Weight percentages of $CaSiO_3$, $MgSiO_3$ and $FeSiO_3$ of the analysed clino-pyroxenes P_I—P_V of the layered series and their parent rocks.

During the formation of the hypersthene-olivine-gabbros the trend of crystallisation was practically constant in direction. Figure 48 indicates clearly that during the crystallisation of the lower layered series the ferrosilite molecule was the only pyroxenic component to increase. It is, however, of some interest that during this early period of the crystallisation no marked decrease in the enstatite molecule takes place but it remains practically constant while at the same time the wollastonite molecule decreases greatly; thus comparing clinopyroxenes I and II the wollastonite molecules decreases from 47 to 32 per cent. while the enstatite molecule decreases only from 34 to 32 per cent. Clinopyroxene II from the olivine free middle gabbro 3662, is still within the limits of clinopyroxene compositions from gabbroic and basaltic rocks. It no longer, however, lies within the field of clinopyroxenes from olivine basalts, but within the field of pyroxenes from tholeiitic basalts (Fig. 54). Clinopyroxene II is in equilibrium with an orthopyroxene, Fs_{57}, richer in the ferrosilite molecule than that in the hypersthene-olivine-gabbros. No comparison can be made with olivine as the middle gabbros are olivine-free. In composition the clinopyroxene approaches the pigeonites as defined by Kuno [1936] and it is noteworthy that it has a smaller optic axial angle than the other analysed pyroxenes of the layered series.

The nomenclature of clinopyroxenes such as occur in the middle gabbros and later layered rocks of the Skaergaard intrusion is ill defined. In the present section, where comparisons are made with clinopyroxenes from other areas, it is necessary to state precisely the way in which names will be used. For pyroxenes, low in sesquioxides, which may, as a first approximation, be regarded as belonging to the diopside, hedenbergite, clinoenstatite, clinoferrosilite solid solution system we consider that the names should be based primarily on chemical composition expressed as molecular percentages of the dominant molecules Wo, En, Fs. We suggest that:—

1) the diopside-hedenbergite series should be defined as solid solutions of diopside and hedenbergite with up to 10 per cent. of extra En and Fs,

2) the clinoenstatite-clinoferrosilite solid solution series should be defined as solid solutions of these two with up to 10 per cent. of Wo,

3) the clinoenstatite-diopside series as solid solutions of these two with up to 10 per cent. of Fs, and

4) the hedenbergite-clinoferrosilite series as solid solutions of these two with up to 10 per cent. of En.

The remaining solid solutions of diopside, hedenbergite, clinoenstatite and clinoferrosilite we shall name pigeonites[1]) in the present section. It will be seen below that these can probably be divided into metastable pigeonites and stable pigeonites which we shall call plutonic pigeonites.

During the formation of the olivine-free middle gabbros the trend of the pyroxene composition changes. The new direction of the trend is marked by a general flattening of the curve of the pyroxene crystallisation,

---

[1]) Since there is a considerable field of pigeonites as defined above it is desirable that they should be further subdivided into varieties. They could be divided into four types by a line, from wollastonite to $En_{50} Fs_{50}$ of the triangular diagram showing molecular percentages of Wo, En, Fs and by a line half-way between the diopside-hedenbergite series and the clinoenstatite-clinoferrosilite series. These might be named diopside, hedenbergite, clinoenstatite and clinoferrosilite pigeonites. This plan is not definitely adopted here because it is felt that the distinction between metastable and plutonic pigeonite may one day provide a better means of subdivision.

representing a more gradual decrease in the wollastonite, and the beginning of a steady decrease in the enstatite, molecule. As in the initial stages of the crystallisation the ferrosilite molecule continues to increase. At the top of the middle gabbros olivine is again found as a primary phase and a little higher in the layered series orthopyroxene ceases to be precipitated as a separate mineral. At the beginning of the ferrogabbros a further change takes place in the trend of the clinopyroxene crystallisation, and it may again be significant that these changes approximate closely to the abrupt changes in the formation of the ferromagnesian phases. At the beginning of the ferrogabbros wollastonite ceases to decrease and there is even a slight tendency for the lime component to increase while at the same time the percentage of enstatite in the clinopyroxene decreases rapidly. Clinopyroxene III, from the hortonolite-ferrogabbro is an unusually iron rich variety containing 50 per cent. of the ferrosilite, and approximately equal percentages of the wollastonite and enstatite molecules; as we have already pointed out (Deer and Wager 1938) it represents an extension of the previously known field of naturally occurring pyroxenes. This pyroxene is in equilibrium with olivine, Fa_{59}, and orthopyroxene is not present.

During the formation of the ferrogabbros the olivine gradually changes from hortonolite to almost pure fayalite, and except for the earlier stages orthopyroxene is not precipitated from the residual magmas. Marked changes in the composition of the pyroxene take place but the trend remains more or less constant. A rapid decrease from 23 to 2 per cent. enstatite is accompanied by an increase from 50 to 69 per cent. ferrosilite and from 27 to 30 per cent. wollastonite. Pyroxene IV occurs in the rock of the purple band close to the top of the main layered series and is in equilibrium with fayalite. A pyroxene close to IV in composition has been prepared by Bowen, Schairer and Posnjak [1933] during their investigation of the melt system $CaO—FeO—SiO_2$ but as far as we are aware this is the first record of the occurrence of a hedenbergite-clinoferrosilite solid solution from natural magma. We have already given reasons for believing that this mineral has formed from the inversion of a β-wollastonite (p. 111 and Deer and Wager 1938, pp. 18 and 19). The trend of the clinopyroxene crystallisation in the main layered series began with normal diopsidic-augite of a type commonly precipitated from olivine basalt magmas and in equilibrium with plagioclase An_{56}, olivine Fa_{37}, and orthopyroxene Fs_{45}. By differentiation the residual magma became gradually richer and richer in iron, and the clinopyroxene crystallising from one of the latest residual magmas, that from which the purple band was formed, was a member of the hedenbergite-clinoferrosilite solid solution series in equilibrium with plagioclase An_{30}, olivine Fa_{98}, and quartz.

16*

In figure 49 the trend of the clinopyroxenes and their parent rocks has been plotted on the basis of the molecular percentages of MgO, CaO and FeO. Compared with the trend of the pyroxene crystallisation the curve showing the progressive change of the rock compositions indicates a smoother variation and although the curves show a general

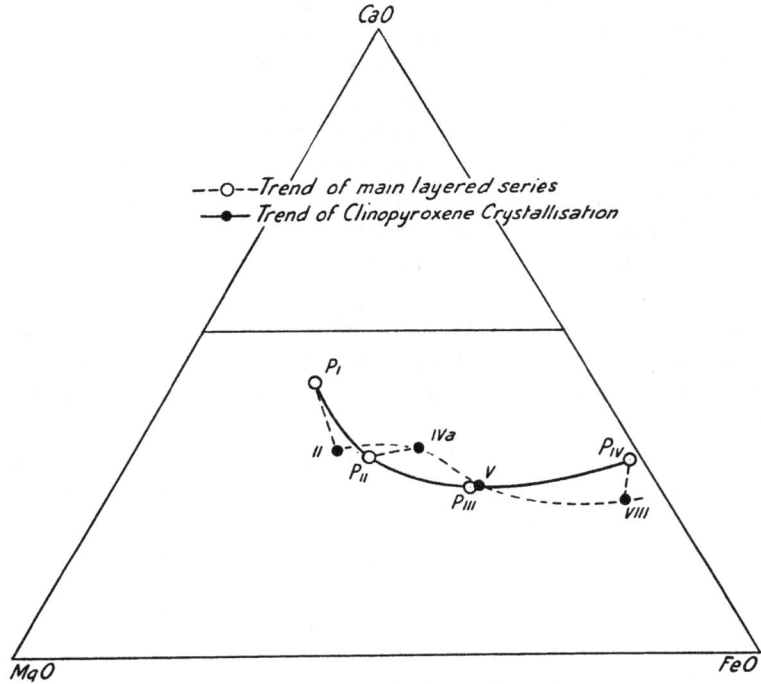

Fig. 49. The trend of the pyroxene crystallization and the trend of differentiation of the parent rocks.

similarity there are, in detail, certain differences. Compared with the initial, rapid decrease in the wollastonite molecule indicated by the curve for the clinopyroxenes, the curve for their parent rocks shows no decrease in the lime component but a rapid fall in enstatite and a correspondingly large increase in the ferrosilite component. This general trend is maintained throughout the formation of the hypersthene-olivine-gabbros and early middle gabbros. During the development of the upper middle gabbros there is a decrease in the lime as well as in the magnesia percentage but throughout the crystallisation of the ferro-gabbros the percentage of lime remains practically constant and the subsequent differentiation involves decreasing enstatite and an increasing ferrosilite percentage. It is of interest to consider the relative amounts of the various components in the individual pyroxenes and their parent rocks relative to the stage reached in the differentiation process. The difference in the ratio of the ferrosilite component of the clinopyroxenes and their parent rocks show a progressive change. It is greatest during the formation of the early pyroxenes of the main layered series. During the crystallisation of the middle gabbros the difference gradually decreases until at the onset of the ferrogabbros the ferrosilite ratio in both clinopyroxene and parent rock is virtually the same. This condition is largely maintained throughout the ferrogabbro stage although in the case of the analysed pyroxene IV, from the purple band, the parent rock is again slightly richer in the ferrosilite component than the pyroxene itself (Fig. 48). It is clear from figure 49 that in the hypersthene-olivine-gabbros the relative percentage of the wollastonite component is greater in the clinopyroxenes than the parent rock. The ratio of the enstatite molecule is approximately the same in the clinopyroxenes and their parent rocks. In these respects they are comparable with clinopyroxenes from olivine basalts. The clinopyroxene from the middle gabbros is comparable in composition with pyroxenes crystallising from tholeiitic basalt magmas (see fig. 54) and like pyroxenes from these rocks it is somewhat poorer in the wollastonite and considerably richer in the enstatite component than the parent rock. A close approximation in the relative percentage of the components in the metasilicate and the parent rock is attained during the early stages of the ferrogabbros and in the case of clinopyroxene III the ratios are almost identical. In pyroxene IV from the fayalite-ferrogabbro there is a slight difference in the ratios of the pyroxene components and their parent rocks shown by a higher percentage of wollastonite and lower percentage of enstatite in the pyroxene.

In figure 50 the trend of the pyroxene crystallisation is compared with the trend of the residual liquids. As the pyroxenes cannot be tied exactly to any of the calculated residual liquids only a general comparison of the trends will be made. The liquid trend shows a much smoother variation than the corresponding trend of rock compositions, a result in keeping with the theoretical calculation of the residual liquids and the difficulty of selecting for analysis a truly average rock at the particular stage of the differentiation, free from either a slight concentration in leucocratic or melanocratic constituents. Certain conclusions may be drawn from figure 50 with regard to the relationship between the clinopyroxenes and the liquid from which it was crystallising at the various stages in the differentiation of the Skaergaard magma. Compared with the composition of the liquid the early pyroxenes are definitely richer in lime and poorer in iron while magnesia is approximately the same in both pyroxene and liquid. This, as we shall show later is the relationship existing between clinopyroxenes and their parent rocks in olivine-basalts. During the middle period of crystallisation a reversal of the ratios takes place, the clinopyroxene becoming somewhat poorer in lime and richer in magnesium than the liquid; the iron ratio of the pyroxene still remains lower than in the liquid. This is the re-lationship existing between the pyroxenes and rocks of the tholeiitic

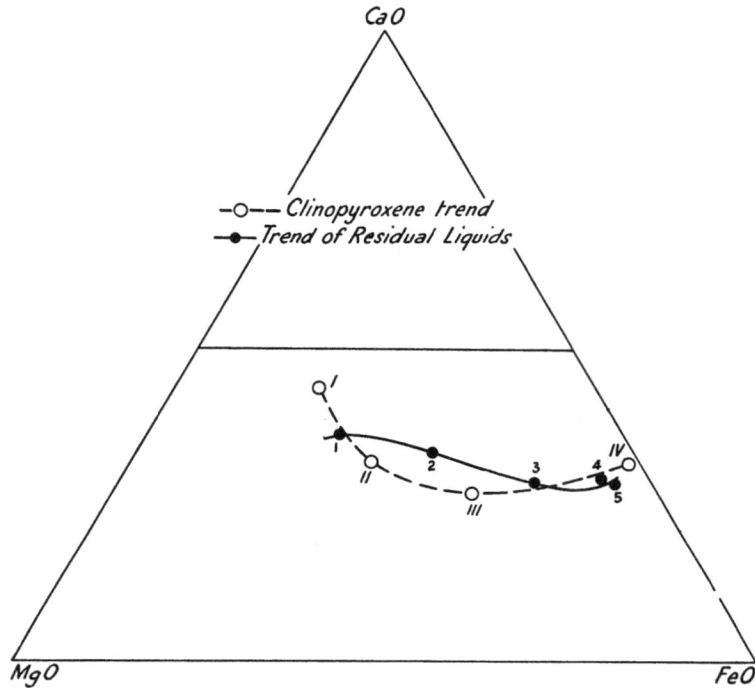

Fig. 50. Molecular percentages of CaO, MgO and FeO of the clinopyroxenes, P_{I-IV}, and the residual liquids.

basalts. A similar relationship between the clinopyroxene and the magma is maintained during a considerable period of the ferrogabbro stage, but later the ratios of MgO, CaO, and FeO in the pyroxenes and residual liquids approach in value, and in the fayalite-ferrogabbro the pyroxene is slightly richer in lime and poorer in magnesia than the liquid from which it was crystallising, while the iron is approximately the same.

In the layered series the higher temperature members of the plagio-clase, olivine and orthopyroxene solid solution series were formed first and were succeeded by successively lower temperature members of the respective solid solution series. As the successive liquid fractions also must have existed at lower and lower temperatures there can be no reasonable doubt that the clinopyroxene series crystallised in a similar way, beginning with higher temperature members which gradually gave place to lower temperature members of the series. The thermal data for the diopside-hedenbergite-clinoenstatite-clinoferrosilite system are not yet completely known. The thermal relationships of the binary systems bordering the major system are known from the work of Bowen and others but no data are available from artificial melts for the interior of the system. The data from the complex natural Skaergaard magma give some indication of the possible thermal relationships within this system.

From the known thermal data relating to the binary systems it is possible to obtain a rough idea of the probable form of the isothermal surface within the four component system. In the binary system clino-enstatite-diopside the addition of the lime magnesia silicate to the clinoenstatite lowers the temperature of the solid solution until at a composition of approximately $En_{10}Di_{90}$ the lowest temperature is reached. From this composition to pure diopside there is a slight rise in temperature from 1381° C. to 1391° C. The $MgSiO_3$—$FeSiO_3$ series is not a simple binary system but over a large part of this metasilicate series the relationship between clinoenstatite and clinoferrosilite is a binary one, in which the temperature decreases from the magnesium to the iron component. The relationship between hedenbergite and clinoferrosilite is also complicated by the incongruent melting of $FeSiO_3$ and by the inversion relations between the higher temperature β-wol-lastonite and the hedenbergite clinoferrosilite series. The sub-solidus relations, however, suggest that the lowest temperature in the series exists at approximately $Wo_{30}Fs_{70}$ which is very close to the composition of the Skaergaard pyroxene IV. In both the diopside-hedenbergite and the clinoenstatite-clinoferrosilite series there is a gradual temperature gradient from the magnesium and magnesium-lime, towards the iron and iron-lime, components respectively. We have already seen that excess $FeSiO_3$ in hedenbergite causes a decrease in the temperature of formation and also that slight excess of $MgSiO_3$ in diopside causes a small decrease in the temperature of the solid solution. The excess of hyper-sthene in the diopside-hedenbergite series must also cause a decrease in the temperature of formation of the resulting clinopyroxene. Any attempt to draw a solidus surface from this data indicates a low tem-perature trough beginning close to the Wo, En, join near to a com-position $En_{50}Wo_{50}$. This trough must then extend towards the heden-bergite-ferrosilite join close to a composition $Wo_{30}Fs_{70}$. The trend of crystallisation of the Skaergaard clinopyroxenes is represented by a continuous solid solution series beginning with a variety close to diopside in composition and extending to a hedenbergite-clinoferrosilite solid solution approximately $Wo_{30}Fs_{70}$ in composition. As this is a true solid solution series the clinopyroxene crystallisation trend should approximate closely to the low temperature trough in the four-component system. Comparison of this low temperature trough and the observed trend of the Skaergaard clinopyroxenes indicates clearly that a close similarity exists between these two directions. The two trends show a parallelism which is especially marked during the crystallisation of the clinopyroxenes of the middle gabbros and maintained during the ferrogabbros.

If this picture of the thermal relationships of the four-component system is approximately correct then it should be possible to estimate the approximate composition of a clinopyroxene crystallising from a basalt magma of the composition of the original Skaergaard liquid. The first pyroxene to crystallise will be a higher temperature member than that derived from the magma crystallised with perfect equilibrium (non-fractional) and will therefore lie away from the liquid composition towards the diopside-enstatite join. The early stage in the crystallisation of a natural magma is usually complicated by the crystallisation of both clino- and orthopyroxene but here we are concerned only with the clinopyroxene. The present conception of the thermal relationships indicates that this clinopyroxene will be richer than the magma in either enstatite or diopside and poorer in ferrosilite. As a magnesium-rich hypersthene is usually precipitated also from the magma this increases the likelihood that the early clinopyroxene will be diopsidic in composition. With the fractional crystallisation of the magma the clinopyroxene will also change in composition. It will, however, bear a higher temperature relationship to the liquid than the pyroxene that would be in equilibrium with the magma if the conditions ceased to be those of fractional crystallisation. The clinopyroxenes should therefore contain a relatively greater ratio of the higher melting components, that is, diopside and enstatite, than the liquid. It has already been pointed out that the Skaergaard clinopyroxenes are richer, relative to the residual liquid from which they have crystallised, in lime and magnesia and poorer in iron throughout the formation of the hypersthene-olivine gabbros, middle gabbros and the major part of the ferrogabbros and that a similar condition probably existed also during the formation of the fayalite ferrogabbros.

This conception of the thermal relationship existing within the four-component system provides an explanation of the large initial decrease in the percentage of the wollastonite compared with the steady value of the enstatite component during the crystallisation of the hypersthene-olivine gabbros. It seems reasonable to consider that the first clinopyroxene to crystallise from the magma will not necessarily lie in the low temperature trough. As the crystallisation continues the clinopyroxene trend will, however, move towards this trough which is pointing towards the En—Fs join and consequently the clinopyroxenes crystallising during this early stage will become poorer in the wollastonite component while showing little or no decrease in the enstatite ratio. As the trough becomes closer to parallelism with the En—Fs join the later pyroxenes will decrease in En and retain a relatively steady percentage of wollastonite.

Kennedy's statement that "the pyroxenes which separate from olivine-basalt magmas are richer in lime than the rock melt from which they separate, while pyroxenes from tholeiitic magmas are poorer in lime than the corresponding magma", is true of the relationship existing between the Skaergaard pyroxenes from the hypersthene-olivine gabbros and middle gabbros (olivine free) respectively. Kennedy's argument, that because pyroxenes of olivine basalt are richer in lime than the parent rock, crystallisation of the pyroxene must necessarily lead to improverishment of lime in the residual liquid, only takes account of one mineral series. Whether lime or any other constituent which enters into a solid solution series decreases in the residual liquid depends on the amount precipitating in all the minerals at any period in the cooling history of the magma. In the case of lime this rests on a balance between the plagioclase and the clinopyroxene series. We have already shown (see fig. 50) that lime decreases very gradually and only to a relatively small extent in the residual liquid. There is no great impoverishment in this constituent in the latter differentiates of the Skaergaard magma as Kennedy postulated for the olivine basalts.

The trend of the clinopyroxene crystallisation outlined above may now be considered in relation to the composition of the other ferro-magnesian phases in equilibrium with the clinopyroxenes. In figure 51 the clinopyroxenes and their parent rocks have been plotted together with the orthopyroxene in equilibrium with the clinopyroxenes. It is evident that the orthopyroxenes are relatively richer in iron and poorer in magnesia than their corresponding clinopyroxenes. Table XXXVIII indicates the gradually diminishing divergence of the enstatite to ferrosilite ratio between the two pyroxenes as the crystallisation continues, and in the middle gabbro 3662, which is the uppermost analysed rock of the layered series containing both clino- and orthopyroxene the difference is slight. It may be significant that the estimated composition $(En_{30}Fs_{70})$ of the orthopyroxene before it ceased to crystallise is almost identical with the enstatite to ferrosilite ratio for clinopyroxene III from a little above the horizon at which orthopyroxene is no longer formed from the magma as a separate phase. In contrast to the lower ferrosilite ratio in the clinopyroxenes compared with their parent rocks, the compositions of the orthopyroxenes from the hypersthene-olivine gabbros and the middle gabbros show that the orthopyroxene is richer in the ferrosilite and poorer in the enstatite component than the cor-

responding parent rock. As the orthopyroxene is one of the minerals
formed mainly from the interprecipitate magma, which has already
been shown to be (p. 127) richer in iron and poorer in magnesia than the
bulk composition of the rock to which it contributes only a part, this
relationship between the orthopyroxenes and their parent rocks is not

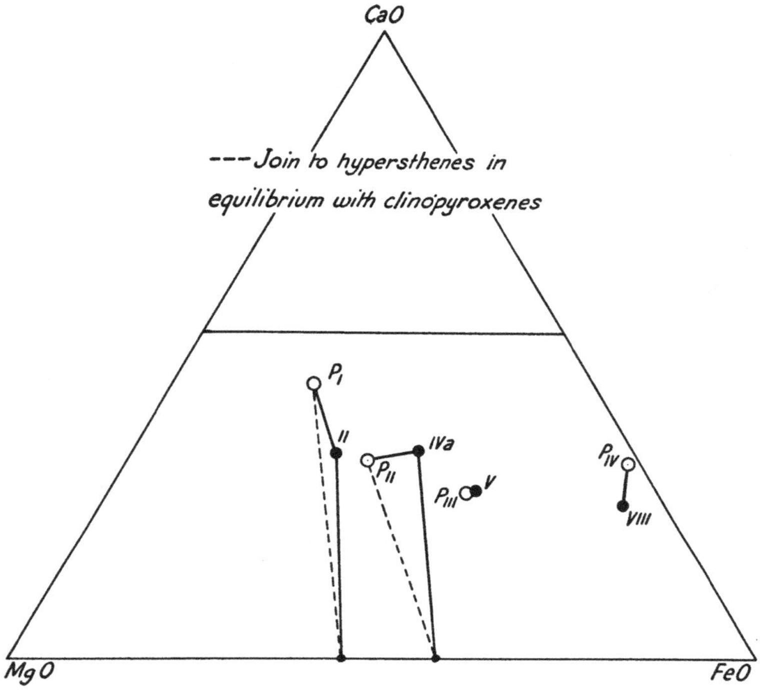

Fig. 51. Molecular percentages of CaO, MgO and FeO of the clinopyroxenes, P_{I-IV},
of the orthopyroxenes in the same rocks and of the parent rocks.

difficult to understand. Since the clinopyroxene is a primary precipitate
mineral and the orthopyroxene is an interprecipitate mineral it cannot,
however, be assumed that their respective compositions are exactly those
of pyroxenes in equilibrium. Throughout almost the whole of the ferrogabbro stage only clinopyroxene is precipitated from the residual magmas.
The crystallisation of a single pyroxene in the ferrogabbros is coincident
with a close approach of the enstatite to ferrosilite ratios in the clinopyroxenes and their parent rocks.

The relationships between clinopyroxenes, orthopyroxenes and
olivines of the same rock are shown diagrammatically in figure 52.
Pyroxenes other than those analysed have been included in the diagram
and their position has been determined by the method already described
(page 79). In the hypersthene-olivine gabbros 4077 and 1690 the ortho-

pyroxene is richer in the iron component than the olivine of the same rock, a condition in accord with the later crystallisation period of the metasilicate and its formation as an interprecipitate mineral. The hypersthene thus bears a similar relationship to the olivine as to the clinopyroxene. In the middle gabbros olivine occurs only as narrow

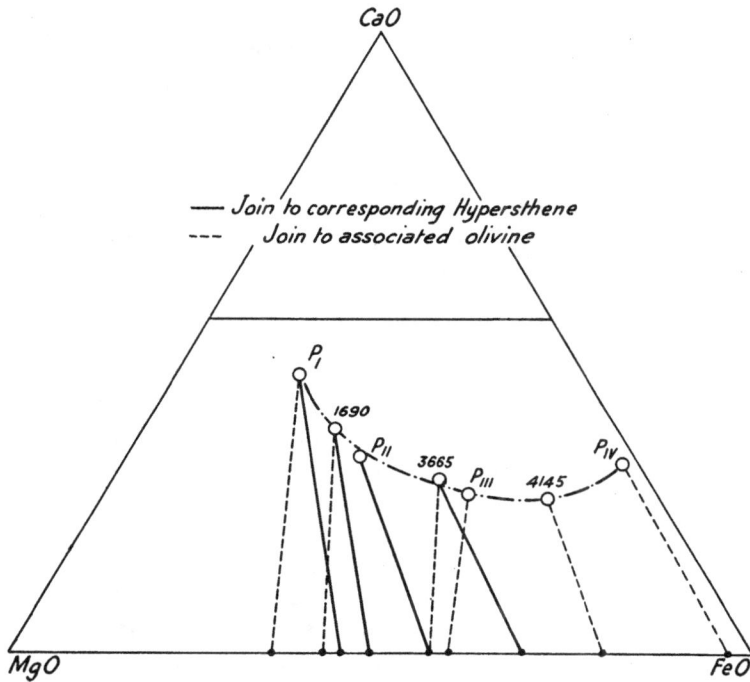

Fig. 52. Molecular percentages of CaO, MgO and FeO of clinopyroxenes and associated orthopyroxenes and olivines.

reaction rims around iron ore and not as a primary phase. Determinations of refractive indices, however, indicate that even as an interprecipitate mineral the olivine is not richer in the iron component than the orthopyroxene. The composition of the olivine from the middle gabbro 3661, has been determined as Fa_{48} whereas the orthorhombic pyroxene from a slightly lower middle gabbro 3662 is Fs_{58}, and when olivine again becomes a primary phase in the lower ferrogabbro 1907, its composition is Fa_{59}. During the earlier stages of the formation of the ferrogabbros olivine is relatively richer in the magnesium component than clinopyroxene. This difference decreases with further differentiation of the residual magma and in the fayalite-ferrogabbros the ratio of the iron to magnesium components is approximately equal in both clinopyroxene and olivine.

Recently a number of authors including Barth, Tsuboi, Kennedy and Kuno have discussed the trend of pyroxene crystallisation from rock magmas. Bowen and his corroborators have provided a considerable amount of data on the crystallisation of the pyroxenes in the melt systems CaO—FeO—SiO_2 and MgO—FeO—SiO_2. With the exception of Bowen the above authors have mainly relied for the determination of the composition of the pyroxenes either on optical data or in the normative pyroxene composition. In other cases analyses of pyroxenes from rocks, which cannot be assumed to be derived from a single differentiating magma have been used.

The work of Wahl and Washington demonstrated the widespread development of enstatite-rich augites in basaltic rocks and the recent work of the authors mentioned above has confirmed these earlier observations as well as attempting to outline the possible crystallisation trends of the later pyroxenes. The investigations of pyroxenes from volcanic rocks have also shown, that whereas an enstatite-augite is the usual pyroxenic phase developed at the effusive stage, both clino- and orthopyroxene are developed during the intratelluric stage.

From a study of the basaltic rocks from various localities, Barth [1929] concluded that the trend of crystallisation within the clinopyroxenes was represented by a progressive enrichment of the early diopsidic pyroxene in hypersthene giving rise on further crystallisation to pigeonitic varieties. Barth summarised his conclusions graphically in a figure showing a number of divergent trends. Part of the diversity of the crystallisation trends is due to including true metastable pigeonites (see p. 255) with more normal pyroxenes from both the olivine-basalts and tholeiitic-basalts. If these metastable varieties are omitted then Barth's conclusions that differentiation takes place from phenocrysts of diopsidic-augite composition to groundmass pyroxenes of a lime-poor enstatite-augite composition is roughly in harmony with the evidence form the Skaergaard layered series. Barth's conclusions based mainly on a study of porphyritic and groundmass pyroxenes and zoned pyroxenes from basaltic rocks are strictly limited to the magnesium-rich pyroxenes and he made no attempt to suggest the form of later differentiation. Daly and Barth [1930] summarised their observations on the clinopyroxenes from the Karroo dolerites in a diagram showing the course of crystallisation of the clinopyroxenes. This optical study indicated that crystallisation began with the formation of clinopyroxenes of diopsidic composition and that, with continued crystallisation, the pyroxene solid solution tended towards more iron-rich, pigeonitic varieties. As in Barth's earlier work there is no suggestion of a directional change towards the hedenbergite-clinoferrosilite boundary at a more advanced stage in the trend of crystallisation.

Tsuboi's [1932] study of the course of crystallisation of pyroxenes from magmas is based mainly on an investigation of Japanese basalts and andesites. In contrast with other Pacific lavas described by Barth in which the porphyritic pyroxenes are mainly of clinopyroxene, Tsuboi found both clino- and orthopyroxenes occurring as phenocrysts. Tsuboi summarised his main conclusions in a figure showing the trend of progressive crystallisation in the composition of both clino- and orthopyroxenes throughout the whole course of intratelluric crystallisation and also within the successive residual liquids. Briefly his conclusions are that porphyritic or plutonic pyroxenes are either clinopyroxenes varying in composition between diopside and hedenbergite or orthopyroxenes varying between enstatite and iron-rich hypersthene. Such trends in the composition of plutonic clino- and orthopyroxenes are considered to result from the limited miscibility at the intratelluric stage existing between the diopside-hedenbergite and the enstatite-orthoferrosilite series. A similar but more limited immiscibility gap had previously been indicated by Asklund's statistical study of pyroxenes and the rocks from which they come [1925]. The pigeonites lie outside the compositional range of these two plutonic pyroxene series and Tsuboi considered that the groundmass pigeonitic pyroxenes result from the complete miscibility of the two solid solution series under effusive conditions.

In order to explain the crystallisation as phenocrysts, sometimes of clinopyroxene, sometimes of orthopyroxene Tsuboi postulated a "two-pyroxene boundary". The trend of crystallisation within the clino- and orthopyroxenes is towards hedenbergite or orthoferrosilite respectively, at the same time the liquid changes until the "two-pyroxene" boundary is reached when the second pyroxene also begins to crystallise. In this way the composition of the original magma controls the earlier crystallisation of either clino- or orthopyroxene. The crystallisation of the Skaergaard clinopyroxenes, as we have shown previously, takes a different trend and although the orthopyroxene follows Tsuboi's trend for the simpler $MgSiO_3$—$FeSiO_3$ series the clinopyroxenes leave the diopside-hedenbergite series and become pigeonitic in composition. Tsuboi's conception of the limited miscibility of the four pyroxenic components under plutonic conditions is not in full agreement with the clinopyroxene crystallisation of the Skaergaard intrusion. The orthopyroxene becomes completely absorbed in the pigeonitic solid solution when an advanced stage in the differentiation of the magma is reached and both clino- and orthopyroxenes have become extreme iron-rich varieties.

Bowen, Schairer and Posnjak's [1933, p. 273] statement "that natural pigeonites are formed under conditions where they are metastable" was put forward with some reserve. They suggested, however, that pigeonite is formed in volcanic rocks by the rapid crystallisation of the pyroxenic components with the precipitation of a single pigeonitic phase, stable only at high temperatures but persisting at lower temperatures in a metastable condition. On the other hand crystallisation of the pyroxenic components with slow cooling in a plutonic environment gives two phases, ortho- and clinopyroxene. In a later paper Bowen and Schairer maintained their earlier opinion that pigeonites are "probably formed metastably at temperatures below their stability range" [1935, p. 203] while crystallisation with slow cooling will lead to the formation of a mixture of orthopyroxene and highly calcic clinopyroxene. Thus they considered that the uniaxial augite "from Mull with a composition of Wo. 9 per cent., En. 35 per cent. and Fs. 56 per cent. would crystallise at low temperatures as an orthopyroxene with a small admixture of lime-rich clinopyroxene". This theory was not put forward, however, as a general explanation for the production of all pigeonites. The difficulty in finding a satisfactory explanation for the formation of pigeonites seems to be mainly due to the lack of data on the possible range of solid solution within the clinopyroxenes. This gap has been reduced by the new data on the determined trend of pyroxene crystallisation within the Skaergaard intrusion. It is now clear that clinopyroxenes of pigeonitic type can crystallise together with orthopyroxene under plutonic conditions, and that both may be relatively rich in the ferrosilite component. With increase in the iron component until there is between 65 and 70 per cent. ferrosilite in both the ortho- and clinopyroxene a single pigeonitic phase crystallises from the residual magma. The recognition of this increasing facility of the clinopyroxene to take the whole of the pyroxene components into solid solution with increasing concentration of the ferrosilite molecule removes many of the earlier difficulties of explaining the occurrence of pigeonites and the apparent anomalies in the way of a simple explanation. The zoned clinopyroxene phenocrysts of the Karroo dolerites possess an outer zone of pigeonite and Tsuboi interpreted these outer margins as the result of crystallisation under effusive conditions while the inner zones he believed to have formed under intratelluric conditions. The probable explanation of these pyroxenes is that with an increase in the ferrosilite molecule and their approach to an iron-rich pigeonite conditions of complete solid solution were attained. Both Barth's and Tsuboi's hypotheses failed to account for the phenocrysts of pigeonite in the Mull andesite, but this mineral is an iron-rich variety containing nearly 60 per cent. of the clinoferrosilite molecule and therefore lies within the one pyroxene field. Combining the earlier work with evidence from the Skaergaard intrusion it appears that pyroxenes of pigeonitic type may be formed in two ways:—

1) Under effusive conditions by complete miscibility of the pyroxenic components due to rapid cooling, the product persisting at lower temperatures in a metastable condition,

2) Under plutonic conditions by the development of a solid solution series with the earlier diopsidic pyroxene as the residual magma becomes rich in iron. The earlier part of this pigeonitic series may be accompanied by a member of the orthopyroxene series but when the concentration of the ferrosilite molecule reaches between 65 and 70 per cent. all the pyroxenic components form a single clinopyroxene of pigeonitic type.

Kennedy [1933] in his study of the trends of differentiation in basaltic magmas has re-emphasised the fact that monoclinic pyroxenes crystallising from olivine-basalt magmas are diopsidic-augites while those precipitated from tholeiitic magmas are lime-poor enstatite-augites. Kennedy based this conclusion on the work of Wahl who was the first to point out the wide occurrence of lime-poor enstatite-augite in basaltic rocks. Washington's later investigations of the Deccan Traps also showed that the difference between the pyroxenes of the plateau basalts (the tholeiitic type of Kennedy) and the cone basalts (Kennedy's olivine basalt type) was expressed chemically in the highly ferromagnesian, hedenbergitic enstatite-augites of the former and the highly calcic or diopsidic-augites of the latter. Kennedy has interpreted these differences as the result of crystallisation of the pyroxenes from two different types of magma, and he considers, following Lehmann, that the original composition of the magma forms the ultimate control of the differentiation. According to Kennedy "in the olivine-basalts the trend of crystallisation is from basaltic augite to diopside and eventually, aegerine-augite" [1933, p. 252], but his own diagram (fig. 1, p. 254) gives no evidence that such a trend is developed during the differentiation of olivine basalt magma. The pyroxenes from olivine basalts show a somewhat wide scattering in his figure but, considered in a general way, as the parent rock becomes richer in iron the clinopyroxenes become richer in iron and also more pigeonitic (e.g. Nos. 1 and 7, Fig. 53). Kennedy plotted the pyroxenes crystallised from tholeiitic magmas together with their parent rocks on a separate diagram (1933, p. 254, fig. 2). Compared with the pyroxenes of the olivine basalts the pyroxenes from tholeiitic basalts lie in a different field and we agree with Kennedy that the two figures "show that the olivine-basalt and tholeiitic magma-types present two very different cases" (p. 253). It seems unnecessary, however, to regard these two groups of clinopyroxenes as products of two distinct and unconnected magmas, and the tholeiitic pyroxenes plotted by Kennedy (p. 254, fig. 2) strongly suggest that the trend developed in the olivine-basalts, is continued by the clinopyroxenes from the tholeiitic basalts, that is from more calcic pyroxenes to types richer in the "pigeonite" component.

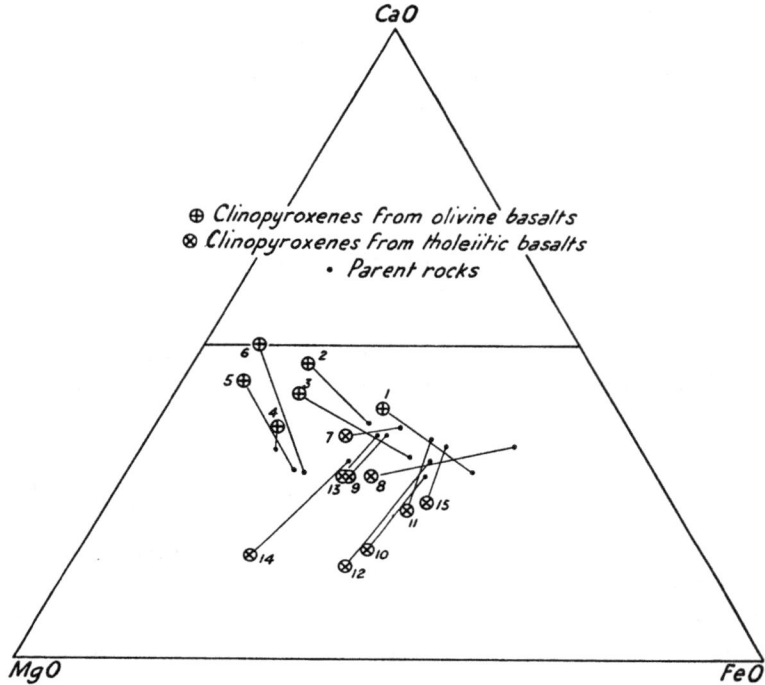

Fig. 53. Molecular percentages of CaO, MgO and FeO of pyroxenes from olivine basalt magmas and tholeiitic magmas in relation to the rocks in which they occur. Composite diagram constructed from a combination of Kennedy's data on olivine-basalt and tholeiitic pyroxenes [Kennedy 1933, p. 254, figs. 1, 2].

With this conception of the relationships we have combined in figure 53 Kennedy's data for pyroxenes from olivine- and tholeiitic basalts, and on a second diagram (Fig. 54) we have shown the approximate limits of the olivine and tholeiitic basalt fields and have indicated in a rough way, the general trend of clinopyroxene crystallisation from the olivine and tholeiitic basalts as we interpret the available data. A similar trend line has been drawn from the points representing the composition of the parent rocks in both the olivine basalt and tholeiitic basalt types. For the sake of clarity and to compare these trends with those taken by the Skaergaard clinopyroxene and the residual liquids, they have been redrawn together with the Skaergaard trends on a separate diagram

(Fig. 55). This diagram shows that there is a close similarity with the general trend taken by the clinopyroxenes of olivine and tholeiitic basalts with the crystallisation trend of the Skaergaard clinopyroxenes and, as we shall show later, with the trend of crystallisation suggested by Kuno [1936]. The basaltic pyroxenes show the same change from diopsidic-

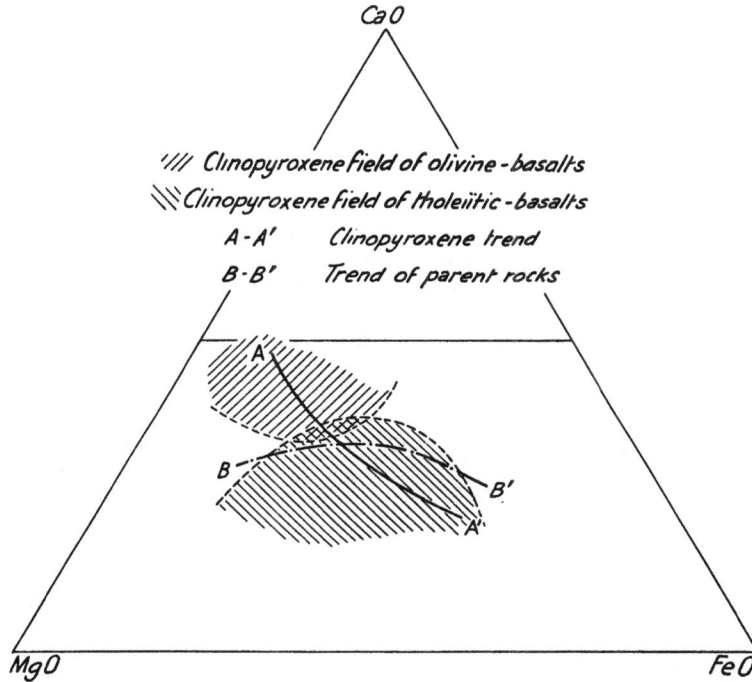

Fig. 54. The fields of pyroxenes from olivine-basalt magmas and tholeiitic magmas, based on Kennedy's data shown in Fig. 53.

augite towards an increasing enstatite-ferrosilite percentage as we have demonstrated for the pyroxenes of the main layered series. The trend begins at an earlier stage than that represented by the clinopyroxenes of the lowest visible rocks of the layered series and continues to approximately the stage in the differentiation series represented by the upper middle gabbros. In the same way the parent rocks show an even closer similarity with the Skaergaard magma trend; they differ only in beginning with more basic types and in showing no examples representing differentiation beyond the upper middle gabbro or lower ferrogabbro horizon of the Skaergaard layered series.

Kuno [1936] observed that intratelluric pigeonites, and pigeonites in the groundmass of basalts with high crystallinity, are always richer in the iron component than the porphyritic clino- and orthopyroxenes.

He also observed that pyroxenes with high ferrosilite and low enstatite percentages crystallise as a single pigeonitic phase even in the intratelluric stage. From this data Kuno concluded that under plutonic as well as effusive conditions the pyroxenic components of rock magmas are capable of forming an unbroken series of solid solutions and this view is shown

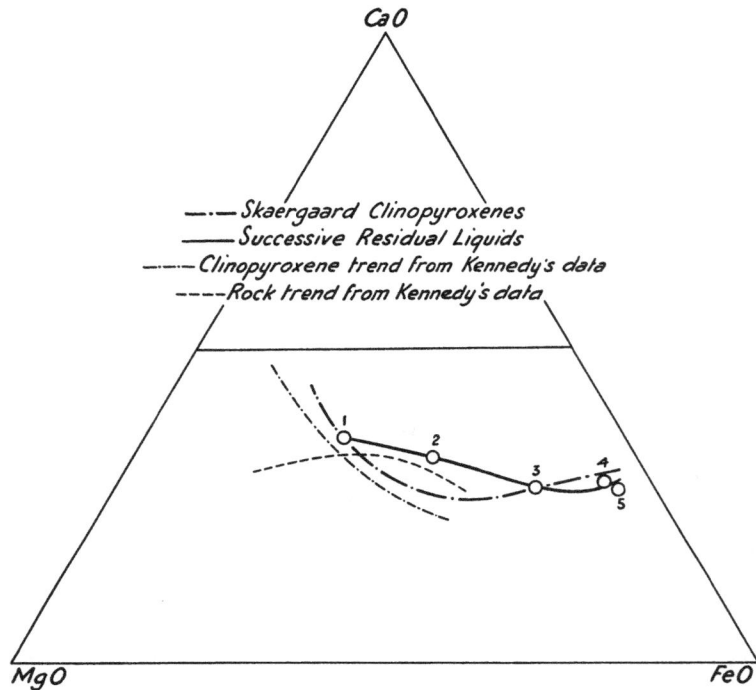

Fig. 55. The trend of crystallization of the Skaergaard clinopyroxenes and residual liquids compared with the clinopyroxene and rock trends suggested by Kennedy's data.

in figure 56. Pyroxene material having a composition such that it lies in Kuno's two pyroxene field (Fig. 56) should crystallise with the formation of both clino- and orthopyroxene. When, by further crystallisation, enrichment in ferrosilite relative to the enstatite molecule is such that the boundary line XY is reached a single pyroxene is formed.

The trend of the Skaergaard pyroxenes is undoubtedly in accord with Kuno's general conclusions and the present investigation makes it possible to define the limits of the "two pyroxene" and plutonic pigeonite[1])

---

[1]) The term plutonic pigeonite is here used to distinguish the metastable types from those, such as found in the ferrogabbros, which were not metastable when formed.

fields with more precision. Kuno draws his division between the two fields from the CaSiO$_3$ apex to the enstatite-ferrosilite join at a composition of En$_{52}$Fs$_{48}$. We have already shown that the orthopyroxene solid solution series in the layered series may continue to be precipitated as a separate phase as far as a ferrosilite content of between 65 and

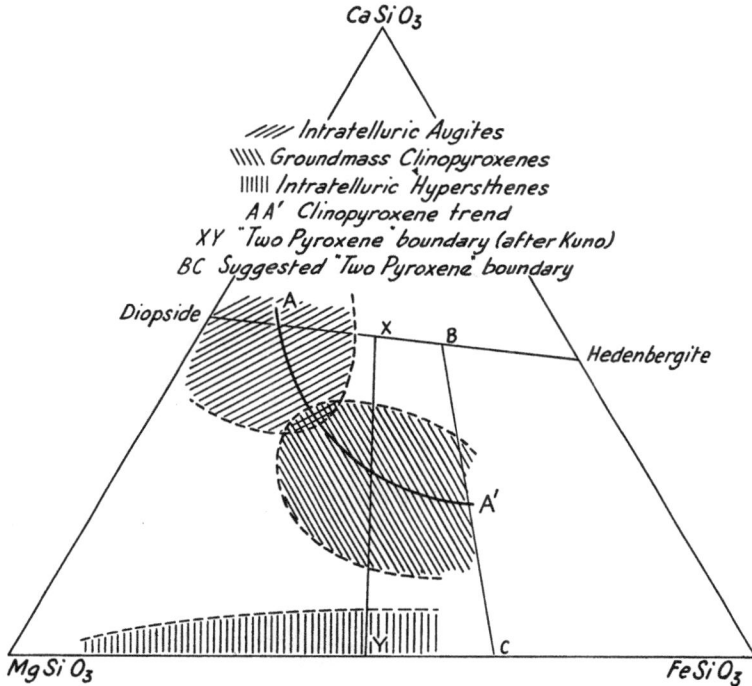

Fig. 56. The fields of intratelluric augites, intratelluric hypersthenes and ground-mass, monoclinic pyroxenes (mostly pigeonites from dolerites or from groundmass of basalts) taken from Kuno's data. [Kuno, 1936, p. 147, fig. 4].

70 per cent. and also that this orthopyroxene is in equilibrium with a clinopyroxene probably close to pyroxene III in composition. On this evidence from the Skaergaard pyroxenes the division between the "two pyroxene" and pigeonite fields under plutonic conditions must be moved towards the hedenbergite-clinoferrosilite join. If this boundary line is drawn to the composition En$_{35}$Fs$_{65}$, pyroxene III still lies within the plutonic pigeonite field, and we consider the data now available indicate this as the more probable limiting position between the "two pyroxene" and single pyroxene fields. From Kuno's data it is possible to draw a line (AA' Fig. 56) which represents the approximate trend of pyroxene crystallisation as he interpreted it from early "intratelluric augites" to plutonic pigeonites. A similar trend has been taken by the clinopyroxenes

17*

of the main layered series of the Skaergaard intrusion and we have also suggested that the data from the pyroxenes of olivine basalt and tholeiitic basalt magmas show the same trend. In figure 57 the trends of clino-pyroxene crystallisation demonstrated by the pyroxenes of the main layered series and the trend deduced from the data of Kuno and Kennedy

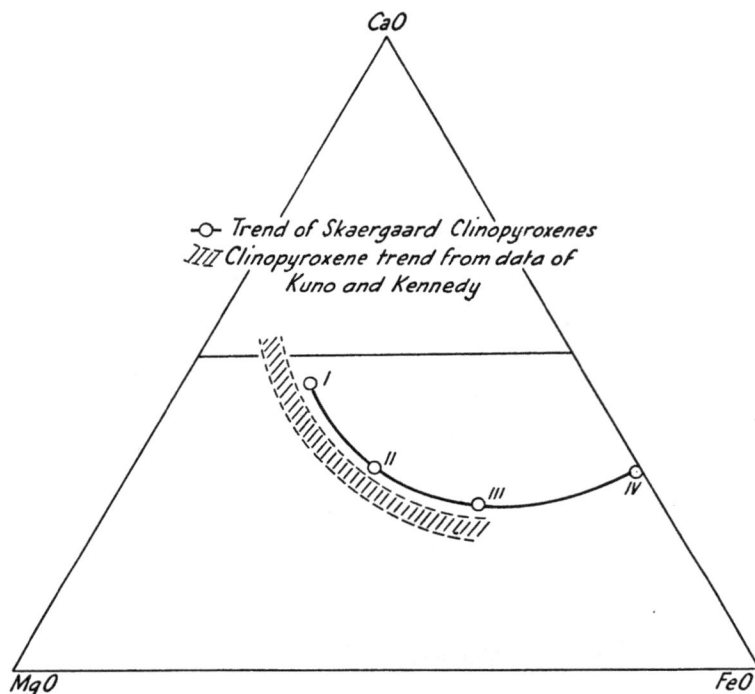

Fig. 57. The trend of crystallization of the Skaergaard clinopyroxenes and the trend of crystallization of clinopyroxene from basalts interpreted from the data of Kuno and Kennedy.

are shown. The combined trend deduced from Kuno and Kennedy shows very similar characteristics to the Skaergaard differentiation except that the crystallisation begins with pyroxenes slightly richer in the wollastonite component. The additional data from Kuno indicates the arrest in the initial trend towards the enstatite-ferrosilite join and the flattening of the crystallisation trend approximately parallel with the enstatite-ferrosilite join, which is characteristic of the later stages of the clinopyroxene crystallisation in the Skaergaard magma.

The normal trend of the clinopyroxene crystallisation within the layered series is shown in several of the preceding diagrams and is sum-marised in figure 57. Throughout the greater part of the crystallisation of the magma the clinopyroxenes separating were not abnormal in composition and many analyses of closely comparable pyroxenes have been made. Moreover throughout the whole of the crystallisation period of the hypersthene-olivine gabbros and the middle gabbros the trend of the crystallisation can apparently be closely matched with the pyroxene trends developed in the basaltic rocks as a whole. It is not until the development of the ferrogabbros that the pyroxenes become unusual iron-rich members of the solid solution series. At this stage in the fractional crystallisation of the Skaergaard magma, the change in the direction of the trend towards the hedenbergite-ferrosilite join has already commenced. Before the development of the, so far, unique ferro-hortonolite-ferrogabbros, with their accompanying iron-rich pyroxenes and olivines, the trend of pyroxene crystallisation is normal a point which we wish to emphasise. In the same way as with the plagioclase series, the clino- and orthopyroxenes separating from the magma are relatively richer in the higher melting components than the liquid from which they crystallised, and as fractional crystallisation proceeds the later members of the series are enriched in the component or components having lower melting temperatures.

XII. MECHANISM OF THE DIFFERENTIATION

(a) Evidence for Currents in the Magma.

Several lines of evidence show that during the slow solidification and differentiation of the Skaergaard intrusion there was a significant circulation of the magma. First there are the flow structures in the border group. In the marginal border group, these are parallel to the contact, as is usual in flow banded plutonic rocks [cf. Grout 1918 B and later papers]. Evidence of flow in the border group is most conspicuously shown by the streaking out of the granophyre inclusion and hybrid material, but some banding is also present due to variation in mineral composition, and there is also slight orientation of the tabular plagioclase crystals. The banding along the east and west margins, where it is best developed, is such that it must have been produced by upward or downward flow parallel to the outer surface of the intrusion (see p. 27 and Pl. 4, fig. 1). In the upper border group, flow structures are sometimes strongly developed and at other times absent; where they occur their direction is such that they must have been approximately horizontal before the regional flexuring and thus they are parallel to the presumed position of the roof of the intrusion. Various hypotheses might be developed to account for the fluxion structures in the border group but the simplest ascribes them to flow of the magma in a direction controlled by the boundary wall.

The layered series, which we have shown accumulated from the bottom upwards, may have been produced either by the vertical sinking of crystals due to their superior specific gravity, or, like most sediments, by a combination of this process and horizontal movement due to currents. Platy minerals, sinking vertically through a liquid, would tend to accumulate with their greatest area parallel to the surface on which they rested. This would be strongly the case, for example, if mica flakes were sinking through water, because the mica has a strongly tabular habit and a considerably greater density than water; also the water has a low viscosity. In the case of roughly tabular, plagioclase crystals, sinking through a viscous magma of nearly the same density as themselves, direct sinking would produce little or no parallelism of the crystals owing to the feebleness of the orientating forces. On the other hand if the magma were undergoing gentle flow there would be an orientating force easily sufficient to lay out the minerals with their flat expanses parallel to the surface of the accumulating pile of crystals, and this direction would also be parallel to the direction of flow of the magma. In considering the origin of the structures shown by the gabbros of the Duluth lopolith, which in our terminology would be called igneous lamination, Grout has likewise come to the conclusion that they could not be produced without flow of the magma [1918 A].

In the trough banding structures, where igneous lamination of the light bands is always well developed, the felspars are rectangular tablets and the long sides of the rectangles are roughly parallel with the length of the trough. If vertical sinking, without any flow of the magma, had taken place, the rectangular tablets of felspar would have no special orientation within the plane of igneous lamination. The fact of their orientation parallel to the axis of the trough banding suggests flow of magma along the trough. This kind of orientation of the rectangular plagioclase tablets is usual in the ferrohortonolite- and fayalite-ferro-gabbros; it is to be regretted that we did not map the direction of the rectangular plagioclase crystals since this would have provided clear evidence for the direction of the magmatic currents.

There are two types of variation in the layered series which are related to the margins, and these point strongly to the action of currents within the magma. The first is the development of the melanocratic layered material for some hundred metres or so adjacent to the inner contact. If the layered rocks accumulated by vertical sinking of crystals, it would require some supplementary hypothesis to account for richness in melanocratic minerals near the margin, but this variation is at once explained if we assume descending currents along the margins which turned along the floor of the intrusion towards the centre, and then rose to complete the circuit. Such currents would tend to deposit the denser material near the margin, and to carry on the less dense to be deposited nearer the centre. The other variation which is related to the margin is the size of the felspar crystals and there is little doubt that other minerals would show the same effect although measurements have not been made to prove it. At any one horizon in the layered series the average size of the felspars increase from the margin towards the centre. On the hypothesis of vertical sinking, this receives no simple explanation, but if centrally directed currents are assumed, from which the felspars were crystallising steadily, the crystals near the centre would be expected to be larger, since they would have had a longer time in which to grow.

The distribution of xenoliths in a magma depends partly on the relative specific gravities of xenolith and magma, and partly on whether there are currents in the magma, and on their strength. A xenolith, lighter than the magma in which it is immersed, would have an upward movement relative to the magma; if the magma was involved in a downward circulation, however, the xenolith would have a reduced upward or even a downward movement. A consideration along these lines of the positions of the acid xenoliths of the Skaergaard intrusion provides further evidence for significant magmatic currents.

According to measurements by Day, Sosman and Hostetter [1914], the specific gravity of the Palisades diabase at 1200° C. is 2.66. It is likely that the initial Skaergaard magma had about the same specific gravity, and that this increased somewhat as differentiation left a more iron-rich residual liquid. The specific gravity of crystalline granites at 1,100° C., based on the observed density and thermal coefficient of expansion, is estimated by Daly [1933, p. 276] as varying from 2.48—2.67, and the estimated range for granite glass at 1200° is 2.26—2.43. It is therefore likely that crystalline granite would have a somewhat lower specific gravity than a basic magma in which it was immersed, and there seems little doubt that melted granite material would have a decidedly lower specific gravity.

In estimating the specific gravity of magmas, appeal to laboratory data is not entirely satisfactory, since the specific gravity of natural magmas, will vary considerably with the amount of volatile constituents present, and the extent to which these are lost during solidification can only be guessed. It has been suggested that a gabbro magma might have a sufficiently low density, due to dissolved volatile constituents, to allow sialic crustal rock to sink into it [Richey 1931—32, p. 132]; if this be the case then the magma might invade the upper crust by stoping. In view of the differing oppinions about the relative specific gravity of gabbro magma and granites, it is useful to have direct field observations proving their relative specific gravities. From East Greenland there comes satisfactory evidence, briefly given before [Wager, 1933, p. 43], to show that the average grey gneiss of the metamorphic complex tends to float in average basalt magma. Fragments of acid gneiss were found abundantly only in four of the basic dykes, out of the many hundreds seen. These were of normal basalt material and in each case there was definite evidence that the exposure was near the top of the dyke. It might be argued that the position of the gneiss inclusions had little to do with relative specific gravities, since they might have been carried up by the rapid flow of the basalt as it filled the fissure, and not by the forces due to relative specific gravities. In all cases, however, the acid gneiss was less abundant, or absent, in the quickly chilled margins of the dykes, and an example of this has been figured [Wager, 1933, fig. 7]. If the gneiss was carried upwards by the rapid upward flow of the basalt, then there is no reason why it should not be present in the quickly cooled margins of the dyke. Since the gneiss inclusions are confined to the central region, they must have reached that position during the tranquil cooling of the dyke, and we must postulate that they collected at the upper limit of the dyke by flotation. The gneiss blocks were unfused and had, therefore, a greater density than if fusion had taken place. This evidence indicates two things: first, that average grey gneiss had a lower specific gravity than basalt magma, and second, that a significant amount of flotation of the gneiss occured during the time taken for a dyke of normal proportions to solidify. Crystalline grey gneiss is likely to have had a lower specific gravity than the original Skaergaard magma and a decidedly lower specific gravity than the later, more iron-rich residual magmas. If the grey gneiss were melted, as we have shown to have been the case in the Skaergaard intrusion, the density would be five to ten per cent. less than that of the crystalline rocks. From these various lines of evidence our final conclusions are that the molten granophyre xenoliths of the Skaergaard intrusion must have had a tendency to float in the Skaergaard magma at all stages of its evolution.

Many of the granophyre xenoliths of the border group form a kind of head to a long tail of hybrid material; they appear as though they had struggled upwards relative to the surrounding material. Whether they were actually moving upwards, or only moving upwards relative to the surrounding magma which was itself involved in a downward flow, cannot be decided. More definite conclusions come from consideration of the granophyre xenoliths in the layered series. Since the layered rocks accumulated from the bottom upwards as a solid mass of sediment, the granophyre patches cannot have floated up from below, but must have been deposited like the crystal precipitate which forms the layered rocks. On the evidence of the relative specific gravities given above, it is most unlikely that the granophyre could have sunk through the Skaergaard magma, especially by the time that the dense hortonolite ferrogabbro was forming. We consider that the granophyre must have been carried down by the magmatic currents, and held there by the simultaneous deposition of the cloud of crystals which formed the layered series and which had a mean density greater than that of the magma.

The discussion of how the granophyre xenoliths came to occupy their present positions is not only important in providing further evidence of circulating currents in the magma, but it seems to show also that material, somewhat lighter than the magma, could be laid down with the crystal precipitate forming the layered series. The labradorite crystals occurring in the lower layered series must have had a density close to that of the magma. Whether the density was a little greater or a little less is difficult to decide. Fermor [1925, pp. 195—97, 206] has given evidence that labradorite crystals sank in some of the Deccan Traps. In dealing with the origin of the anorthosites, Bowen [1917] postulated that bytownite remains suspended in normal gabbro magma while labradorite sinks in a magma of dioritic composition and he considered that field observation on the anorthosites supported this view. Vogt [1924] held that labradorite would sink in a basic magma. The majority of observers, however, have sought to explain the origin of anorthosites by flotation of labradorite crystals. As Daly has remarked [1933, p. 416] the density relations existing between basic plagioclase and basic magma are probably delicate. In the Skaergaard intrusion plagioclase, varying from basic labradorite to andesine, has been laid down, with other minerals, at the bottom of the magma. It may be that in the early stages when labradorite was forming from the scarcely modified Skaergaard magma that the plagioclase would sink, but it seems unlikely that andesine would sink in the later iron-rich magmas. The mechanism suggested above whereby granophyre inclusions were laid down with the pile of crystals forming the layered series may also be applied to the felspar, if the felspar had a tendency to float. The process suggested is that a cloud of felspar, pyroxene and olivine crystals was formed having a mean specific gravity considerably above that of the magma, and of sufficient numerical density for the felspars, even if they had a lower specific gravity than the magma, to be carried down as a result of the interference of the sinking ferromagnesian minerals. Bowen, in an early paper [1915, p. 15] tentatively suggested this kind of mechanism. Evidence is given below that, in the later stages of the formation of the layered series, felspar, not held down by some such process, floated upwards to give the hedenbergite-andesinite (p. 283).

(b) Theory of the Convective Circulation of the Skaergaard Magma.

The flow structures, the igneous lamination, the variation of the layered series in relation to the margin and the distribution of the granophyre xenoliths all point to the existence of significant currents in the Skaergaard magma. These currents cannot be those set up during the formation of intrusion, for they had obvious regularity and persisted throughout nearly all the protracted period of solidification. It has been suggested that currents may be produced in a magma by the separation of a gaseous phase as separate bubbles which stream upwards; this process has been called two phase convection by Daly [see for instance 1933, pp. 367—74]. The possibility that this process might have contributed to the currents developed in the Skaergaard magma must be briefly examined. In the Skaergaard intrusion there is evidence of a greater concentration of volatile constituents in the upper border rock than in the contemporaneous layered series. Thus the upper border rocks show more late stage alteration—involving hydration—than the contemporaneous layered rocks and the analysed upper border rocks are richer in $H_2O +$ than the average layered rocks (see p. 182). They also have a higher ratio of ferric to total iron (Table XXXXII), and this may have been brought about by water, which reacted with the ferrous iron to give ferric iron and hydrogen, the latter escaping through the overlying cover (see pp. 183—84). These two features of the upper border rocks suggest that, at some stage, there was a concentration of water in them but it does not indicate that this occurred by upward streaming of a gas, a process which would be expected to produce a vesicular, upper border rock. Since the Skaergaard intrusion was relatively deep-seated at the time of its formation it is likely, on general grounds, that the water would not form as a gaseous phase during the main differentiation although it might when water became strongly concentrated in the late stage residual magmas. In the acid granophyres there are minute drusy cavities but even in this case it is not safe to deduce that a gaseous phase was present when this rock was completely molten because of the irregular shape of the drusy cavities which suggest that the vapour phase only developed during the late stages of the crystallization of the rock. We consider that two phase convection involving a gaseous phase is unlikely to have happened in the Skaergaard intrusion.

The hypothesis which we wish to put forward to account for the circulation in the Skaergaard intrusion is based on the idea of convection, due to increase in density of the magma on cooling. Without the separation of a solid phase, slight increase in density would take place on slight cooling, but if some crystalisation also took place there would be a much more marked increase in the mean density of the liquid plus crystals (except in certain peculiar circumstances which need not be considered [cf. Osborne and Roberts, 1931]). Grout [1918, A and B] has extended the term, two phase convection, to cover the case of convection due to formation of solid phases in the liquid; it is a two phase convection rather like that visualised by him which we postulate. Cooling of the Skaergaard magma must have taken place from the top and sides of the intrusion, but mainly from the former. The cooled and partly crystallised magma along the top and walls would have a higher specific gravity than the rest and would sink. Thus a con-

vective circulation would be established in which currents descended
along the walls, crossed the floor to about the centre, then rose to the
top where they would spread out towards the walls again. While passing
along the top of the intrusion the magma would be sufficiently cooled
and increased in density to sink again and repeat the process. Such a
circulation must tend to be established in any liquid which increases
in specific gravity on cooling, and which is cooled from the upper surface.
Whether a convective circulation will actually be established, in any
give case, depends on such factors as the compressibility, conductivity,
thermal coefficient of expansion and viscosity of the liquid. Our knowledge of these properties for natural magma is not sufficient to allow us
to deduce whether a convective circulation would be established or not,
but in the particular case of the Skaergaard intrusion the existance of
magmatic currents, such as would result from convection, are shown
by direct observation. Other features of the intrusion, besides those
used in the previous section to prove the existence of magmatic currents, also find a satisfactory explanation on the theory of a convective
circulation[1]) and are discussed below.

According to Jeffreys [1929, pp. 138—39] convection in a basalt
magma should establish a thermal gradient of approximately 0.3° C.
per kilometre. Using the best available data, Bowen [1935, p. 208]
has estimated that the melting point of forsterite increases 4.7° C. per
thousand atmospheres increase in pressure. This means an increase in
melting point of about 1.5° C. per kilometre of descent in the crust.
A direct determination of the increase in melting point of fayalite with
pressure [Bowen and Schairer 1935, p. 208] gives 7° per 1000 atmospheres and this corresponds to approximately 2° C. per kilometre of
descent. The equivalent figure for diopside [Daly 437, p. 66] is about
4.6° C. and for anorthite [Alling 1936, p. 9] is about 1.0° C. If the upper
part of the Skaergaard magma is regarded as on the point of precipitating
crystals of the primary phases, plagioclase, olivine and pyroxene, then the
increase in pressure as the magma is carried down by the convective cir-

---

[1]) It would be of interest to know the approximate velocity of the currents
in the Skaergaard magma. If an estimate of the viscosity of the magma and the relative
densities of the inclusions and the basic magma could be made, a minimum velocity for
the descending currents could be obtained from the fact that molten acid gneiss
was carried down against its tendency to float. Another line of approach could be
based on the difference in size of the crystals at the margin and at the centre of the
intrusion combined with experimental evidence of the rate of growth of crystals.
Vogt [1921—1923, pp. 240—42] estimates that the rate of growth of diopside is
about a millimetre per minute in a melt of its own composition and he suggests
that the rate is a function of the percentage of the substances in the melt. A little
more experimental data of this kind would allow a rough estimate of the speed
of the currents to be made.

culation should cause some crystallisation, since the thermal gradient is about 0.3° centigrade per kilometre and the raising of the melting point for these minerals lies between 1° and 5° centigrade per kilometre of descent. The amount of crystallisation will be slight because the latent heat evolved on crystallisation will tend to reverse the processes. Nevertheless, the slight crystallisation will increase the mean density of liquid plus crystals and tend to increase the strength of the convection currents. The relation between the temperature gradient in a magma undergoing convection and the increase in melting point of the minerals with depth also shows that the conditions at the bottom of such a magma will not result in any remelting of the solid phases as has often been suggested for crystals sinking through a stationary magma. There is not sufficient thermal data to make a more exact analysis of what would be likely to happen during such a convective circulation, and in any case the mathematical treatment would be complex. However, the net result of the process should be the formation of crystals towards the bottom of the circulating magma, and here they should accumulate if their collective density be greater than that of the liquid, while the liquid, disentangled from among the crystals, should rise again as a central upward current.

While in Greenland we concluded that the layered series was formed, as a precipitate, from the bottom upwards, and, from considerations of probable thermal and melting point gradients, we reached the general view of the mechanism of formation of the layered series which, in rather more detail, has just been given. Later we found that Adams [1924, pp. 462—65] and Jeffreys [1929, p. 139—40] had inferred, from theoretical considerations and the known properties of basalt magma, that in a convecting basalt magma, at moderate depths in the crust, solidification should take place from the bottom upwards. Our field observations and early inferences were not affected by previous knowledge of the theoretical possibility of such a mechanism.

It is not necessary to discuss whether a convective circulation in magmas, as we have postulated, is possible; given certain conditions, it is inevitable. What is required is to show, from a study of the rocks themselves, that convection has taken place, and this we consider has been shown for the Skaergaard intrusion in the previous section. As evidence for convection in plutonic intrusions has previously not been definitely found, it is of interest to consider whether there are any known factors present in the case of the Skaergaard intrusion, which would favour the establishment of such a circulation. First, there is the shape of the intrusion. If the form of a cooling liquid is that of a sheet, then a series of convection systems tends to be established, bounded by vertical divisions and giving a series of polygonal cells, having a diameter about four times the thickness [Jeffreys 1926]. Convection is at its optimum in a mass of liquid having a diameter about four times its thickness and, during cooling, there was a considerable period of time when the Skaergaard magma had a form approximating to this. Another factor favouring convection is low viscosity of the magma. From the rate of flow of a basalt lava, Becker [1897, p. 29] estimated that the viscosity of basalt is about sixty times that of water, that is, about the viscosity of olive oil at room temperature. Bowen [1915, p. 185] obtained an estimate of the viscosity of an artificial melt by the rate of sinking of crystals forming in it; although the melt was of rather different composition from natural basalt, Bowen's results support Becker's estimate. A viscosity of this order should allow vigorous convection if other factors are also favourable. Harker [1909, p. 221—22] gives in the following words the results of some experiments by Vogt on the effect of various oxides on the viscosity of slags. "Ferrous oxide promotes fluidity strongly, and magnesium to a less degree. Lime and probably soda also tends to promote fluidity; but even a small content of potash, in an acid or moderately acid magma imparts a noteworthy degree of viscosity". In a later publication Vogt summarised the results of his experience with melts of various compositions and these support his early conclusions [1921—1923, pp. 233—39]. The initial Skaergaard magma had a general basaltic composition and should therefore have had considerable fluidity. Certain peculiarities of its composition, namely richness in ferrous iron and poverty in potash, should have given it greater fluidity than many basalts, and changes in composition during the early and intermediate stages of the differentiation, resulting in especial richness in ferrous iron, should further have increased the fluidity. Hydrostatic pressure probably increases the viscosity of a magma [Daly, 1933, pp. 73 and 192—93; Bridgmann 1926]; and it has been suggested that this is the factor tending to prevent convection in the lower substrata of the crust. However, in an intrusion into the upper crust such as the Skaergaard intrusion, the increase in viscosity due to this cause would probably be slight. This brief inquiry suggests that the initial Skaergaard magma had a viscosity less than that of normal basalt, and that the residual liquid, when about half was crystallised (at which stage evidence for convection is most definite) had a still lower viscosity as a result of the particular trend in composition taken by the magma. The high ratio of ferrous iron to ferric and the low potash content, by causing unusually low viscosity, may have been important factors in the production of convection currents in the Skaergaard intrusion.

(c) Origin of the Igneous Lamination, Rhythmic Layering and other Structural Features of the Layered Series.[1])

The igneous lamination is due, in the first place, to the platy habit assumed by the plagioclase as it crystallised from the Skaergaard magma, a habit which is usual for bytownite and labradorite in basic igneous rocks. From the initial Skaergaard magma and from the early residual magmas the plagioclase crystallised in roughly square, thin tablets. The more acid plagioclase, crystallising from the later residual magmas, usually formed less thin tablets; these were no longer square, but rectangular, and they became increasingly so with increasing acidity. The bladed form assumed by some of the pyroxene, and even sometimes by the iron-rich olivine, contributes slightly to the igneous lamination but where these minerals are granular they have had no effect.

The other factor involved in the formation of the igneous lamination was magmatic flow; this caused the tabular crystals to have their greatest area parallel to the direction of flow, a direction which was also parallel to the surface on which the crystals were accumulating. As differentiation proceeded and the habit of the plagioclase crystals gradually changed—one direction in the plane of the tablet becoming longer than the other—there was rough orientation of the long direction of the tablets parallel to the direction of flow of the magma. While

[1]) There is a close analogy between the structural features of the layered series and of sedimentary rocks. The igneous lamination of the layered series is similar to the lamination of a micaceous silt; in both cases it is due to the lamellar form of the material deposited. The unlaminated material of melanocratic bands in the layered series, like sandstone is composed of fragments with a general rounded shape and in both there is no fissility. In some ways, though perhaps not in origin, rhythmic layering is like graded bedding, there being, in both cases, upward grading from heavy to light constituents. In the igneous rock, however, the heavy constituents are of approximately the same size as the light but have a greater specific gravity, while in sediment the heavy grains have roughly the same specific gravity, their greater weight being due to their greater size. In other ways the rhythmic layering is comparable with rhythmic variation of sandstones and shales, the light bands of the layered series being comparable with the shale because they represent material carried by feeble currents, while the dark bands are comparable with sandstone composed of heavy grains only transported by vigorous currents.

Throughout the varying structural features of the layered series there is a constant trend in the composition of certain minerals which finds no close analogy a among sediments but is similar to the changes taking place in the fossil content of sedimentary series and in a like manner may be used for establishing a time sequence. Thus in the layered series the stage of evolution reached by a mineral belonging to a solid solution series may be used to establish the chronological order of the rocks just as the stage of evolution of a fauna or flora. Furthermore, the abrupt appearance or disappearance of a mineral, as with fossils, provides datum lines for subdivision.

igneous lamination is a structure which would not occur without flow
of the magma, it should be distinguished from the usual fluxion structures
of igneous gneisses which are due to the flow of an almost crystallised
magma so that orientation is brought about by one crystal pressing on
another. In the case of the igneous lamination of the Skaergaard layered
series, the orientating force was produced by friction between the
crystals and flowing magma. It is because of this difference in origin
that the term igneous lamination has been used instead of fluxion
structure.

The rhythmic layering, the most striking feature of the layered
series, can only have been produced by rhythmically repeated variation
in the conditions of deposition. Workers on the better known layered
intrusions, the Ilimausak batholith [Ussing 1912], the Duluth lopolith
[Grout, 1918], the Bushveld Complex [Hall, 1932, Wagner etc.], the
Bay of Islands Complex [Ingerson, 1935, Cooper, 1936] and the Stillwater Complex [Peoples, 1936], have reached no consensus of opinion
as to the origin of the layering. The explanation usually put forward
turns on the idea of intermittent crystallisation due to gas emanation,
or to relief of pressure, as a result of earth movement or extrusion of
lava. In the case of the layering of the Skaergaard intrusion a hypothesis,
based on variation in the velocity of the convection currents, seems to
provide an adequate explanation. Such a hypothesis has two merits;
first, it appeals to a variation in the conditions of formation which
would be expected to be as varied as the layering is actually found
to be; secondly, variation in velocity seems to be a usual feature
of convection currents. To explain our view, let it be imagined that
a cloud of crystals of the primary phases, including heavy crystals such
as olivine and pyroxene, and light, such as the felspar, were being carried
along by the convection currents near the bottom of the intrusion, from
the margins towards the centre. At the same time let it be imagined that
the crystals were sinking through the magma at different rates due to
their different specific gravities. If the current and supply of crystals
remained constant, then the proportions of heavy and light constituents
deposited at any one place would also be constant, but there would
be more of the heavy constituents near the margin than further in,
a feature which is observed in the Skaergaard intrusion. If the strength
of the currents increased, the heavy minerals would be carried further
than they were before, and they would be deposited in abundance where
they were formerly scarce because the weaker current had not carried
them so far. Conversely, slowing up of the current will cause light constituents to be deposited in abundance where, with a more powerful
current, they were rare. Variation in velocity is apparently a usual
feature of convection currents. Such variation combined with different

rates of sinking of the crystals, because of their different specific gravities, should produce a winnowing effect on the crystals forming in the Skaergaard magma, and gravity stratified layering should result.

Detailed consideration of the mechanism here suggested for the production of layering will not be attempted; the problem is complex since the crystals would be forming over a range of positions and the rate of sinking would depend on their size and shape. Furthermore it is possible that some of the layering, especially in the case of the sharply and strongly differentiated trough banding structures, may have been produced by the lighter constituents being whisked and rolled along the surface of the accumulating heavy constituents. The matter is considered a little further in the following section, dealing with the structures of the trough banding horizon.

The way in which convection currents would produce marginal enrichment of the layered series in heavy constituents, and also increase in the size of the felspar crystals from the margin towards the centre, has already been explained (p. 263). These features of the layered series provide two of the best arguments for the existence of convection currents in the magma. The saucer form of the layered series and the banking up of the layered series against the border group are also features to be expected in a precipitate accumulating at the bottom of a convecting liquid. For this reason the saucer form of the layering is regarded as a primary feature (see pp. 62—63) and not the result of subsequent bending which was the suggestion offered by Ussing for the rather similar case of the Ilimausak batholith [1912, pp. 321—24].

(d) The Origin of the Trough Banding.

When originally formed the trough banding structures were symmetrical about a vertical plane, and were directed towards a point a little south of the centre of the intrusion. Trough banding is found in the hortonolite-ferrogabbros in which the plagioclase consists of decidedly rectangular tablets. In the strongly banded rocks of the trough bands, e. g. 2572, there is a tendency for the long direction of the felspar tablets to be parallel to the general trend of the trough structure, indicating that the currents were directed along the troughs. The currents responsible for the trough banding were thus directed roughly centrally and conform to the convection current theory.

Well developed trough banding structures were only formed at one horizon. Below there is even layering extending over wide areas, and there is often good rhythmic layering and igneous lamination. These features would be produced by broad and fairly gentle convection currents with periodic variations in velocity. At the horizon of the trough banding structures, the convection currents were apparently broken up into narrow and powerful streams which flowed along the trough bands, while separating them, there were zones with less powerful and more irregular currents. Between the trough banding structures, where the currents were irregular and less powerful, the precipitate of minerals would accumulate more easily and therefore stand up higher than along the zones of strong currents. This accounts for the trough banding structures being banked up against the irregular pile of material between the troughs. The strong differentiation into melanocratic and leucocratic layers in the trough banding zones indicates strongly varying velocities, while the absence of banding in the rock between the troughs indicates irregular eddy currents allowing no separation of light and heavy crystals. There was apparently some variation in the width of the currents, since banding, at one horizon, may spread out a little over the intervening massive rock, and at another horizon may narrow.

The various types of layering and banding in the area round the houses, shown on the map, figure 11 and the section, figure 12, may be interpreted in terms of varying convection currents. In doing this it must be remembered, that the currents would probably vary in character with the distance from the margin, and that the most north-easterly group of trough banding structures shown in the figures are far from the margin, and the south-westerly group near. Differences in the nature of the trough bands as they are traced from the north-east to south-west along the present line of outcrops are believed to be due to variation in the nature of the convection currents at different distances from the margin, rather than to persistent differences along the whole length of the troughs. The lowest, widespread, rhythmic layering of figures 11 and 12, that running through the northern house, is considered to represent a broad current of variable velocity which, since the banding dies out to the north-east, is considered to have become steady in velocity in that region. Then, over the whole area, the currents apparently became steady and slow, giving an unbanded and only feebly laminated horizon. This was succeeded in the region round the Main House, by powerful and variable currents, giving good rhythmic layering and igneous lamination, but further north-east, the steady currents apparently continued as rhythmic layering is lacking. In the Main House area the violent and variable currents broke up into separate currents along the trough bands, the break up occurring rather earlier towards the south-west than towards the north-east. Once the narrow currents had been established, they were usually fairly persistent, but an exception is the current producing trough banding F which, besides starting late, died out early. To the north-east there are cases of separate currents, at a low level, joining together at higher levels into a single broader one;

for example, the separate trough bandings B and C unite upwards into a single broad trough banding structure. On the neck of the Skaergaard, which is nearer the margin, banding is on a smaller scale and this is presumably due to more rapid changes of velocity. Near the margin, where the vertical currents are changing to an almost horizontal direction, rapid changes of velocity may reasonably be expected. Further south-west, that is still nearer the margin, the trough banding structures disappear and continuous small scale banding occurs which gradually gives place to the false-bedded, irregular type found everywhere close to the inner contact. Here the currents were apparently not divided into definite streams, and they had highly variable velocities.

If it is accepted that the more north-easterly examples of trough banding represent the general conditions some way from the margins and the south-westerly the general conditions near the margin, it is possible to get a still more complete picture of the nature of the currents at this horizon. The descending currents were apparently not divided into definite streams but only became so as they crossed the floor of the intrusion. Near the contact they were apparently powerful since much of the light material was carried on leaving the more melanocratic marginal zone. Here they were also highly irregular since the rhythmic layering is strongly developed and of the false-bedded type. The most definite division of the currents to give trough banding occurred at an intermediate distance from the margin represented by the position of the Main House. Further from the margin the definite lines of current became less abundant, and there was also a tendency at higher horizons for two or three lines of current to amalgamate into one broader current.

Trough banding is not found higher in the series on Basistoppen and at first we considered that this indicated the disappearance of the narrow lines of current as cooling and solidification continued. On Basistoppen, however, the rocks are far from the margin and this may be the reason for the absence of trough banding structures rather than that they have died out in the later layered rocks. At the Base Houses horizon where the trough banding is best developed there is evidence that the structures die out towards the centre of the intrusion. This would be expected on the convection hypothesis as, near the centre of the intrusion, currents from all directions must be meeting and turning upwards. It may well be that the trough banding structures would be found at higher horizons, equivalent to the horizon of the upper-ferrogabbros of Basistoppen, if rocks at a suitable distance from the margin were still preserved.

(e) Convection and the Major Differentiation.

The effect of gravity is evident to the eye in the gravity-stratified, rhythmic layering of the Skaergaard intrusion. Although not so immediately obvious, it has been shown that gravity controlled the cryptic layering and, therefore, the major differentiation; both have arisen as a result of the accumulation, at the bottom of the intrusion, of the successive solid fractions which separated from the magma. Cooling of the magma would mainly take place from the top surface, and the cooling might have occurred either with or without convection currents. If without convection currents then crystallisation must have taken place in the upper part of the magma where the heat was being lost, and to give the layered series, which accumulated from the bottom upwards, it is necessary to postulate that vertical sinking of the crystals took place. On the other hand, if convection currents accompanied the cooling of the magma, the process would be as described in a previous section (pp. 266—270). Considering the major variation only, the important difference between vertical sinking and the somewhat more complex process involving convection, is that, in the latter, the solid phases would be formed gradually in the descending current due to increase in hydrostatic pressure, and they would be carried by bodily movement of the magma to the neighbourhood of the bottom of the intrusion, while, in the former, the solid phases have to sink long distances through the magma as a result of specific gravity differences between the solid phases and the liquid. Even on the convection theory there must have been some sinking of the crystals relative to the magma in order that they should come to rest at the bottom of the liquid. The amount of settling, however, is of the order of a few metres instead of a few kilometres, as is required by the direct sinking hypothesis.

Direct sinking of crystals has apparently taken place in many sills and lava flows and has been observed by Bowen [1915] in laboratory melts approaching the composition of basalt. In appealing to the hypothesis of direct sinking to explain the Skaergaard intrusion we should therefore have the support of analogy with other described cases. The convection theory, however, is only a modification of the direct sinking hypothesis in which the sinking is aided by the convection currents, and also the formation of the solid phase takes place at lower levels in the liquid. As we have pointed out, even with the intervention of convection, some direct sinking must have taken place. The main evidence for convection comes from the structural features of the layered series and not from the major variation which would result from direct sinking, if such took place. Taking into consideration the probable differences between the densities of the liquid and the various solid phases it is difficult to imagine vertical sinking of crystals from the top of the magma to the bottom without more marked separation into light and heavy constituents; for this reason the convection theory provides a preferable explanation of the major variation. Whether we postulate direct sinking of crystals or sinking aided by convection currents, gravity is equally the cause of the movement.

(f) Origin of the Border Group.

The border group is material solidified along the top and sides of the magma chamber where loss of heat took place From a consideration of the composition of the minerals of the border group and of certain structural features, the order of solidification has been shown to be from outside, inwards, as would be expected. The layered series is made up of two parts, the primary precipitate of unzoned crystals and the interstitial material which represents the magma surrounding the collection of primary crystals. In the same way the border rocks may be made up of a certain proportions of primary crystals and of the magma in existence at the time, or it may consist simply of the magma without primary crystals. From the chemical composition, a rough estimate may be made of the proportion of primary crystals and magma. If the composition of a border rock approaches that of the contemporaneous layered material, it must be supposed that the rock, like the layered material, includes about eighty per cent. of primary crystals. If the composition of the border rock is that of the contemporaneous magma, it must be assumed that it is the result of solidification of the magma without any primary crystals; intermediate compositions will indicate an intermediate proportion of primary crystals and magma. The extent of the zoning, especially of the felspars, also gives an indication of the amount of the primary precipitate. Where about eighty per cent. of the rock is primary precipitate, as in the layered series, zoning, due to solidification of the twenty per cent. of interstitial material, is slight. Stronger zoning will suggest a greater proportion of magma relative to primary precipitate.

The analysed outer border rock from Ivnarmiut, 1837, was formed before any of the layered rocks now visible, and exact comparisons are not possible. As a rough estimate this rock may be considered to have been formed when 30 per cent. of the magma had solidified. The mean between the composition of the probable layered rock and magma at this stage (Pl. 27) is very close to the composition of this outer border rock, and this provides tentative evidence that the rock was formed of forty or fifty per cent. of primary crystals, the remainder being the magma in existence at the time. The considerably greater amount of zoning of

TABLE XXXIX.

	A	4298	B
SiO$_2$	46	45.51	45.5
Al$_2$O$_3$	15.5	14.36	13.5
FeO + Fe$_2$O$_3$	16.5	17.69	21
MgO	6	5.38	5
CaO	9	9.91	8.5
Na$_2$O	3.2	2.43	3.2
K$_2$O	0.26	0.35	0.4
TiO$_2$	2.6	3.96	2.3
MnO	0.15	0.20	0.25
P$_2$O$_5$	0.07	0.32	1.2

A. Layered rock comtemporaneous with 4298.
B. Liquid contemporaneous with 4298.

the plagioclase compared with that of the layered series supports this estimate.

For the inner part of the border group on Ivnarmiut, the evidence is more reliable, and points to roughly the same conclusion. The time of formation of the olivine-free border rock, 4298, was the same as that of the uppermost middle gabbros whose composition can be accurately estimated. The composition of the contemporaneous magma may also be estimated with moderate accuracy, and it is a good deal different from that of the layered rock (Table XXXIX). For most of the constituents the percentage composition of the border rock lies between that of the contemporaneous magma and the contemporaneous layered rock, suggesting an intermediate proportion of primary crystals and liquid (Table XXXIX). In this particular border rock the amount of lime is greater, and the amount of soda less, than in either the contemporaneous liquid or layered rock. This suggests the presence of more basic plagioclase, the reason for which is not clear. Another unexplained difference is, that the iron is in a more oxidised state than the contemporaneous layered rock. The major constituents may be expected to vary somewhat because of variation in the relative amounts of the primary, felspar and ferromagnesian crystals, and more interest is attached to certain minor constituents, especially K$_2$O, and P$_2$O$_5$, which do not occur appreciably in the primary minerals, and which are present in very different amounts in the liquid and contemporaneous layered rock. The amount of K$_2$O present in the border rock is roughly the mean between that in the contemporaneous liquid and layered rock, suggesting that the amount of primary crystals in the border rock is about forty or fifty per cent. compared with about eighty per cent. in the layered rock. The relative amounts of P_2O_5 should be more significant, as it does not occur appreciably in solid solution in any of the primary precipitate phases. On the other hand we do not care to put too much reliance on the estimated amount of P_2O_5 in the later residual magmas which, for reasons already given (p. 219), is probably too high. However, assuming the estimated value of P_2O_5 in the liquid is correct, then the proportion of magma to crystals appears to be about 1:4. If the estimated value of P_2O_5 in the liquid is high then the proportion of magma would have been greater. As in the outer border rock previously discussed, the greater amount of zoning suggests a greater proportion of magma to primary crystals than in the layered series.

The upper border rocks are regarded as partly the result of contamination with acid gneiss (pp. 180—82). The composition of the material which added to acid gneiss would give certain of the upper border rocks is given in Table XXVIII. The material which would give the low, upper border rock, 4163, is sufficiently close to the layered rock at ninety per cent. solidified (see Pl. 27) to indicate that it must have been formed at this stage. Similarly the high upper border rocks 3050 and 3052 must have been formed when about seventy per cent. of the total magma was solidified. The estimate of the composition of the materials before contamination is also close to that of the estimated contemporaneous liquid but neither set of estimates is sufficiently exact to enable the proportion of primary crystals to magma to be deduced.

The low marginal border group, excluding the chilled margin and the perpendicular felspar rock which are considered below, was apparently formed from a certain amount (about 50 per cent.) of primary crystals, the remainder being the contemporary magma. Like the layered rocks, although to a less extent, the low marginal border rocks have a composition which does not correspond with that of any magma but represents a magma plus a concentration of the solid phases in equilibrium with it. To get a rock composed of homogeneous, early crystals, surrounded by magma of the average composition then in existence, some relative movement of magma and crystals must be postulated. The low marginal border group is only exposed where the contact is dipping inwards; here, so far as the slope allowed, primary crystals could collect, as in the layered series. On this view of the origin of the border group, the proportion of primary crystals to magma should depend on the steepness of dip of the margins, and if the margins were dipping outwards there should be no opportunity for the primary phases to collect. It is unfortunate that low border rocks are not exposed where the margin is outward dipping, and the high marginal border rocks found on Tinden and Kilen are probably different for another cause. It is likely that another mechanism produced the perpendicular felspar rock and that this also contributed to a varying extent to the production of other low border rocks.

On the convection theory there should be no primary crystals in the magma at the top of the chamber, and the upper border rocks and also the high marginal border rocks should therefore have formed from the successive magmas without addition of primary crystals. It is unfortunate that the hybridisation with acid gneiss makes it difficult to decide whether this is the case or not, and more detailed work would be of value. It may be asked how there could be any solidification at the top surface of the liquid as, on the convection theory as so far developed, it has been postulated that the temperature is there too high for solidification to take place. The temperature is, however, only too high for solidification where the magma is actively taking part in the circulation. Along the boundaries of a circulating liquid any convection currents will be slowed down due to friction with the walls [cf. Grout 1918 B, p. 495—96], and from the stationary or nearly stationary magma, heat would be lost so that solidification could take place [see also Adams 1924, p. 467]. Although the differences between the upper border group and high marginal border group, on the one hand, and the low border group, on the other, may be mainly due to hybridisation with acid gneiss, some of the differences may also be due to the higher rocks having formed by solidification of successive residual magmas in which there were no primary phases.

The width of the border group should depend, among other things, on the rate of heat loss and the time available for its formation. Thus the marginal border group at a constant level in the intrusion should be thicker along the hanging wall, where the heat loss would be greater, than along the foot wall; and comparing different levels in the same part of the intrusion, the greater thickness should be opposite the highest horizon of the layered series, for, in such positions, it would have had the longest time in which to form. The mapping shows that these expectations are realised. The conditions at low levels in the intrusion are shown on Uttentals Plateau and here the border group is narrow. This seems to be partly because the northern boundary is a foot wall where heat loss should be slight and partly because the layered rocks were banked against the border group at an early stage, but conditions here are complicated by the abnormal transitional layered rock whose origin is uncertain. On Mellemö the marginal border group has a greater width than along the northern border, and this is to be correlated, first with the steeper margin from which heat would be more rapidly conducted away, and secondly with the longer time available for growth, as the hortonolite-ferrogabbros banked against the border group were formed at a relatively late stage in the cooling history of the intrusion.

The remarkable perpendicular felspar rock, found in the outer part of the low marginal border group, seems to have had a special manner of origin. In describing this rock on an early page (p. 150) it has been suggested that the concentration of high temperature plagioclase which it shows may be the result, either of diffusion, or magmatic flow. Since there is much evidence for convection currents in the Skaergaard intrusion, the hypothesis of magmatic flow is preferable. The rock seems to be an example of fractional crystallisation by the method originally suggested by Becker (see p. 284). In a circulating magma Becker considered that the less fusible material would crystallise attached to the cooling surface, leaving the residual magma enriched in the more fusible constituents. Examples of this mechanism do not previously seem to have been described, but it provides an apparently satisfactory explanation of the perpendicular felspar rock, and the process has probably also produced those outer border rocks which have wavy pyroxene and felspar crystals set at right angles to the margin. It is possible that this mechanism has contributed to the formation of other low border rocks which have apparently contained some primary precipitate, but without textural evidence it is not possible to distinguish between the mechanism suggested by Becker and the collecting together of primary crystals, by gravity, on the sloping wall of the intrusion. A sequence of rocks in passing inwards, from high to low temperature types such as found in the border group, is to be expected on Becker's hypothesis just as on the hypothesis we have put forward. In explaining the border group we have preferred a modification of Becker's views, partly because it is only in certain cases that there is textural evidence in harmony with Becker's original hypothesis, and partly because there seems to be a difference, other than due to hybridisation, between the high marginal border rocks such as those on Tinden and Kilen and the low border rocks of the northern two-thirds of the intrusion. On Becker's hypothesis there should not be any difference while on the proposed modification of his hypothesis a difference would be expected.

Although the present evidence is by no means conclusive because of complications due to hybridisation, it seems that the upper border, and high marginal border groups are primarily the result of the solidification of the magma in existence at the time. Such solidification of the existing magma could not lead to any change in composition of the remaining magma. In our view the accumulation of successive crystal fractions to form the layered series controlled the change in composition of the Skaergaard magma and, therefore, also controlled the composition of the upper border rocks.

(g) The Earliest and Latest Stages in the Differentiation.

When the magma was first injected to form the Skaergaard intrusion, there would only be a few irregular currents due to the act of injection; the regular system of convection currents would only be slowly established. It is possible that the gabbro-picrite which is found in the northern border group immediately succeeding the chilled marginal gabbro, represents the result of vertical sinking of olivine before the convective circulation was established. The available evidence suggests that from the original Skaergaard magma plagioclase and perhaps pyroxene were formed contemporaneously with olivine. If our view of the origin of the gabbro-picrite is correct then the absence of primary crystals of plagioclase and perhaps of pyroxene in the gabbro-picrite must be due to the slowness with which these crystals settled compared with olivine.

As the intrusion solidified the thickness of material remaining liquid became gradually less and at the same time there was probably not much reduction in area. In the latest stages the conditions for a single system of convection currents must have become progressively less favourable. We have not obtained evidence that the circulation of the Skaergaard magma ever disintegrated into a series of independent convection systems, but this might be shown by careful mapping of the directions of the longest axes of the plagioclase crystals in the upper ferrogabbros. What we do know is that igneous lamination continues up to the purple band but is entirely absent above. By the time the thickness of the remaining liquid was reduced to about three hundred metres, and its composition had changed, so that it was probably more viscous, no igneous lamination was produced; this indicates that convection currents had ceased. In the unlaminated layered rocks the mechanism of differentiation must have been somewhat different from that in the main layered series. In the formation of the unlaminated fayalite-ferrogabbro, with its large olivine crystals up to one centimetres in diameter, we seem to see the effect of direct sinking of heavy olivine crystals leaving an olivine-free basic granophyre above. Differentiation by such a process, like that which has been postulated for the layered series, is the result of crystal fractionation controlled by gravity, but it differs from the case shown by the layered series, and it is desirable that there should be two distinct names. We propose to use the name non-selective, gravitational fractionation when, as in the layered series, all the solid phases have settled out of the liquid, and to use the name selective gravitational fractionation when only certain of the solid phases are precipitated, while others remain suspended or even float to upper levels.

Under the Basistoppen raft there is a variable thickness of hedenbergite-andesinite forming the highest horizon of the layered series. The andesinite does not occur further east where the Basistoppen raft is missing and its formation seems dependent on the presence of the raft. Relatively large, zoned plagioclase crystals, whose inner half is about An_{40}, make up the bulk of the pyroxene andesinite, and these cannot have formed from the magma depositing the upper ferrogabbros from which only plagioclase more acid than An_{35} could form. It seems necessary to postulate that the plagioclase of the hedenbergite-andesinite formed from the magma depositing the middle gabbros or lowest ferrogabbros, and that they accumulated under the Basistoppen raft because they had a lower specific gravity than the contemporaneous magma. It has already been pointed out (p. 266) that the plagioclase crystals separating from the Skaergaard magma, at any rate at the later stages, may have had a lower specific gravity than the liquid, and that, if this be the case, they must have been held down by the dense minerals associated with them. It is not, therefore, contrary to our hypothesis on the origin of the layered series, if we also postulate that a little plagioclase escaped being laid down with the heavy minerals at the bottom of the intrusion and collected by flotation at the local top of the liquid. In our view the origin of the hedenbergite-andesinite is another example of differentiation by selective gravitational fractionation. It is interesting to note that the size of the plagioclase crystals in the andesinite is comparable with those of the central middle gabbros and not with the upper ferrogabbros (Table IV). Where the top of the liquid was forming the upper border rocks, a few plagioclase crystals might well have floated up into the stationary magma and contributed to the formation of the upper border group, but, if this happened, it has escaped notice because of the similarity between the plagioclase which had floated upwards and the inner part of the plagioclase being directly precipitated. Only under an inclusion such as the Basistoppen raft, below which little or no cooling would be taking place and typical upper border rocks would not be forming, would plagioclase-rich rocks be expected. The amount of hedenbergite-andesinite is very small and the process which gave rise to it, whatever it may be, was relatively unimportant in the differentiation of the Skaergaard magma.

The remaining rocks of the Skaergaard intrusion to be considered are the hedenbergite-granophyres of Brödretoppen and the acid granophyres. The hedenbergite-granophyres are closely related in composition to the basic hedenbergite-granophyre of the unlaminated layered series, and perhaps they represent drafts of a late residual magma carried above the Basistoppen raft as the raft sank lower. The acid granophyres probably represent the very last part of the Skaergaard magma to solidify.

Considerable pressures must have been exerted to drive the acid granophyre magma into irregular veins, far from the place where the last
residual magma would naturally occur, and differentiation by the filter-
press method may, therefore, be expected to have contributed to its
formation.

(h) Former Views on Convection in Magmas and other Comparisons.

Becker in 1897 first made the suggestion that fractional crystallisation as a result of convection might be responsible for some cases
of differentiation in magmas (1897 A and B). In his approach to the
problem he first considered some of the previously suggested methods
of differentiation and in particular gave evidence, for which there is
now still more support [Bowen, 1921] that diffusion is too slow a process,
even in basic magmas with low viscosity, to allow any considerable
concentration of the less fusible constituents by crystallisation at the
cooling surface. But if diffusion be inadequate, he suggested, in the
following words, that convection currents might perform roughly the
same function:—

"If we suppose a dyke in cold rock filled with mobile lava . . . . . the mass will
be subjected to convection currents because the liquid near the walls will be cooler
than that near the median plane of the dyke. A circulation of lava will take place,
the descending flow at the sides being compensated by ascending flow near the
central surface . . . . . If the lava is a homogeneous mixture of two liquids of different
fusibility, then the crusts which first form upon the walls will have nearly the same
composition as the less fusible partial magma. If we follow mentally a small portion
of the liquid in its circulation, it will clearly deposit at each of its early contacts
with the growing walls a part of its less fusible component, and at each complete
revolution it will have a different composition. This composition will always tend
towards that which represents the most fusible mixture of the component compounds."

Becker gave no examples of igneous complexes in which this process
had taken place and, indeed, no certain example has yet been found.
Nevertheless, if Becker's hypothesis is slightly modified and the early
crystallising minerals are considered to separate along the floor of the
intrusion, due to gravity, instead of attached to the walls, the explanation
is similar to the one adopted for the Skaergaard intrusion. The importance of Becker's papers is that they first clearly linked the process
which the chemist calls fractional crystallisation with differentiation of
igneous rocks, but it is also of interest that the mechanism he suggested,
and of which no example has so far been recorded, is practically the
same mechanism as we believe to have acted in the Skaergaard intrusion.

In seeking an explanation of the symmetrically-arranged rock types
in the Shonkin Sag, Palisade Butte and Square Butte laccoliths, Pirsson

[1905, pp. 181—97] rejected on field evidence the hypothese of successive injections and of hybridisation with material from the walls, and he rejected the hypothesis of molecular flow or diffusion, on the theoretical grounds put forward by Becker. He considered that the differentiation resulted from establishment of an ordinary convective circulation [pp. 187—88] and that as the magma slowly crystallised:

....."part of the material solidified would remain attached to the outer wall, and part would be in the form of free crystals swimming in the liquid and carried on the currents. As the crystals are solid objects and of greater specific gravity than the liquid, there might be a tendency for the crystals to drag behind and accumulate on the floor of the chambers..... In this manner it may be possible to understand how these would form a femic marginal crust and a great thickness of femic material at the bottom of the laccolith."

Daly [1914, p. 224] has criticised Pirsson's views considering that convection currents would not form in a sheet-like laccolith such as Shonkin Sag, because there would be little difference between the rate of heat loss from the bottom and top surfaces, and that the time available between injection and solidification would not be sufficient to allow convection currents to be effectively established. These arguments seem valid for relatively thin laccoliths such as Pirsson was considering. Osborne and Roberts [1931] have given another reason why convection should not have taken place in the Shonkin Sag laccolith, although this does not seem to be a necessary inference from the data if Grout's idea of two phase convection is used. It seems that Pirsson did not sufficiently explore the idea of direct sinking of the femic constituents, an idea favoured by Daly and further supported by the work of Osborne and Roberts. It is disturbing to find it suggested in a subsequent paper [Barkinsdale 1936] that the Shonkin Sag laccolith is due to successive injection and hybridisation. The convection hypothesis for the differentiation of the Shonkin Sag laccolith is clearly of such doubtful probability that it cannot be used in support of the convection hypothesis for the Skaergaard intrusion.

After working on the rocks of the Duluth lopolith Grout also came to believe in the importance of convection currents in magmas. From a consideration of the properties of basic magmas he sought to prove mathematically that convection currents would be established in a large body of basic magma. As already stated (p. 268) the data at present available are not sufficient to provide a convincing proof that convection currents would be established in natural magmas although they seem to show that it is not an unreasonable postulate for certain magmas. By means of convection currents Grout satisfactorily explained certain structural features of the Duluth gabbro which are similar to the igneous lamination of the Skaergaard intrusion [1918, A, pp. 452—57]. His explanation of the phenomenon has been adopted for the Skaergaard examples (pp. 263 and 271). In explaining the banding of the Duluth lopolith which is apparently similar to the rhythmic layering of the Skaergaard intrusion Grout [1918, A, p. 457] only offers the very general suggestion that convection currents combined with:

"....rhythmic effects in the way of cooling, intrusive action, or gas emanation (all of which are known to be rhythmic), might rhythmically change the mineral composition of the crystals growing along the walls, and thus result in banding."

In our view the banding of the layered series is not the effect of rhythmic changes of these kinds but of rhythmic variation in the velocity of the convection currents. Grout's conception of the conditions along the sides of a crystallising magma undergoing convection is rather similar to ours. Thus he suggests [1918, B, p. 495], that the descending marginal current would be slowed up near the wall, especially as early formed crystals became abundant, and eventually movement would cease, and the magma plus crystals would solidify to a solid rock. Fluxion structures parallel to the border he considers would result from slight flow of the viscous magma and its included crystals. We consider that the fluxion structures in the low marginal border group of the Skaergaard intrusion were formed in a rather similar way to the layering of the layered series. The flow foliation in the upper border group is highly variable in its development and it may have formed as suggested by Grout.

Grout also states [1918, B, p. 496]:

"Most of the crystals, being heavier than the magma, would tend to settle; and though formed during the cooling along the top and sides of the chamber, they would probably lodge at the sides and especially along the bottom..... It will be expected that whichever (crystals) form first will segregate towards the bottom. It is only the coincidence of high gravity and early crystallisation that results in a strict gravitative arrangement."

Grout realised that the lower layers of an igneous complex formed as he postulated would consist of early crystals irrespective of specific gravity, providing this was greater than that of the liquid. Unfortunately when Grout applied these clearly expressed conceptions to the case of the Duluth gabbro (a study of which had presumably largely given rise to them) he had to modify them, and in so doing he weakened the arguments for convection. On Grout's hypothesis, the lower layers of the Duluth gabbro should contain the earlier felspars, that is, the more anorthite-rich. Grout believed that the whole three miles of the banded gabbros of the Duluth lopolith contained plagioclase of approximately the same composition (basic labradorite or acid bytownite) and he concluded [1918, p. 639] that the early basic crystals:

.... "must have remained in contact with the mother liquor until equilibrium was established and they became average in composition. The crystals may have settled a little, but the viscous magma more than likely moved with them in convection most of the way. The end of crystallisation came when the crystals lodged in the more viscous wall or floor and there, slowly maintaining equilibrium, the crystals adjusted their composition to that of the magma around them, some bytownite, some labradorite."

If there was sufficient magma about the crystal to effect this change, then the result is as if the crystals grew in the magma without convection currents. This critical evidence is opposed to the hypothesis of differentiation by convection currents which Grout, largely on theoretical grounds, had previously put forward. In the layered series of the Skaergaard intrusion this difficulty does not occur since the felspars become more acid with increasing height. At first the plagioclase in the gabbros and related rocks of the Bushveld Complex was believed to have a roughly constant composition, but recent work by Lombard [1935] has shown a steady upward variation from basic to more acid felspars which, however, is interrupted once. Perhaps detailed work will prove the same to be the case in the Duluth lopolith, thus removing a big ('ifficulty in the application of Grout's clearly stated theoretical conceptions of convection to the case of the Duluth gabbro.

Since Pirsson's and Grout's attempts to explain certain types of differentiation by convection, the idea has not been further developed, so far as we know, for any specific examples, although it has occasionally been mentioned as a possibility [e.g. Grout 1925, p. 486, Daly 1933, p. 354]. The nature of convection in magmas has been discussed from the point of view of the Earth's thermal history by Adams [1924, pp. 462—68] and Jeffreys [1929, p. 140]. The conception of convection currents in magma reservoirs has been used in a suggestive paper by Holmes [1931] in which he seeks to explain the close association of acid and basic rock in the British Tertiary province, and Backlund [1932, pp. 15—18], has extended Holmes' ideas to the case of certain alkali complexes. In a later more speculative paper, Holmes [1935] has considered deep convection currents as the cause of many of the major features of the Earth, but such applications do not concern us here.

From published descriptions there seems to be a strong similarity between the types of differentiation in the layered series of the Skaergaard intrusion, and other layered plutonic masses. The similarity suggests the probability of similar origin, but the suggestions so far put forward to account for the other layered intrusions are, so far as we know, frankly, preliminary in character. In the case of the Bushveld layered rocks, there is a general unanimity among geologists who have studied them in the field, that the layering results from post-emplacement differentiation of a homogeneous magma [Hall 1932, p. 279]. The dominant, but not exclusive, action of gravity has been considered the underlying cause of the differentiation, and Hall [pp. 279—80] has sought to show that, on the average, the density of the rocks decreases upwards. The evidence for this is not convincing and in any case is not a necessary result of gravitational control of differentiation, as Grout has pointed out [see p. 286]. Just before his death Wagner [1929, p. 46] summarised his views which involved the idea of intermittent, crystallisation differentiation due to checks in the cooling, and also Vogt's notion of re-fusion of crystals on sinking down into hotter levels of the intrusion. The recent work of Lombard [1935] has shown that in the Bushveld Complex there is, on the whole, a steady change in composition of the felspar and pyroxene, as in the Skaergaard layered series, but that there is at least one abrupt reversal, and he has suggested [p. 32], that sinking under the influence of gravity, together with successive waves of injection, may account for the phenomenon. The action of convection currents has not been postulated, except briefly by Daly [1933, p. 354]. Work on the other known layered intrusions of basic composition, such as the Bay of Islands Complex [Cooper 1936, Ingerson 1934, Buddington and Hess 1937, Ingerson 1937], the Stillwater Complex [Peoples 1933], The Great Dyke of Southern Rhodesia [Lightfoot 1927], Sierra Leone Complex [Dixey 1922, Junner 1930], is also not sufficiently advanced to contribute to the interpretation of the Skaergaard rocks.

Although the rock types in the Ilimausak batholith of south-west Greenland are very different, there seems some similarity in their distribution and structural features to those of the Skaergaard intrusion [Ussing, 1912]. The Ilimausak mass is probably a batholith formed by stoping. In the southern half, which is probably the lower part, there is a border phase of augite syenite which has a relation to the central layered rocks similar to the relations between the border group and layered series of the Skaergaard intrusion (see, for example, Ussing's figures 4, 5 and 10). The layered Ilimausak rocks are saucer-shaped and, in places, are either banked against the margin (see Ussing, fig. 12), or give place to unbanded rocks near the margin. The evidence points clearly to the varied layered rocks having formed from a homogeneous magma by differentiation, and Ussing believed that the mechanism involved was crystallisation differentiation combined with gravitative effects. He explained the upper naujaite by flotation of the early and light sodalite, combined with the tendency for the sodalite, which is an early mineral, to form as a result of convection or diffusion where cooling was taking place (pp. 349—50). The lujavrite which underlies the naujaite, is poor in sodalite because this mineral has been concentrated above. The banded kakortokites which are closely related in composition to the lujavrite, are different in structure, being in the form of markedly gravity-stratified sheets (pp. 355—62) showing exactly the same features as the Skaergaard layered series. The hypothesis which Ussing adopted to explain the banded kakortokite is: —first, recurrent crystallisation due to outside effects causing reduction in pressure; second, sinking of the crystals so formed under the influence of gravity at various rates. Ussing (pp. 361—62) points out, as we have done for the Skaergaard intrusion, that the magma responsible for the Ilimausak batholith was probably unusually mobile. It seems from Ussing's description of the Ilimausak batholith that, despite differences in composition, there are many striking similarities with the Skaergaard intrusion. In view of Ussing's magnificient, detailed study of the Ilimausak batholith, it is surprising to find that Wegmann [1935, 1938, pp. 74—83], has recently put forward a view which has no similarity with Ussing's original interpretation but which is based on an idea that the layered Ilimausak rocks are the results of transformations, by a process akin to migmatisation, of a horizontal series of earlier rocks. We hope shortly to visit the Ilimausak batholith and make detailed comparisons with the Skaergaard intrusion.

XIII. DISCUSSION OF THE TREND
OF DIFFERENTIATION

(a) The Original Skaergaard Magma compared with other Magma Types.

From the evidence of the chilled margins, the Skaergaard magma had apparently a fairly usual basaltic composition before differentiation (p. 140). It was close to a member of the Porphyritic Central Magma series of Mull and Ardnamurchan, and a still closer approach to its composition is obtained by averaging two parts of the Porphyritic Central type of silica percentage 48 and one part of Non-porphyritic Central type of silica percentage 47 (see p. 142). Bowen [1928, pp. 134—39] and others have suggested that the Porphyritic Central magma is not on the line of liquid descent because it usually contains porphyritic plagioclase even in chilled examples; they have considered that it was produced by addition of plagioclase crystals to another, more fundamental and completely liquid, magma. After this suggestion had been made Thomas [Richey, Thomas etc. 1930, p. 98] wrote:—

"Perhaps the statement in the 'Tertiary Mull Memoir', that a porphyritic structure is 'always' to be observed in the more quickly cooled rocks of the Porphyritic Central type, is a little too general..... While admitting the possibility of the Porphyritic Central magma never having been quite free from solid phases, there appears to be evidence that these, in certain cases, have been reduced to mere nuclei, and that the liquid, in which they later became re-grown without serious modification of their composition, was capable of yielding its own suite of differentiates."

The chilled Skaergaard magma does not show any porphyritic felspars and we consider that at the time of its intrusion the magma was completely liquid. The authors of the Ardnamurchan Memoir consider it possible that the Porphyritic Central magma may have originated as Bowen suggested but that:—

"....there is nothing inherently impossible in the re-solution of accumulated early formed plagioclase, and that this condition was approximated to, even if not completely attained, is suggested by the great textural differences exhibited by marginal and internal portions of the same rock-mass, and between plutonic and hypabyssal representatives of the Porphyritic Central type."

It is of interest to note that in a convecting magma crystals should not melt as a result of sinking per se but various other processes may be postulated allowing re-melting to take place. It is not our intention here to consider the origin of the Skaergaard magma—the differentiation of the magma, after emplacement, being our main interest—but it should be emphasised, that the Skaergaard magma was free from porphyritic plagioclase crystals when intruded.

There are two respects in which the original Skaergaard magma was rather different from a mixture of the Porphyritic and Non-porphyritic Central magmas of the British Tertiary province (see p. 142) —first, the iron was in a more reduced state and secondly, there was rather less potash. Other widespread basalt magmas have, however, a comparable state of reduction of the iron as can be seen from analyses of basalts from Oregon [cf. Kennedy 1931, p. 64, and the average of 6 given by Daly 1933, p. 201] the Deccan [Washington 1922, p. 797], Iceland, Spitzbergen and Franz Joseph Land [see Wolf 1931, pp. 934—35] and dolerites from the Karroo [Daly and Barth 1930, p. 101]. The amount of potash present in basalts which in other ways are similar to the original Skaergaard magma is usually two or three times that in the Skaergaard magma. However, among the analyses of Hebridean basalts there is one having a lower potash value and another only slightly higher [Bailey, Thomas, etc. 1924, p. 24, Anals. III and V] and from the rest of the North Atlantic Tertiary Igneous province there are also basalts with a low potash content, for example, in the Faroes [see Walker 1936, p. 890] Iceland [Holmes and Harwood 1918, p. 196] and West Greenland [see Wolf 1931, pp. 932]. As with the degree of reduction of the iron, so with the amount of potash, examples are to be found which are similar to the original Skaergaard magma, although a general review of the literature shows them to be rare.

The trend of differentiation in the Skaergaard magma is apparently in an unusual direction. In trying to understand the reason for this it will be useful to inquire first, to what extent the trend occurs in other supposed cases of crystallisation differentiation of basalt and, secondly, whether the Skaergaard trend, which undoubtedly results from strong crystal fractionation, is peculiar because of the small differences in composition which have just been pointed out, or whether other basalt magmas, of more average composition, undergoing the same strong crystal fractionation, would give the same results.

(b) The Effect of Assimilation of Acid Gneiss on the Composition of the Skaergaard Differentiates.

The layered series has been formed by the accumulation, at the bottom of the magma, of the minerals successively separating during cooling and it has been shown that this process has dominantly controlled the changes in composition of the successive residual magmas, at any rate until the latest stages of solidification. In the upper border group and to a less extent elsewhere the rocks are locally heterogeneous due to hybridisation with acid gneiss inclusions. This hybridisation process took place approximately in the position in which the rocks are now found, and it can not have affected significantly the composition of the changing Skaergaard magma. It is possible and even likely, that there was another type of mixing involving complete assimilation of acid gneiss to give a homogeneous mixed magma. If this took place then the composition of later crystal fractions must have been affected by the composition of the assimilated gneiss. Definite evidence for this is difficult to obtain but we can be certain that it has taken place to some extent as acid gneiss inclusions, in various stages of hybridisation, are found in all the Skaergaard rocks except the upper ferrogabbros and later differentiates. It has been shown that these inclusions became completely liquid, and that the acid magma so formed was extremely viscous (p. 197). The available experimental evidence on silicate melts (Greig 1927) shows, with reasonable certainty, that acid magma is not immiscible with basic and thus some small part, at least, of the re-formed acid magma would become mixed with the basic to give a homogeneous liquid. The field evidence shows that mixing of the two was a slow process due to the high viscosity of the regenerated acid magma. If convection currents existed in the Skaergaard intrusion they should have aided the mixing process.

The amount of complete solution of acid gneiss in the Skaergaard magma can only be estimated very roughly. One method of doing this is to compare the total quantity of the various constituents in the various differentiates with the amounts estimated to have been present in the original magma. The total amount of P_2O_5 in the various differentiates is estimated as being greater than that in the original magma (see Table XXXV) and it has been suggested that the excess may have come from the incorporation of acid gneiss (see p. 220). Our knowledge of the total bulk of the layered rocks with a high P_2O_5 content is, however, too slight for any great weight to be attached to this deduction. The amount of the Skaergaard granophyric differentiates probably gives an upper limit to the amount of acid gneiss completely incorporated. Some of this, as we shall suggest below (p. 308), may well have come from the original Skaergaard magma as a result of strong fractionation, but it is possible that the whole has come from the solution of acid gneiss. The total amount of acid differentiates from the Skaergaard intrusion has been roughly estimated as of the order of one per cent. The amount of acid gneiss which has been completely incorporated into the magma is probably of the same order or less, and this value will be used as a working hypothesis. This estimate does not include the considerable amount of acid gneiss now present as inclusions and as hybrid material, as this never became completely dissolved in the magma.

The complete incorporation of acid material into the Skaergaard magma during the protracted period of cooling complicates the interpretation of the late stages of the differentiation but it may safely be assumed that it did not significantly modify the early and middle stages. Thus the observed differentiation of the Skaergaard magma while ninety-five per cent. solidified is considered to be essentially the result of strong crystal fractionation acting on a basic magma of the composition of the chilled marginal gabbro.

(c) Former Views on the Course of Fractional Crystallisation of Basalt Magma.

Evidence for the course of fractional crystallisation of basalt magma comes from direct observation of the interstitial residuum of basalts. C. N. Fenner has maintained that this is the only reliable evidence [1929, pp. 243 and 249—50] but others have considered that the segregation veins of basalt and hypabyssal basic rocks also indicate the trend of crystallisation differentiation. Evidence is also provided by microscopic observation of the sequence of crystallisation and trend of zoning in basic plutonic rocks. The field relations and mineral composition of the normal series of calc-alkaline igneous rocks, when interpreted in the light of experimental investigations of simple silicate systems in the laboratory, has also led many petrologists—foremost among whom is Bowen—to consider that the series is the result of fractional crystallisation of basalt. A misuse of Bowen's inspiring work seems to have crept into much petrological thought. Thus the abundance of basalt magma and the importance of the calc-alkaline series among igneous rocks, combined with a strong predilection for the theory that the differentiation of igneous rocks is due to fractional crystallisation, has led to the calc-alkaline series of igneous rocks being regarded as evidence for the course of fractional crystallisation of basalt. This step cannot be regarded as legitimate.

In discussing the composition of the interstitial residuum of basalts Fenner states [1929, p. 242]:

"Out of a large number of sections of plateau basalts that I have studied, the great majority show features indicating that the residual liquid, up to complete crystallisation, was highly ferriferous, and that it deposited magnetite, pyroxene and felspar."

Somewhat similar observations have also been made by others, especially Newton and Teall [1897, p. 487—88] and Krokström [1932, p. 000]. Fenner concludes that the result of fractional crystallisation of basalt is to produce "a final liquid rich in certain constituents, particularly iron and alkalies, and poor in silica, magnesia, and lime".

These observations were considered by Fenner in the light of what was then known of the thermal behaviour of the minerals concerned, and he wrote [1929, p. 227]:—

"The same principle that indicates to us that in the crystallisation of the felspars anorthite enters in high ratio in the solid phase and albite and orthoclase in the liquid, likewise indicates that in the crystallisation of pyroxenes the ratio of one or more of the refractory members (pre-eminently magnesium metasilicate) is increased in the solid phase, and that of the more easily fusible members (especially ferrous metasilicate, probably with some excess of Fe_2O_3, and accompanied by the soda-iron pyroxene acmite) is increased in the residual liquid.

....In the scheme of differentiation by crystallisation one of the requirements was to formulate an acceptable method by which anorthite diminished until it finally became almost nil, while alkali felspars correspondingly increased. To meet this requirement the principle of the crystallisation of solid solutions has been appealed to, and this explanation has been generally hailed as most successful. It is also requisite, however, that the silicates of magnesia, lime and iron should be eliminated almost entirely in order to attain the composition of the most siliceous rocks by this process, and we find that the same principle that indicates the alkaline felspars should increase in ratio also indicates that iron compounds should increase."

By combining the experimental results obtained up to 1928 with the evidence from the rocks themselves, Bowen [1928, pp. 79—80] reached the conclusion that fractional crystallisation of basalt magma did not produce enrichment in iron. It is understandable that the actual composition of the interstitial residuum of basalts should be in doubt since it is difficult to examine microscopically and almost impossible to analyse chemically. Bowen [1930, p. 453] stated clearly his agreement with the idea that, in a series of rocks resulting from fractional crystallisation, there should be enrichment in iron relative to magnesium, and he points out that almost any calc-alkaline series of rock shows this effect. Thus the ratio $\frac{FeO}{MgO}$ in the Katmai series of rocks is 1.2 at fifty-four per cent. SiO_2; 1.6 at fifty-eight per cent. SiO_2; 3.0 at sixty-six per cent. SiO_2 and approaches infinity at seventy-six per cent. SiO_2. Commenting on this he writes [1930, p. 453]:

"The concentration of FeO relative to MgO in rocks that are by hypothesis successive mother liquors from crystallisation is plain enough. Fenner points out that rock series such as the Katmai series show that concentration of alkaline felspar relative to calcic felspar that the doctrine of successive mother liquors demands. In both the felspathic crystallisation series and the ferromagnesian crystallisation series, the requirements of crystallisation-differentiation are met in the Katmai rocks and in any other sub-alkaline rock series. The crystallisation theory requires the progressive concentration noted within each of these two crystallisation series taken individually, but it specifies nothing regarding the concentration of the two series relative to each other. In other words, theory has nothing to say as to whether the net result will be an increase in the absolute concentration in residual liquors of the iron-rich molecules of the one series or the alkalic-siliceous molecules of the other series. Indeed, the one or the other is bound to dominate. Which is it? In urging that residual liquors from crystallisation should suffer absolute enrichment in iron Fenner tacitly assumes that the effect shown by the ferromagnesian series would dominate the effect shown by the felspathic series."

In reply to this Fenner [1931, p. 549] gave an argument which he considered indicates that there should be enrichment in iron as fractional crystallisation of basalt proceeds:—

"As bearing upon the general relations of iron compounds to other components of magmas, and the possibility of their removal by crystallisation to give residues poor in iron, the authors (Bowen, Schairer and Williams) argue that in the competition in a crystallising magma between the two principal solid solution series— plagioclase felspars and iron-bearing metasilicates (e.g. pyroxenes)—it is not possible to tell with our present knowledge which side will prevail in the diminishing residual liquid. Strictly, this is true, but there are certain general principles which control the result and give a suggestion of what is likely to occur.

The competition is chiefly between the lowest melting plagioclase felspar— albite—and the lowest melting pyroxene, which may be either ferrous metasilicate, ferrous calcium metasilicate, or some still more complex mineral compound. The final result, with respect to the relative amounts of plagioclase and iron-pyroxene left in the liquid, is dependent upon two physical properties of the two competing minerals— the temperature of melting and the molar heat of melting (or solution). For albite these properties are almost exactly known from Dr. Bowen's determinations; for the iron-pyroxenes experimental investigations encounter the difficulty that is always present in dealing with iron minerals—the difficulty of preserving the state of oxidation unaltered; therefore, our knowledge of these properties is not at all exact. What little evidence we have hardly indicates that the physical properties of the iron-bearing pyroxene are so different in these respects from those of albite that we should expect to obtain as a result of crystallisation the overwhelmingly large amounts of albite and the vanishingly small amounts of iron-pyroxene (or its equivalent) that we find in many rhyolites and granites.

This has a bearing of some importance upon the theory of crystallisation-differentiation. The view adopted in the theory is that iron metasilicates are almost completely eliminated by crystallisation, as this assumption is necessary to obtain a magma of rhyolitic composition, but this can hardly be said to rest upon either experimental or theoretical evidence except of the vague sort explained above, which, as far as it goes, seems to point in the contrary direction."

The views of Bowen and his collaborators on this matter have not been changed significantly by their recent investigation of several

systems containing iron. They continue to maintain [e.g. 1938, pp. 407—
11] that the normal calc-alkaline series of rocks which does not show
absolute enrichment in iron but only enrichment of iron relative to
magnesium, is due to fractional crystallisation because the series is
in harmony with the results of fractionation in experimentally determined systems. These views have been widely accepted. Fenner has
continued to question Bowen's views and to present petrological evidence
opposed to it [1938, pp. 367—400].

The evidence provided by segregation veins for the course of
fractional crystallisation of basalts and dolerites, recently summarised
by Kennedy [1933, pp. 242—47], gives no indication of an iron-rich
residuum. The quartz dolerites tend to have sporadic segregation veins
of granophyric composition while the olivine-rich basalts and dolerites
have quartz-free veins of a general trachyte or syenite composition.
Fenner accepted the usual view that these veins were derived from
the basic magma, but considered that they were not produced by crystal
fractionation since they differed in composition from the observed iron-
rich residuum of basalts. He considered that the latter must have resulted
from fractional crystallisation as there could be no opportunity, during
the cooling of a basalt, for other processes to intervene. The problem
of the segregation veins will be considered later when an attempt is
made to explain why they apparently give evidence conflicting with
that from the residuum of basalts and the iron-rich rocks of the Skaergaard intrusion.

There remains to be discussed the evidence provided by the composition of the late crystallising material of coarse dolerites and gabbros
which has been to some extent estimated by microscopic methods. From
a study of two tholeiitic dolerites of the Midland Valley of Scotland,
Walker [1930, pp. 368—76] came to the conclusion, although on slender
evidence, that:—

"Both tholeiites support Dr. Fenner's contention that iron is concentrated in
the residuum of basaltic rocks; for the chlorophaeite—a mineral rich in iron—is
of late formation in both rocks and so is the ilmenite in the northern exposures of
Kinkell quarry. The author believes, however, that this late crystallisation of minerals
rich in iron left a still later residue much poorer in that constituent—a constituent
whose presence in any abundance would raise the refractive index of the glass far
above the recorded values."

To set against this observation there is the more usual conclusion
which is the one, for example, which Bowen reaches on the basis of
Asklund's careful work on a noritic gabbro from the Stavjö region [Bowen
1928, p. 80—83], that, in a gabbro which has undergone crystal fractionation due to zoning, the late, and last, materials to crystallise are
of dioritic and granitic composition respectively.

Now that Bowen and his collaborators have devised a method of dealing with melts containing iron, it should be possible to make rough experiments with average basalt to determine whether, in fact, there is enrichment in iron in successive residual liquids when fractional crystallisation takes place, but, until this is done, indirect evidence is all that is available. From the foregoing review of this evidence it is apparent that much of it is conflicting while the quotations from papers by Bowen and Fenner show that theoretical deductions, based on our present knowledge of the thermal behaviour of silicate melts, also leads to no consensus of opinion.

(d) The Early Stages of the Fractional Crystallisation of the Skaergaard Magma.

The formation of the layered series has resulted in very strong fractional crystallisation of the Skaergaard magma. The fact that a certain amount of the contemporaneous liquid was entrapped in the interstices of the primary precipitate should not have affected the trend in composition resulting from fractionation but should only have reduced the amounts of the later liquids. For descriptive purposes, the continuous process of fractional crystallisation of the Skaergaard magma has been divided, into three parts, the early, middle and late stages (see pp. 235—36). During the early stage seventy-five per cent. of the magma solidified but the result of the strong fractionation only produced rocks and magmas which belong to the eucrites and olivine-gabbros as these are normally defined. Theoretical considerations indicate that there should be much less change in the composition of rocks and magmas during the early stages of fractionation than during the late, but it is of interest that the current definitions of eucrites and olivine-gabbros are such that the rocks produced by strong fractionation of the Skaergaard olivine-gabbro magma, while seventy-five per cent. of the original liquid solidified, are to be included under these names. With less complete fractionation eighty or ninety per cent. of the Skaergaard magma would have solidified before the rock types or residual liquids moved out of the range of normal eucrite and olivine-gabbro. The evidence from the Skaergaard intrusion suggests that some of the variation in normal eucrites, gabbros and basalts is due to fractional crystallisation. This will be accepted in a general way by many petrologists but it is evident that the trend of fractional crystallisation, as deduced from the Skaergaard case, is not one of increasing richness in silica; indeed the silica content decreases slightly as differentiation proceeds. The main changes during the early stage of the Skaergaard differentiation are increase in the amounts of ferrous and ferric iron, soda, potash, TiO_2 and P_2O_5 and decrease in the amounts of MgO and Al_2O_3, while CaO remains approximately constant. The nature of the original basalt will presumably have an effect on some of these trends. The trends shown by the Skaergaard rocks which are most likely to be true for the majority of basalts undergoing similar fractionation seem to us to be the increase in FeO, Fe_2O_3, Na_2O, K_2O, TiO_2, P_2O_5 and the decrease in MgO. The behaviour of Al_2O_3 and CaO in the Skaergaard case, and the slow rise of Na_2O may be connected with the tendency for basic plagioclase to be abundant in the early fractions, a feature which has given the Skaergaard magma a similarity to the Porphyritic Central type of the British Isles.

(e) The Middle Stages of the Fractional Crystallisation of the Skaergaard Magma and Comparisons.

The results of strong fractionation of the Skaergaard magma are unambiguous throughout the early and middle stages of the solidification, that is while approximately 95 per cent. of the original magma solidified. It is only after this that the field evidence is more difficult to interpret and doubt introduced by the solution of acid gneiss in the magma. During the middle stages the successive fractions show the same trend as in the early stages but the changes are more rapid. During this stage the amount of total iron in the solid fraction eventually exceeds thirty per cent. while MgO falls to below one per cent. There is also a sudden increase in the amounts of TiO_2, Fe_2O_3 and P_2O_5 which has been described and explained (pp. 227—30). The deduced composition of the successive residual magmas follows roughly the same course as the solid fraction but is without the abrupt changes which take place in the latter. After about ninety-five per cent. has solidified, the magma contains its maximum of total iron-oxides—about twenty-five per cent.—which is less than the maximum in the rocks. The middle stage ends abruptly with the residual magma changing its trend towards a granophyric composition (Fig. 46). The trend in composition of the solid fraction persists in the direction of enrichment in iron a little longer than the liquid, then the layered series ends and the distinction between solid fraction and residual magma cannot be made with precision.

The increase in iron and decrease in magnesium during the Skaergaard differentiation is mainly due to gradual enrichment of the ferromagnesian minerals in iron although the amount of ilmenite and magnetite has a subsidiary effect. For olivines and pyroxenes it has been shown from a study of zoning, and comparisons between phenocrysts and groundmass minerals that the later and lower temperature fractions are richer in iron (For olivines see references in Vogt [1921], and Tomkieff [1939]. For pyroxenes see Barth [1929], Tsuboi [1932], Krokström

[1932], Kuno [1936], Fenner [1938] and discussion on pp. 255—61). Less direct evidence of the enrichment of the ferromagnesian minerals in iron as fractionation proceeds is also afforded by certain petrogenetic studies such as those by Foslie [1921] and Lombaard [1935]. Experimental evidence for the olivines and for the $MgSiO_3$—$FeSiO_3$ clinopyroxenes [Bowen and Schairer 1935] likewise shows that there is enrichment in the iron compound with fractionation. This point can be said to be definitely established. On the other hand, that fractional crystallisation of basic magma produces absolute enrichment in iron, is not at present widely believed.

During fractionation of a basic magma it is to be expected that the plagioclase will become more albite-rich; and it has long been realised that the composition of the modal plagioclase (and to some extent the composition of the normative plagioclase as well) is of prime importance in the classification of igneous rocks, and in attempts to trace their petrogenesis. However, it is not usually recognised that the composition of the ferromagnesian constituents may be used in the same way as that of the plagioclase and, failing determinations of the composition of the ferromagnesian minerals in the rock itself, that the ratio of iron to magnesium given by the bulk analysis of the rock may be used to serve roughly the same purpose. In figure 58 the iron-magnesium ratios of all recent analyses of Greenland basalts and all analysed Mull basalts (except one showing abnormally high K_2O) are plotted against the percentage of albite in the normative plagioclase. The graph would have been more significant if the actual rather than normative composition of the plagioclase had been plotted but normative values have been used as the actual composition has not usually been assessed accurately, no doubt because the crystals are zoned. The distribution of points for these basalt analyses shows that, with increasing acidity of the plagioclase, the iron-magnesium ratio rises fairly steadily. It seems highly likely that those basalts represented by points towards the top right hand part of the graph are late differentiates produced by crystal fractionation. On the same graph has been plotted the values for the successive Skaergaard liquids, while the first ninety-per cent. solidified. The general position of the points representing Greenland and Mull basalts is obviously related to the curve for the Skaergaard magma, and this suggests that, in the magma reservoir from which the basalts were derived, differentiation had proceeded as in the Skaergaard intrusion. Since ratios and not actual amounts have been plotted in this figure the deduction may fairly be made that fractional crystallisation has been operative, but this evidence does not preclude the possibility of the simultaneous dilution of the magma series by assimilation of rocks, having a composition which would not materially affect these ratios.

The trend of differentiation by crystal fractionation should be shown by increasing iron content of the ferromagnesian minerals just as it is shown by the increasing albite content of the plagioclase. If both these changes indicate the same direction of differentiation then the chances of this being the real trend, and due to fractional crystallisation, are much increased. Petrologists have usually relied on increasing silica percentage as the best guide to the sequence of rocks in a dif-

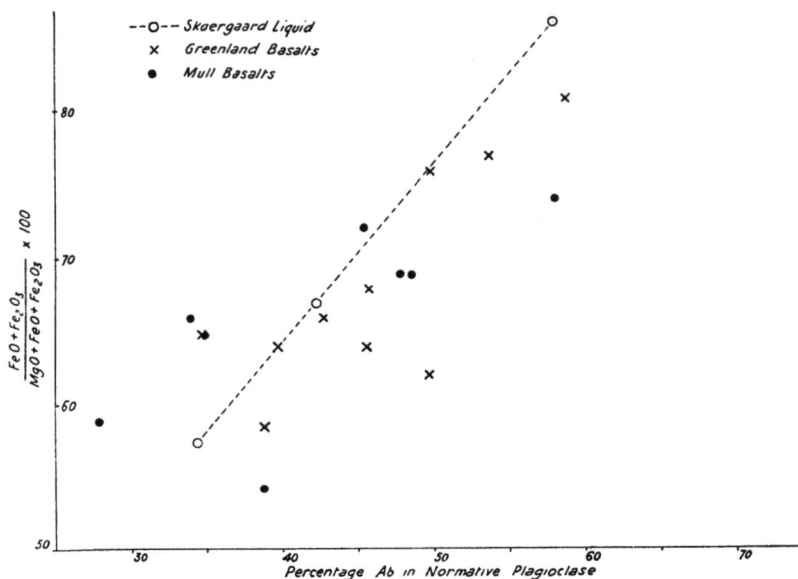

Fig. 58. Iron-magnesium ratios of all recent analyses of Greenland basalts and all analyses of Mull basalts (except one with high K_2O) plotted against the composition of the normative plagioclase.

ferentiation series, although various authors, e. g. Brammall [1933, p. 100] and Krokström [1937, p. 271—72] have expressed doubts of the value of this procedure and it has already been suggested from the evidence of the Skaergaard intrusion that the practice should be critically reconsidered (p. 226). In tracing the differentiation of normal basic magma it would seem, from the preceding discussion, that the composition of the plagioclase and the iron-magnesium ratios should be a better guide than the SiO_2 percentage, and that they should provide better values against which to plot variation diagrams. The fact of a rock series having a steady and correlated increase in iron-magnesium ratios and in the albite content of the plagioclase does not preclude the possibility of extensive assimilation of material while the differentiation was in progress, a point which should be strongly emphasised. In attempting to distinguish between the effect of assimilation. and differentiation the absolute amounts and not ratios should be our guide.

The difficulty of finding analyses of rocks with which to compare the more iron-rich differentiates of the Skaergaard intrusion has already been pointed out, and it is clear that rocks of these compositions are rare among accessible crustal rocks. A scrutiny of basalt analyses, however, shows that with increasing iron content the amount of magnesium usually falls. This is the effect to be expected if the basalts are derived from a source undergoing crystal fractionation such as occurred

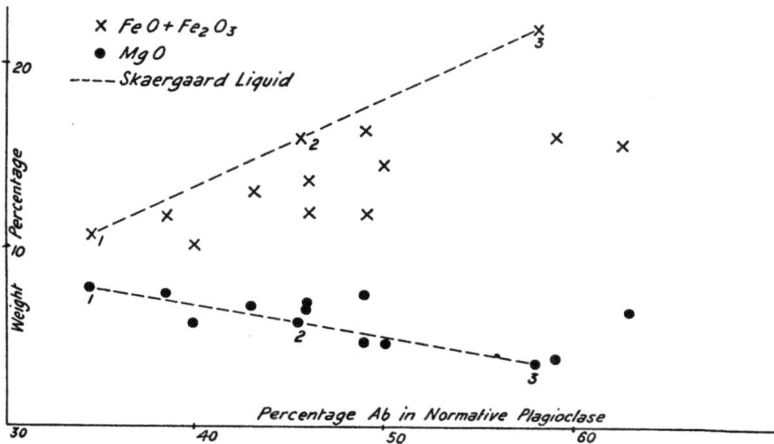

Fig. 59. Percentages of MgO and FeO + Fe$_2$O$_3$ for all recent analyses of Greenland basalts, and for the Skaergaard liquid, plotted against composition of the normative plagioclase.

in the Skaergaard intrusion. To illustrate this point more fully the iron and magnesium contents of the analysed Greenland and Mull basalts have been plotted against the proportion of albite in the normative plagioclase (Figs. 59 and 60). In calculating the composition of the normative plagioclase it has been assumed that all the soda is present as albite even if calculation of the standard norm shows a little as nepheline. On the same graphs the amounts of ferrous oxide and magnesia in the Skaergaard liquid during the early and middle stages of the differentiation have been plotted. The information given in figure 58 proves that roughly the same graphs as figures 59 and 60 would have been obtained had the analyses been plotted against the iron-magnesium ratio. The Greenland basalts show a fairly definite increase in total iron oxides and decrease in magnesia with increasing content of albite in the normative plagioclase. The changes are approximately the same as for the estimated successive residual magmas of the Skaergaard intrusion while the first ninety-five per cent. was solidifying. The same thing is shown for the three Plateau basalts of Mull and also for the Porphyritic Central basalts if allowance is made for general reduction in the amounts of both iron and magnesium due to abundant plagioclase. The Non-Porphyritic Central basalts depart far from the curve for iron in the Skaergaard rocks. The same trends are also indicated, though less clearly, by the usual variation diagram of the Mull Normal magma series over the range of the silica percentages of the typical basalts,

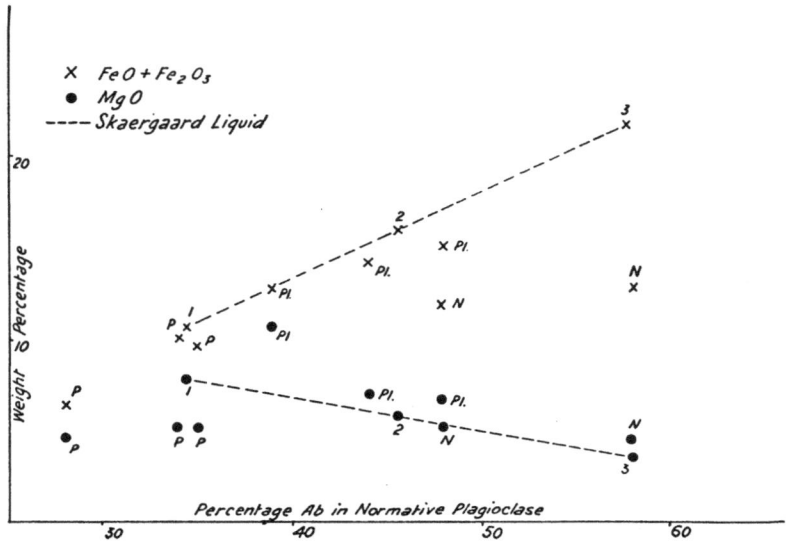

Fig. 60. Percentages of MgO and FeO + Fe₂O₃ for Mull basalts and Skaergaard liquid plotted against composition of normative plagioclase (P = Porphyritic Central Basalt; Pl = Plateau Basalts; N = Non-Porphyritic Central Basalts).

i.e. from forty-five to fifty per cent. [Bailey, Thomas, etc., 1924, fig. 2]. The total amount of iron remains steady or increases slightly while that of magnesium falls. The same is shown (where the abundance of plagioclase is not so great as to mask the effect) on the variation diagram of the Porphyritic Central magma series [Richey, Thomas, etc., 1930, fig. 7]. This suggests that, over this limited range, increasing silica percentage does roughly indicate successive stages of differentiation. We consider, however, that for basic rocks variation diagrams plotted against normative plagioclase, are a more reliable indication of the trend during fractionation.

Other Plateau basalts have not yet been considered along these lines but they would probably show the same thing as the Greenland and Mull examples, namely, moderate enrichment in total iron, and strong decrease in magnesium, as differentiation proceeded. If it be accepted that the sequence of differentiation of basalts may be deduced from the composition of the plagioclase and from the iron-magnesium ratios, then it appears that differentiation of the Greenland and Mull basalts, which may be regarded as typical of those of the North Atlantic Tertiary Province, was in the same direction as that of the Skaergaard magma, although not reaching such extreme types. The more iron-rich basalts of the province correspond to the early middle stages of the fractional crystallisation of the Skaergaard magma.

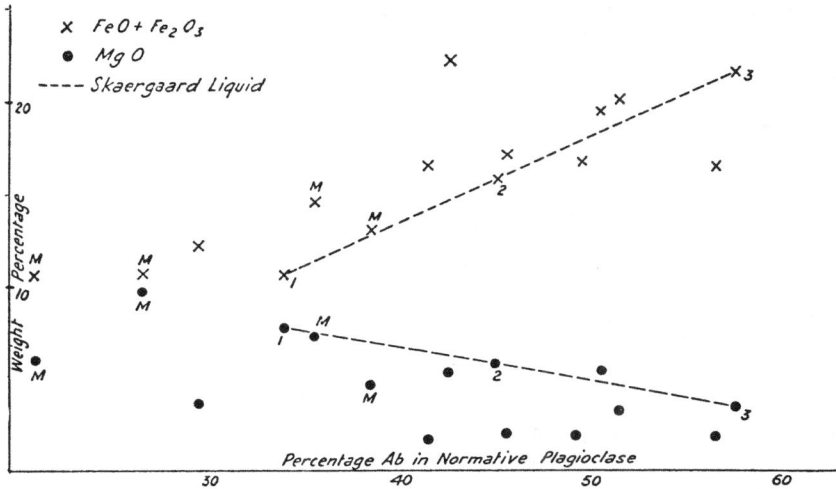

Fig. 61. Percentages of MgO and FeO + Fe_2O_3 for the analyses of Breven and Hällefors dyke rocks [Krokström, 1932 and 1936], and for the Skaergaard liquid, plotted against the composition of the plagioclase (M = rocks classed by Krokström as marginal).

Among basic hypabyssal and plutonic rocks there is also evidence that the Skaergaard trend of differentiation has been followed to some extent. Fenner [1929, pp. 228—31] draws attention to two sills, the Palisades Sill of New Jersey and the differentiated sheet of Electric Peak, Yellowstone Park, which show decided increases in the iron-magnesium ratio in the presumed later differentiates, and also a slight absolute enrichment in iron oxides. The differentiated basic mass described by Foslie [1921] also shows, at a late stage in the differentiation, a slight absolute enrichment in iron. The pyroxenes of the Bushveld complex show a tendency to become more iron-rich as the associated plagioclase becomes richer in albite [see Lombaard 1935] and, although existing bulk analyses of the rocks do not show absolute enrichment in the presumed later differentiates, yet it seems likely that they would do so if a fair sampling could be carried out to arrive at the composition of the average rocks.

In the Breven and Hällefors dykes which have been described by
Krokström [1932 and 1936], there are types closely similar to the more
iron-rich Skaergaard rocks. If the same rocks that Krokström has plotted
against silica percentage [1936, pp. 229—30], are plotted against the
composition of the normative felspar, the variation diagram (Fig. 61)
shows striking similarities with that of the Skaergaard liquid, and
it may be that the rocks, in order of increasing acidity of the plagioclase,
represent successive differentiates due to crystal fractionation. With
this interpretation the rocks which Krokström classifies as of marginal
type are the early differentiates and those which he classifies as central

TABLE XXXX.

	Gabbro excluding olivine gabbro	Olivine Gabbro	All Gabbro	All Basalt	Basalt as named by authors	Plateau Basalts
SiO$_2$...............	49.5	46.5	48.2	49.1	48.8	48.8
Total Iron oxides....	8.6	9.8	9.1	11.8	11.7	13.4
MgO	6.6	8.9	7.5	6.2	6.0	6.7

are the later. Among plutonic and hypabyssal rocks, the only examples,
so far as we are aware, which are closely comparable with the Skaergaard iron-rich rocks are these from the Hällefors and Breven dykes.

Consideration of the average compositions of certain basic rocks
which have been computed by Daly shows an interesting fact which
seems to be significant in this connection. The amount of iron in the
averages of various types of basalt is greater than that in the averages
of various types of gabbro of approximately the same silica percentage
(see Table XXXX showing information taken from Daly's book [1933,
p. 17]). It seems likely that many gabbros are, partly at any rate, early
crystal fractions mechanically separated from residual liquids, while
basalts on the other hand represent the result of sudden chilling of
magma (with of course some crystals if the basalt be porphyritic). If
this is so then the higher iron, and lower magnesium, content of basalts
compared with gabbros suggests that basic magma, on the average,
has undergone the same trend of differentiation as is shown by the
early stages of the Skaergaard magma.

A complete review of the basic igneous rocks has not been attempted,
but it seems safe to conclude that, among plutonic and hypabyssal basic
masses, rocks similar to the middle-stage differentiates of the Skaergaard intrusion are extremely rare while among basalts they are fairly
common, although not reaching such an extreme limit. The evidence

from the Plateau basalts suggests that differentiation along the same lines as in the Skaergaard intrusion, and due to similar fractional crystallisation, has taken place in the magma reservoirs from which they were derived. Differentiation of basalt magma by fractional crystallisation seems therefore to have been a significant petrogenetic process, although the resulting changes in composition have not been in the direction commonly believed but towards enrichment in iron, as suggested by Newton, Teall, and Krokström, and as firmly believed by Fenner.

By analogy with the behaviour of relatively simple silicate melts in the laboratory, and by appeal to the rocks themselves, Bowen has sought to prove that the normal calc-alkaline series of igneous rocks has been derived from basalt by crystal fractionation. However, fractional crystallisation acting on the Skaergaard magma, which was close to normal basalt in composition, has led, in the middle stages, to iron-rich rocks whose composition precludes them from belonging to the normal calc-alkaline series; and the same trend occurs among the North Atlantic Tertiary basalts and probably elsewhere. Thus it seems necessary to postulate either:—

1) that the slight differences in the composition of the Skaergaard magma from average Plateau basalt were sufficient to cause an unusual trend of differentiation and that similar effects have been concerned in the production of some of the Plateau basalts of the North Atlantic Tertiary province, or

2) that the trend of fractional crystallisation of basic magma which is shown by the Skaergaard intrusion is the normal one.

If the latter view is accepted then it must be assumed that in the production of the normal calc-alkaline series of igneous rocks some other process as well as, or in place of, straightforward crystal fractionation has been at work. These two alternative possibilities will be dealt with after the late stages of the Skaergaard differentiation have been briefly discussed in the next section.

(f). The Late Stages of the Fractional Crystallisation of the Skaergaard Magma.

After passing through an iron-rich stage, the differentiation of the Skaergaard magma suddenly takes a new direction. This is the stage represented by the unlaminated layered rocks and the transgressive basic, and acid, granophyres. In the formation of these rocks the method by which the crystal fraction separated from the residual liquid was not the straightforward sedimentation process which gave the main layered series. Selective gravitative fractionation, (p. 282), as has happened in various well known sills, seems to have been responsible for the formation of the unlaminated layered series. Still later, there seems to have been filter-press action which has produced a final, highly-acid residuum now found as sills and veins cutting all the earlier differentiates. The mechanisms involved in the production of these later differentiates cannot be demonstrated so clearly as those which have produced the layered series but they are of kinds not uncommonly encountered in other intrusions.

The trend of differentiation of the Skaergaard intrusion is best shown by the triangular diagrams already given (Figs. 45 and 46), the three corners of which represent normative amounts of the more refractory, medium refractory and less refractory components. During the early and middle stages of the differentiation, both the composition of the rocks, and the deduced composition of the successive residual magmas, trend away from the more refractory components mainly towards the medium refractory components—the iron compounds—but also slightly towards the less refractory—the acid felspars. Ninety-five per cent. of the whole magma was crystallised during these two stages. While the remaining five per cent. solidified the magma takes a different trend towards the acid-felspar corner; iron-rich basic granophyres are first formed and these later give place to acid granophyres of fairly common type. The final acid granophyre is soda-rich and potash-poor, like the original Skaergaard magma. The change in trend between the middle and the late stages is shown in the diagram as very abrupt. This is because the total amount of felspar in the upper ferrogabbros is reduced relative to the ferromagnesian constituents, an effect perhaps due to the high density of the liquid preventing some of the light felspar crystals collecting at the bottom of the liquid as layered rock. It seems likely that such a selective gravitational effect took place but it has not been possible to prove it. Even if some process of this sort is exaggerating somewhat the abruptness of the change from the middle to the late stages it seems certain from the trend shown by liquids 1 to 3 and 5 onwards, that the change is sharp. Although the method by which fractionation has taken place in the late stages is less clear, and although some of the effects may be the result of assimilation of acid gneiss, it appears that fractionation ultimately produced a magma of general granitic composition. This is the result expected by Bowen and others for the strong fractionation of basalt magma.

The sequence of compositions during the differentiation of the Skaergaard magma cannot be explained in terms of the thermal diagrams at present available. Certain features, such as the disappearance of olivine in the middle gabbro stage and the formation of the hedenbergite-clinoferrosilite solid solutions by inversion from a higher temperature mineral, find close analogies in the investigated systems; these suggest that soon a more complete explanation, based on the experimental work, will be possible. The trend of the differentiation may, however, be explained roughly in terms of the solubilities or melting point ranges of the different phases. The available evidence suggests that iron-olivines and hedenbergite pyroxenes separate from silicate melts approaching basic magma in composition at a higher temperature than albite, orthoclase and quartz, and at a lower temperature than the magnesium-rich pyroxenes and olivines. The trend of fractional crystallisation of the Skaergaard magma is in harmony with this, since the sequence is: first, rocks consisting mainly of basic plagioclase and magnesium-rich ferromagnesian minerals, then, rocks consisting of medium plagioclase and iron-rich ferromagnesian minerals, and finally, acid rocks made up of acid-plagioclase, quartz and orthoclase with but little ferromagnesian minerals.

The occurrence of granophyric segregation veins in many dolerite sills and dykes seems to be analogous to the late stage formation of granophyre in the Skaergaard intrusion. However, there is one important difference; the granophyric segregation veins are derived from quartz-dolerites, whereas the original Skaergaard magma was moderately rich in olivine. From olivine-basalt the usual segregation veins are quartz-free and related to syenites in composition.

Fenner considered that granophyric differentiates from basic magma could not be the result of straightforward fractional crystallisation as this process should give the iron-rich material which he had found in basalts [1929, p. 243]. Walker [1930, pp. 368—76] found evidence that both types of differentiate occurred in succession, as in the Skaergaard intrusion, and he considered that both resulted from fractional crystallisation. Among the products of fractionation of the Skaergaard magma there is a conspicuous iron-rich stage and, if it be conceded that the granophyric and syenitic segregation veins of dolerites and basalts also result from fractional crystallisation, it is of importance to enquire, why in these cases the iron-rich stage is usually missing.

A partial explanation of these difficulties may be found in Bowen's suggestion [1928, p. 91] that "the products of crystallisation at intermediate stages will be represented by certain layers of the crystals". In the case of plagioclase it is well known that the intermediate layers are frequently medium to acid felspar and, although less conspicuous, it is established that the zoning of pyroxenes and olivines is such that more iron-rich and lower temperature solid solutions are formed marginally. Bowen has suggested that the layers representing the intermediate stages of crystal fractionation should be dioritic in composition,

20*

while we should expect them to be iron-rich. It should not be difficult
to decide the matter by careful observation and measurement under
the microscope. If the intermediate stages of fractional crystallisation
are present in the body of the rock, then the segregation veins should
only represent the latest stages, and should be comparable with the
one or two per cent. of granophyric material of the Skaergaard intrusion.
The total bulk of the segregation veins in any particular dolerite or
basalt is relatively small and is roughly comparable with the proportion
of the late stage granophyre material found in the Skaergaard intrusion.

Another possibility, also dealt with by Bowen [1928, p. 100], is
that rapid cooling may largely prevent fractionation by zoning. Instead
of zoned plagioclase, pyroxene and olivine, the crystals may be more
or less homogeneous and of mean composition. If we postulate that
rapid cooling produced crystals whose mean composition corresponded
roughly with the composition of the rocks formed during the early
and middle stages of the Skaergaard differentiation, the iron-rich stage
would be eliminated and only the latest fraction, consisting of granophyric or syenitic material, would be found.

The original Skaergaard magma was of olivine-basalt composition
and yet it has given rise to a small amount of granophyric differentiate.
Some of this, but probably not all, may be derived from the complete
assimilation of acid gneiss into the magma from whence it was later
re-developed, with a modified composition, by fractional crystallisation.
Bowen has shown [1928, pp. 70—74] how early crystallisation of olivine
and its removal from the liquid should result in enrichment of the
residual liquid in silica so that a quartz-bearing, late residuum might
develop. In the Skaergaard intrusion there has certainly been the
strong fractionation postulated by Bowen and it seems likely that
some of the late granophyre material has been produced in this way.
The low state of oxidation of the iron should, however, produce the
reverse effect, for ferrous silicates contained in the olivine and pyroxenes
are produced instead of magnetite, and this reduces the amount of
silica in the liquid. If the iron had entered entirely into the pyroxene,
more SiO_2 would have been used up than is actually the case in the
Skaergaard intrusion, where some enters the orthosilicate, olivine.
Whatever the detailed causes may be, strong fractionation of olivine
basalt magma seems to have produced, in the Skaergaard intrusion, a
final granophyric differentiate as Bowen has maintained it should.
Bowen has also suggested that rapid cooling and weak fractionation
may produce a late syenitic residuum as in the segregation veins of
olivine-basalts. The Skaergaard intrusion, which was slowly cooled,
gives no definite evidence on this point but it seems a likely postulate
providing the separate existence of the iron-rich stage is suppressed by

the rapid cooling. The late residuum of the Skaergaard magma is soda-rich and if it had contained less quartz, as a result of less strong fractionation or of less assimilation of acid gneiss, the resulting rock might have had a composition similar to the fayalite-quartz-syenites which occur in considerable abundance elsewhere in the Kangerdlugssuaq region.

In summing up our views on the origin of the late differentiates of the Skaergaard intrusion it must be admitted that they have not the same pretence to certainty as those on the main layered series owing to doubt about the amount of solution of acid gneiss in the basic magma. Up to the end of the middle stage it is unlikely that assimilation of acid gneiss had any significant effect, yet there is a small amount of interstitial granophyre present in the rocks. At this relatively early stage, filter-press action could have produced a granophyre. We consider it likely that granophyre would have formed by strong fractionation without assimilation of acid gneiss and that such assimilation did not radically affect the nature of the late differentiates but only the amount. The late acid granophyre of the Skaergaard intrusion cannot be considered as exactly homologous with the granophyric segregation veins of quartz-dolerites as the original Skaergaard magma was of olivine-dolerite composition. We prefer to regard it as due, at any rate in part, to the strong fractionation of the olivine-gabbro magma, a view in harmony with Bowen's work. The syenitic segregation veins of olivine-basalts and the granophyric segregation veins of quartz-dolerites are probably not closely comparable with the late Skaergaard differentiates, as they are the result of cooling without fractionation (except by zoning), while 90 per cent. solidified; only after this did a sudden separation of the residual liquid from the solid phases occur.

(g) Reasons for regarding the Trend of the Skaergaard Differentiation as typical of the Fractional Crystallisation of Average Basalt.

Strong fractionation of the original Skaergaard magma in the early and middle stages leads from a fairly common type of olivine-gabbro to iron-rich types and then to ferrogabbros. The exact mechanism by which the later rocks, amounting probably to less than five per cent. of the whole, were produced is open to doubt, although evidence has been given which suggests that they also are mainly the result of crystal fractionation. The early and middle stages of the fractionation are, however, of greater interest partly because it is clear that they are the result of strong fractionation, and partly because the trend is in an unexpected direction. An attempt will now be made to decide whether the trend during the early and middle stages is the one which average basalt magma would take as a result of strong fractionation. The evidence already given on this point would be reasonably conclusive if the Skaergaard magma had had, in every way, the composition of average basalt; actually it differs from this in the more reduced state of the iron and in the lower amount of potash.

There seems to be no likelihood that the low potash content of the Skaergaard magma influenced the fractionation so that ferrogabbros were produced, but the low state of oxidation of the iron, on first consideration, might well have had this effect. In a magma rich in ferric iron, magnetite would be likely to be a solid phase formed in considerable quantities and if this separated in the early stages it might be expected that the middle stage of enrichment in iron would be inhibited. In the Ardnamurchan Memoir, Thomas definitely inclines to the view that the Mull normal magma series is the result of fractional crystallisation and he writes [1930, pp. 94—95]:

"Such information as is furnished by those rocks with low silica percentages from the Tertiary province may be summarised by the statement that olivine, iron-ores, and basic plagioclases are often the least soluble magmatic constituents, and that with falling temperature they begin to crystallise at a very early stage....

Viewed in this light and assuming that the Plateau magma is the parent stock, the Normal Mull Magma Series presents no great difficulties. It is obvious that if we initially abstract iron-ores, olivine and basic-plagioclase, and follow this by the abstraction of other crystalline phases such as augite and less basic plagioclase, we can produce just such a progressive variation in magmatic composition as is represented by the sequence of increasingly siliceous magma-types."

Among basic igneous rocks early crystallisation of iron ore is frequently to be deduced from the textures but, quite as frequently, it appears to have crystallised during the middle and late stages of the solidification. Even if iron ores begin separating early it does not necessarily follow that they will form in sufficient quantities to reduce the iron in the successive residual liquids and prevent the formation of ferrogabbro magma. It has already been shown (pp. 228—230) that ilmenite and magnetite both became primary phases at about the middle gabbro horizon, when the concentration of Fe_2O_3 in the magma was about three and a half per cent., and the concentration of TiO_2 about two and a half per cent. This is interpreted to mean that a fairly normal basic magma is saturated with magnetite when the Fe_2O_3 content is about three and a half per cent. and saturated with ilmenite when the TiO_2 content is about two and a half per cent. With lowering temperature and changing composition of the Skaergaard liquid due to fractionation, the amount of TiO_2 in the liquid when saturated with ilmenite decreased slowly at first but, in the late stages, the decrease was fast. Under similar conditions the amount of Fe_2O_3 in the magma when saturated with magnetite, increased steadily, reaching as much as five and a half per cent. towards the end of the middle stage but falling rapidly in the latest, acid granophyre stage.

This behaviour of the magnetite and ilmenite suggests that in a basic magma these minerals will be among the first to crystallise if the amounts of Fe_2O_3 and TiO_2 exceed about three and a half and two and a half per cent. respectively; if under these amounts there will be no primary precipitation until these values are attained. Supposing a basic magma were rich in Fe_2O_3 and TiO_2, then magnetite and ilmenite should be precipitated early. This will cause a reduction in the amount of ferrous iron in the magma since this is a constituent of both these minerals. The general trend of the crystallisation differentiation should not, however, be materially affected, as early precipitation of ferrous iron as magnetite and ilmenite will only cause the olivines and pyroxenes to be more magnesium-rich and therefore to be abundant in the early solid fractions. In this way the composition of the early solid fractions will be changed but the composition of the residual liquid should trend towards, and eventually reach, the trend taken by the Skaergaard magma.

For three reasons the above argument must be regarded as much simplified: first, magnetite and ilmenite are both solid solutions; secondly, the amount of ferric iron and titanium occurring in other minerals of the rocks, especially the clinopyroxene, will effect the result; and thirdly, it appears, from the Skaergaard example, that the solubility of magnetite and ilmenite is sensitive to variation in the composition of the magma. It is not maintained that the results of fractional crystallisation will not be modified to some extent by the state of oxidation of the iron. Greater oxidation of the iron should produce early, solid fractions relatively enriched in iron ore and magnesium-rich ferromagnesian minerals, and this would probably affect the proportion of felspar in the magma in the middle and late stages. The lime content of late residual magmas might also be expected to be modified. Nevertheless a state of greater oxidation of the iron should not greatly modify the trend in composition of the residual magma, and an iron-rich middle stage, having the general composition of the ferrogabbros, would be expected. Thus we are strongly of the opinion that average basalt magma, subjected to the same degree of crystal fractionation as took place in the Skaergaard intrusion, will follow the same trend of fractional crystallisation, and give rise to solid and liquid fractions of ferrogabbro composition.

It is widely believed among petrologists that fractional crystallisation of basalt magma is an important petrogenetic process. Rocks of ferrogabbro composition are, however, rare and it is necessary to

account for this if our views on the trend of fractionation of basalt
are to be acceptable. In the first place the amount of residual magma of
ferrogabbro composition, even with strong fractionation, is relatively
small; therefore, material of this composition would not be expected
to be abundant. Furthermore strong fractionation of basic magma,
from the first stages of crystallisation, as occurred in the Skaergaard
intrusion, is probably rare. Such fractionation requires slow cooling of
a considerable mass under tranquil external conditions. Whether the
process is due to convection currents or vertical sinking of crystals, it
is likely that strong fractionation from the earliest stages of crystallisation,
will produce layering of some kind. The number of basic masses showing
layering is small, and for this reason we believe strong fractionation of
basic magma at high levels in the crust has been rare. Fractionation of
basic magma by filter-press action, which is probably an important
process in some situations, can only take place when about three-
quarters of the rock has solidified, and it will therefore give a different
result from fractionation of the Skaergaard type. Fractionation began
in the Skaergaard intrusion when the first high temperature crystals
formed and, from then onwards, it continued uninterrupted. The ferrogabbros were produced by fractionation of this kind and it seems likely
that filter-press action would not produce a somewhat different result,
a matter which will be considered in a later paper.

The two respects in which the Skaergaard magma differed from
average basalt, namely, the low potash, and the low state of oxidation
of the iron, probably had the important effect of reducing the viscosity
of the magma (see pp. 269—70). Although these differences in composition, in our view, would not affect greatly the trend in composition
of the residual magmas if fractionation took place, yet they may have
been the chief factors making the fractionation possible. Thus strong
fractionation, by the mechanism which occurred in the Skaergaard
intrusion, requires low viscosity of the magma and it may be that the
low potash and high ferrous-iron content were the factors which reduced
the viscosity so that the process could occur. Without these constituents
the magma might have been unable to develop convection currents or
allow direct sinking of crystals, and cooling would have taken place
without the strong fractionation of the Skaergaard kind. The composition of Plateau basalts indicates that differentiation along the Skaergaard trend has occurred in the deep magma reservoirs from which they
have come. The large size and slow cooling of such magma reservoirs
would probably allow differentiation of the Skaergaard type even if the
viscosity were not quite so low as that of the Skaergaard magma.

The evidence presented suggests that strong fractionation of average
basalt magma should lead to iron-rich types and then to rocks of ferro-

gabbro composition. In our view the rarity of ferrogabbros is not due to their production from a magma of unusual composition but because uninterrupted, strong fractionation of a basic magma, such as took place in the Skaergaard intrusion, is rare. The normal calc-alkaline series of igneous rocks, which is often regarded as resulting from fractional crystallisation of basalt, is not the sequence which is shown by the Skaergaard differentiation; thus we consider that some other process as well as, or in place of, fractional crystallisation has been responsible for it.

(f) The Origin of the Normal Calc-Alkaline Series of Igneous Rocks.

The most commonly accepted alternative to the fractional crystallisation theory for the origin of the normal calc-alkaline series of igneous rocks is some form of hybridisation between basalt and granite material with a limited amount of fractional crystallisation of any hybrid magma that may be formed. In the triangular diagram, figure 62, the values of MgO, FeO and total alkalies have been plotted for:- (1) the successive Skaergaard liquids, (2) all analysed hypabyssal and extrusive rocks of Mull and (3) Daly's average basalt, andesite, dacite and rhyolite [1933, pp. 9—28]. The diagram resembles the earlier triangular diagrams of the Skaergaard rocks, figures 45 and 46, in that MgO, the significant constituent of the more refractory minerals is at one corner, while FeO, the significant constituent of the medium refractory minerals is at another corner and total alkalies, the significant constituent of the less refractory minerals is at the third corner. This modification of the earlier type of diagram has been used since it necessitated less calculation and it is desirable to plot all examples rather than selected ones. The new diagram shows the same kind of trend for the Skaergaard magma as the earlier. For the Mull rocks there is considerable scattering but the general trend is represented by the shaded area. The almost straight line obtained from Daly's averages of basalt, andesite, dacite and rhyolite, is taken as defining the trend of the normal calc-alkaline series. All the Breven and Hällefors analyses [Krokström 1932 and 1936] have also been plotted in a similar way (Fig. 63); on the whole they show less scattering than the Mull analyses. The figures show the result of fractional crystallisation of the Skaergaard magma in terms of these three constituents and, on the basis of the arguments just given, this is regarded as roughly the trend of strong fractionation of average basic magma. The rocks of the Hällefors and Breven dykes show a trend comparable with that of the Skaergaard magma and therefore, they are considered to represent the results of strong fractionation. Some of the Mull rocks show a tendency to enrichment in iron and this is also regarded as the result of crystal fractionation; others show an approach to the trend of the normal calc-alkaline series. In terms of the constituents considered, the linear relationships shown by the averages of normal calc-alkaline rocks may be interpreted as due to mixtures of granite and basalt in various proportions. Such a

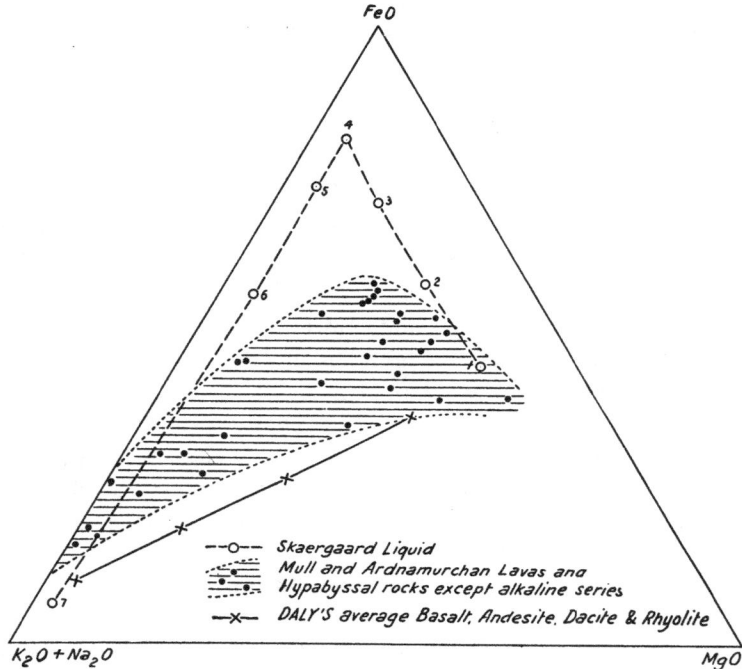

Fig. 62. Analyses of Mull and Ardnamurchan, lavas and hypabyssal rocks (except alkaline series) together with the Skaergaard liquids, and Daly's averages for basalt, andesite, dacite and rhyolite, plotted on a triangular diagram showing MgO, FeO and total K_2O and Na_2O.

linear relationship for three particular constituents of the rocks might, by chance, have been the result of crystal fractionation; therefore this curve, in itself does not indicate that the rocks are mixtures although it is in harmony with that view. On the other hand evidence has been given that the trend of fractional crystallisation of basalt on this particular diagram is approximately along the two sides of the triangle followed by the Skaergaard magma, and this is markedly different from the course shown by the normal calc-alkaline rocks which is along the third side of the triangle. We consider the fact that the normal calc-alkaline trend is markedly different from the one resulting from fractional crystallisation of basalt, gives considerable support to the usual alternative theory, that the calc-alkaline series is essentially the result of hybridisation between the two end materials, granite and basalt.

With this conception of the significance of the graphs, the area representing the hypabyssal and volcanic rocks of Mull may be considered. The points distributed near the top right-hand side of the area

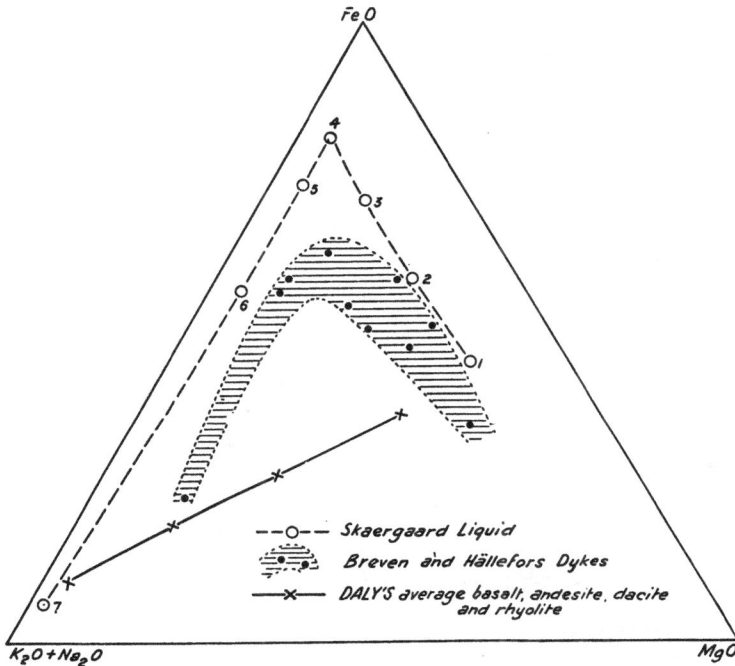

Fig. 63. Analyses of Breven and Hällefors dyke rocks plotted on similar diagram to figure 62.

suggest that in the magma basin from which the rocks were derived, basalt magma had been fractionated as in the Skaergaard intrusion but that a less extreme limit was reached. The points near the lower side suggest that the rocks which they represent were produced by mixing of granitic material with differentiates of the basalt at various stages. The points near the top left hand side of the area suggest that the rocks represented were produced, either by mixing of granite material with middle differentiates of the basic magma, or as late differentiates of the basic magma comparable with the basic and acid granophyres of the Skaergaard intrusion. In the Mull Memoir, Bailey and Thomas tentatively suggested that the normal magma series might have resulted from assimilation [1924, p. 31] but Bowen [1928, pp. 75—78] gave evidence for rejecting this view and replacing it by the idea of fractional crystallisation. In the Ardnamurchan Memoir, Thomas definitely inclined

to Bowen's view that the normal magma series was produced by crystal
fractionation and not by assimilation of granitic material, believing, as
we have mentioned (p. 310) that iron ores as well as olivine and basic
plagioclase were the minerals of the early solid fraction. The variation
diagram of the Tonalite-Monzonite series of Ardnamurchan differs in

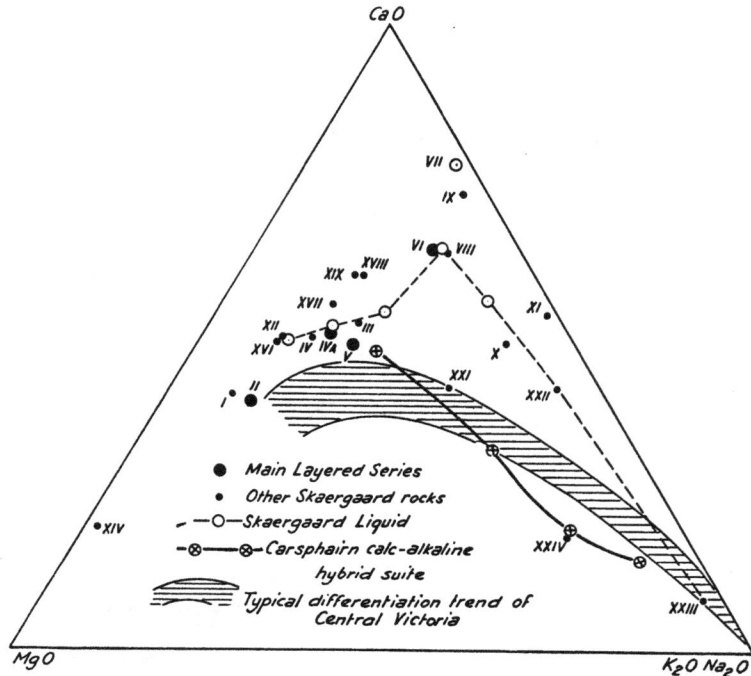

Fig. 64. Skaergaard analyses, and other rock suites, plotted on diagram showing
CaO, MgO and total K₂O and Na₂O.

tertain ways from that of the normal magma series although the general
crend of the curves is the same. Thomas considered that there were
good reasons for believing that this magma series was produced by
hybridisation. Our view that hybridisation as well as crystal fractionation
has contributed to the production of many of the Mull rocks is thus not
directly opposed to the views of the authors of the Mull and Ardnamurchan Memoirs, who, on the whole, have envisaged both possibilities
and have inclined only tentatively to one or the other. It is mainly
those who have considered these rocks from a theoretical point of view
who have maintained strongly that they are the result of crystal fractionation.

    Other triangular diagrams which have been used to show the trend
of differentiation have also been tried for the Skaergaard rocks. Thus in
figure 64 the Skaergaard rocks have been plotted on a diagram showing

CaO, MgO and total alkalies and on the same diagram are given plots of the Carsphairn calc-alkaline hybrid suite [Deer, 1935] and the typical differentiation trend of Central Victoria [Edwards, 1938]. Again the Skaergaard trend differs from the other two while the Carsphairn calc-alkaline hybrids have a resemblance to the differentiation trend of

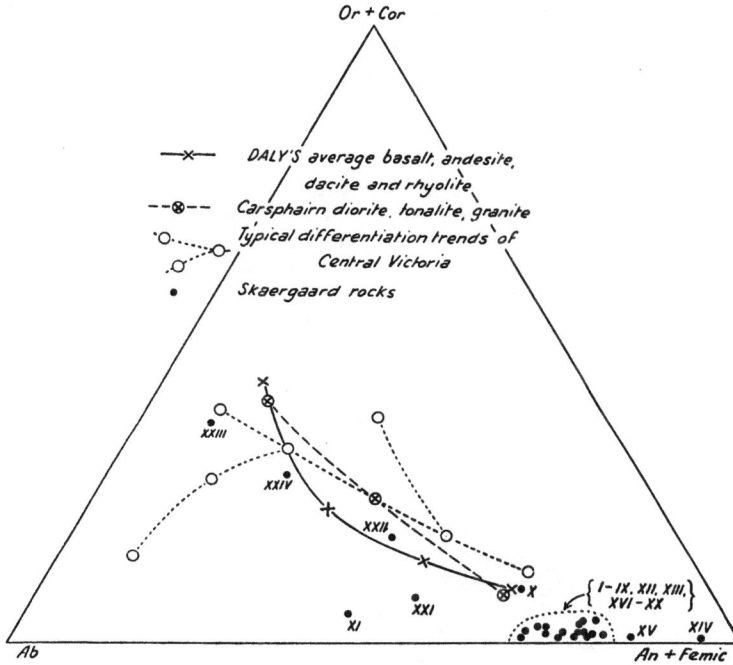

Fig. 65. Skaergaard analyses and trends of other rock suites on a triangular diagram of normative compounds.

Central Victoria. In figure 65 normative constituents are plotted in a way that has been widely used. On this diagram the points resulting from the fractional crystallisation of the Skaergaard magma, while ninety per cent. solidified, are all clustered together, and only the late stages show similarities with the trend of Daly's average calc-alkaline rocks and the other two series plotted.

The diagrams just given are not as useful for the present purposes as those especially devised to show enrichment in iron compounds. In figure 66 the same results as are presented in figures 62 and 63 are shown by plotting normative constituents. In this case only selected analyses of the Mull, Breven and Hällefors rocks have been used but the justification for this procedure is given in the previous figures where no arbitrary selection was made. On a similar diagram, figure 67, analyses of three differentiated sills are presented. The Palisades Sill of New

Jersey shows a similarity with the Skaergaard case but the Lugar Sill [Tyrrell, 1916] and the Shiant Sills [Walker, 1930] do not. In the two latter cases there is slight enrichment in the ratio of the iron compounds to the magnesium and lime compounds but no absolute enrichment. In these two examples the differentiation is apparently a selective gravitational effect due mainly to sinking of olivine. This should tend

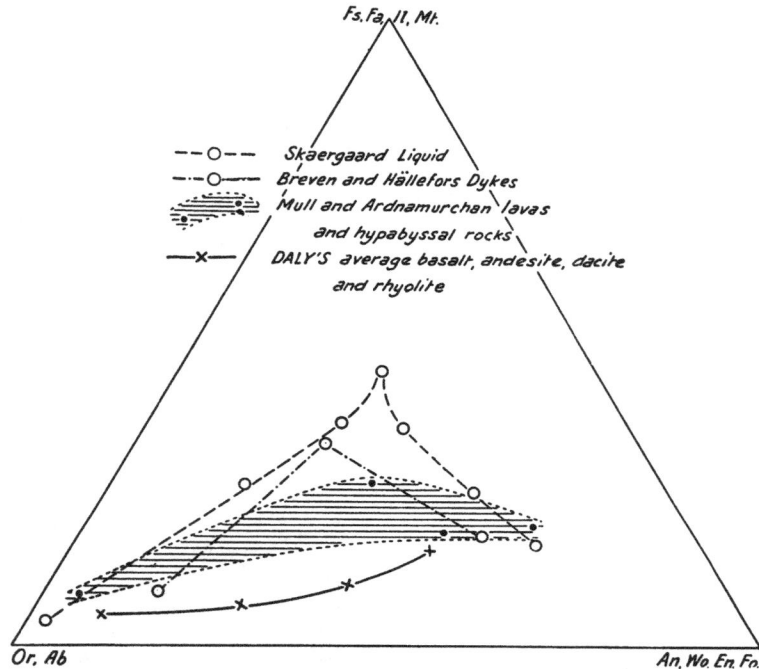

Fig. 66. Mull and Ardnamurchan, lavas and hypabyssal rocks, compared with the trend of the Skaergaard liquid and other rock series.

to produce a straight line differentiation but other factors, which we will not attempt to evaluate, are no doubt involved.

In the last diagram of the series (Fig. 68) has been plotted the Carsphairn calc-alkaline suite which has been described by one of us [Deer 1935] and regarded as the result of hybridisation. It follows closely the trend of Daly's average basalt, andesite, dacite and rhyolite and thus, as far as the grouping of the constituents in this diagram is concerned, the normal calc-alkaline series is the same as a suite believed to be hybrid. Analyses of the Glen More ring dyke have also been plotted. These do not follow the trend which we expect for the crystal fractionation of basalt but rather the trend which is believed to be the result of hybridisation. In view of the careful work done on the Glen More ring dyke [Bailey etc. 1924 and Koomans and Kuenen 1938], and in view

of the straight line graphs obtained by plotting the Lugar and Shiant
Isles sills, we shall not do more than suggest that this way of considering
the analyses favours, on the whole, the hybridisation hypothesis for
the Glen More ring dyke rocks.

Hybridisation by incorporation of basic rock in granite or grano-
diorite magma can frequently be observed and the mechanism of the

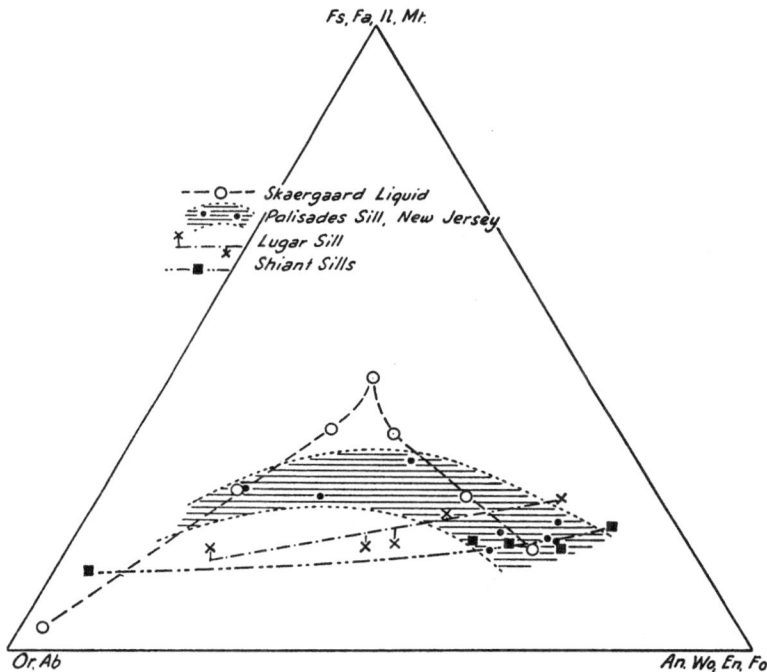

Fig. 67. Analyses of Pallisades, Lugar and Shiant sills compared with the trend of
the Skaergaard liquid.

process has been discussed in detail by Bowen [1928, pp. 175—223].
Such a process has been suggested by many authors as a method by
which certain diorites and other intermediate rocks have been produced.
In a general paper on the origin of rocks by contamination Nockolds
[1934] has expressed his belief that hybridisation between a granite
magma and a basic rock has been the most important process by which
the intermediate rocks were formed. His reasons for considering the
process of primary importance in the production of the intermediate
rock seem to be based on the fact that evidence for this process, at the
present stage of our knowledge, is more clearly to be seen, and also that
the process fits into his idea of contrasted differentiation. There are
many difficulties in the way of the latter theory and neither of these
lines of evidence is convincing. Intermediate rocks might also be produced

by assimilation of acid rocks by basic magma, a view advocated by Daly and others. The process has occasionally been seen in an arrested stage among the rocks available to observation as, for instance, in the Skaergaard intrusion (pp. 185—199) and in cases quoted by Daly [1933, pp. 428—36] although in some of the latter the interpretation is open to doubt. The physico-chemical aspects of the process whereby basic magma

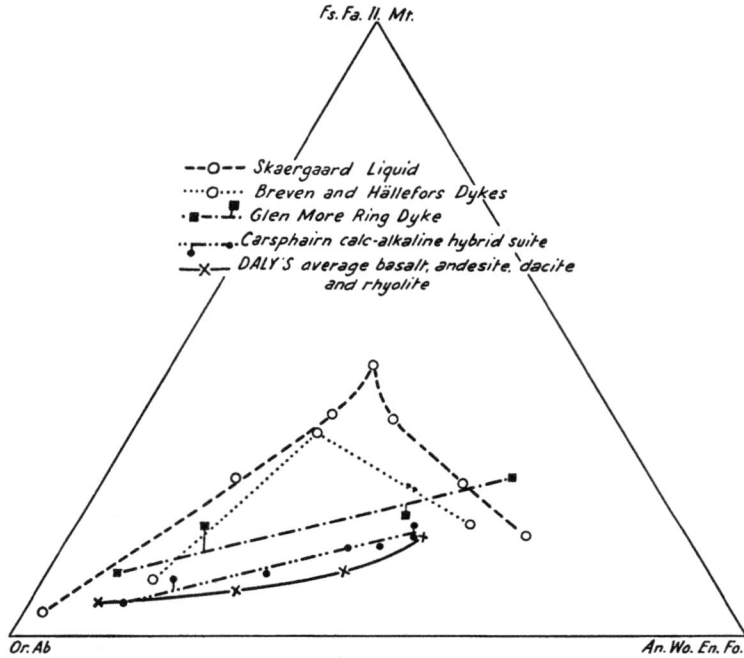

Fig. 68. Carsphairn calc-alkaline hybrid suite and other rock series compared with the trend of the Skaergaard liquid.

might incorporate acid rocks has been considered in some detail by Bowen [1928, pp. 187—88 and 199—201] and the possibility of this being able to take place cannot be doubted. The problem is rather whether such mixing of acid rocks and basic magma is an important process in the Earth's crust. It will probably be always difficult to prove this view since textural evidence of the former acid rock is completely destroyed by the process. There is, however, a line of attack on this problem, besides that based on bulk chemical analyses of rocks: the intermediate rocks, formed by the incorporation of solid basic rock in acid magma, must usually have been produced at the low temperatures of crystallisation of the acid magma, while the intermediate rocks formed by incorporation of acid rock in basic magma must usually have formed at the higher temperatures at which basic rock is liquid; it is possible

that some of the minerals developed in these two different environ-
ments will differ, and afford a clue to the conditions of the hybridisation
process. For instance, augite-bearing intermediate rocks as found in
Mull and in the Skaergaard intrusion may be the result of the incorporation of acid crustal rocks in basic magma, while the dominantly
hornblende-bearing intermediate rocks may represent the results of the
lower temperature hybridisation of solid basic rock by granite magma.
This is a matter we shall take up in another place.

Most petrologists will agree that from a physico-chemical point of
view the incorporation of acid rocks in a basic magma is a possibility,
but a difficulty in the way of accepting it as a significant petrogenetic
process in the earth's crust is to understand how the mechanical mixing
could be brought about. If convection currents may exist in a relatively
small intrusion like that of the Skaergaard, then with greater probability
they may exist in larger masses of basic rock, in existence for longer
periods of time. If convection currents are a feature of deep reservoirs
of basic magma they provide a method by which effective mixing of
basic magma and sialic crustal rocks may take place. At the same
time it should be pointed out that the high viscosity of the re-
formed acid magma (which is proved by field observation in the
Skaergaard case) will in all circumstances hinder the process. In another way convection may be important; as Holmes has clearly shown
[1931] it provides a means whereby heat, liberated by the crystallisation of basic rock, can be transported to the upper part of the
magma reservoir and be made available there for melting of sial
material. In this way he has envisaged the production of uncontaminated granite magma.

Great quantities of olivine-basalt, allied to the Plateau basalt magma
of Mull, have been available for extrusion at certain times and places
in the igneous history of the world. At other times and other places
great quantities of a more acid tholeiitic basalt, comparable with the
Non-porphyritic Central basalts of Mull, have also been available. In
Mull these two types have been regarded as giving rise to two distinct
magma series. The authors of the Mull Memoir [1924, pp. 30—31] considered that the Plateau magma type holds a parental position in Mull
petrology and they have considered how the change from this to the
Non-porphyritic Central type could be brought about. They did not
accept the hypothesis that straightforward crystal fractionation could
effect the change, basing this conclusion on the alkaline character of the
last products of consolidation which occur as segregation veins in the
basalts. Although this evidence does not now seem to us adequate, since
fractionation in deep magma reservoirs may well have been different
from that taking place in a rapidly cooled lava, it led them to suggest

tentatively that the Non-porphyritic Central type was developed from the Plateau magma type by assimilation of acid rocks.

At a later date Bowen [1928, pp. 175—77] discussed this problem and graphically determined the possible compositions of material that would have to be added or subtracted to change average Plateau basalt to the Non-porphyritic Central type. The material to be subtracted has a composition which could be calculated into basic plagioclase, 53 per cent., olivine, 27 per cent., diopside, 7.5 per cent., and magnetite 4.5 per cent. According to Bowen these are the minerals which observation shows to be the early crystallising constituents of Plateau basalt. It should, however, be noted that the olivine to be subtracted consists of forsterite and fayalite in the proportions of sixteen to eleven, which is more iron-rich that would have been expected for early olivine crystals. The material which would have to be added has a very remarkable composition, and as Bowen says, it is doubtful whether any mixture of known igneous and sedimentary rocks would be of this composition. One of its remarkable features is the high percentage of FeO. On these grounds Bowen favoured the hypothesis that the change of Plateau magma into Non-Porphyritic Central magma was brought about by crystal fractionation and not by assimilation; this is the view adopted in the Ardnamurchan Memoir [1930, pp. 95—96].

For the reason that "there are not sufficient amounts of ultra-basic rock known to represent the complementary magmas of the more acid types" Kennedy [1931, pp. 67—83] subsequently rejected Bowen's suggestion and put in its place the hypothesis that "early separation and concentration of olivine and basic plagioclase would lead on one hand to an under-saturated olivine-rich basaltic magma (Hebridean Plateau type) and on the other hand to a more acid differentiate, the degree of acidity depending on the stage of the differentiation attained". Kennedy's conception seems to be that early olivine and basic plagioclase, seperating in one part of the magma, might be mechanically transferred to another part; in this way part of the magma would be impoverished in these constituents while another part would be enriched. In this paper he also maintained that basalts which are close to the Non-porphyritic magma type in composition have a more universal distribution over the world than the Plateau basalt magma [1931, p. 67]. In a later paper Kennedy [1933, p. 255—56] again considered the relation between the olivine-basalt magma type (= Plateau Magma type) and the tholeiitic magma type (= the Non-porphyritic Central type) and concluded, in view of the results of further investigation, that: "there is no evidence to show that either magma is a derivative of the other, nor is it possible to relate them both to some common parent. ......
we are justified, therefore, in regarding them, for the purpose of petro-

genesis, as primary magmas". In his latest paper [1938] much the same views are expressed but there is also a mention of the possibility of hybridisation [1938, p. 38].

The relationships between the Plateau magma type and the Non-porphyritic Central magma type, using these terms in their wide sense, is one of the important problems of petrology towards the solution of which the study of the Skaergaard intrusion seems able to provide critical evidence. The result of strong fractionation of the Skaergaard olivine-basalt magma was to produce iron-rich types while the greater part of the magma solidified, and, during this stage, there was not an increase but a slight decrease in silica content. Thus crystal fractionation of olivine-basalt apparently does not by itself lead to the Non-porphyritic Central magma type but a magma of that composition would be produced if hybridisation with granitic crustal rocks took place simultaneously with, or after, the crystal fractionation. Direct dilution of typical olivine-basalt by re-fused granitic material would reduce the amount of iron as well as magnesium in the mixture. However, if the fractionation of the olivine-basalt is postulated to have taken place to some extent in the direction of enrichment in iron, as has taken place in the Skaergaard case, then hybridisation with granitic material should be capable of yielding the tholeiitic magma type. The conception of convection currents in the deep-seated reservoirs of basic magma provides a plausible mechanism whereby mixing of the basic magma and the re-formed acid magma might take place. The significant world distribution of these two magma types—olivine-basalt magma appearing both in the continental and oceanic regions although more frequently in the latter, while the tholeiitic magma is practically restricted to the continental areas where a sial crust presumably exists—is also in favour of the idea that hybridisation with sialic rocks is a factor in the derivation of one type from the other. The foregoing argument has been developed using the grouping of basalts into the types which were adopted for Mull. If these were further subdivided, taking into consideration the iron-magnesium ratio, which the differentiation trend of the Skaergaard magma indicates to be of such importance, then the hypothesis just brought forward to account for the relationship between the Plateau magma type and Non-porphyritic magma type (using these terms in the broad sense) should be capable of being stated more precisely. Further discussion will therefore be postponed until this can be done.

In the preceding paragraphs of this section new evidence from the Skaergaard intrusion has been presented which indicates that in general the normal calc-alkaline series is the result of hybridisation between granite and basalt and not the result of crystal fractionation of basalt magma. This must not be taken to mean that we believe that rock types belonging to the normal calc-alkaline series are not produced occasionally, and in small amounts, by crystal fractionation of basic magma. As Daly has frequently insisted, there are probably more ways than one in which a rock, or even rock series, may be formed. In particular we do not maintain that granite has never resulted from fractional crystallisation of basalt. In fact, in the Skaergaard intrusion a small amount of granitic material, occurring as late transgressive veins of acid granophyre, has probably been partly formed by strong fractionation of the Skaergaard magma. The conclusion to which the present work leads is rather that granites and the intermediate rocks of the normal calc-alkaline suite are not usually produced by fractional crystallisation of basalt, acting alone. It has always been difficult to explain by the latter hypothesis how it happens that the known bulk of intermediate rocks is relatively small, and there now seems to be the additional difficulty that the intermediate rocks have not the appropriate composition. Since we do not believe in the origin of the bulk of post-Cambrian granites by fractional crystallisation of basalt magma, we are drawn to believe, with Daly, Holmes and others, that they are due to remelting of the sial crust.

SUMMARY

Sections II and III. Topography and General Geology of the Region.

The Skaergaard intrusion which occupies fifty square kilometres of high mountain country on the north side of the large fjord of Kangerdlugssuaq cuts transgressively through the metamorphic complex of Pre-Cambrian age, the overlying Kangerdlugssuaq Sedimentary Series and the Tertiary volcanic rocks. The metamorphic complex consists dominantly of a somewhat granulitised acid gneiss which is poor in potash. The Kangerdlugssuaq Sedimentary Series, about five hundred feet thick in the neighbourhood of the intrusion, consists of shallow water silts, sandstones and conglomerates, the age of which, from fossil evidence, has been shown to be late Cretaceous and early Eocene. Succeeding the sediments without any significant time interval is a series of basalts with subordinate tuffs; of these 4,500 metres are still preserved in the neighbourhood of the intrusion and, when the intrusion was formed, the total thickness was probably 7,000 metres. Certain sills or thin laccolites, varying from peridotite at the base, through well-characterised spotted olivine-gabbros, to olivine-free gabbros, were injected into the lavas before the Skaergaard intrusion was produced; these are included in the estimates of thicknesses of the volcanic material. Later than the Skaergaard intrusion, and later than the middle Eocene sediments preserved further north, the coastal region of this part of East Greenland was bent into a sharp flexure, and simultaneously a dense dyke swarm was injected. The Skaergaard intrusion was affected by the flexuring process in the same way as the surrounding volcanic and metamorphic rocks, and in the southern part of the intrusion the dyke swarm is dense. The estimated age of the Skaergaard intrusion is middle Eocene.

Section IV. Field Observations on the Intrusion.

On the basis of field observations the intrusion has been divided into a layered series, forming the central part of the complex, and a border group, which forms a narrow envelope about the layered rocks

(Fig. 9). The latter is sub-divided into a marginal border group, adjacent to the steeply dipping margins of the intrusion, and an upper border group which once formed a continuous capping to the layered series, but which has now been largely removed by erosion. The marginal border group is chilled close to the contact, then follows a complex series of coarse, basic rocks showing feeble fluxion structures, indicating either upward or downward flow parallel to the walls of the intrusion. Inclusions of acid material, now granophyre or hybrid quartz-gabbros, are abundant throughout the marginal border group, while in the northern marginal border group masses of gabbro-picrite are common. The upper border group is composed of coarse quartz-gabbros, sometimes with, and sometimes without, fluxion structures, and often containing reefs or schlieren of acid material. The fluxion structures are parallel to the presumed position of the roof of the intrusion, and the reefs of acid material have the same disposition.

The junction between the border group and the layered series, called the inner contact, can be defined within a metre along the east and west sides of the intrusion. It is a surface parallel to the flow structures of the border group which, in turn, are parallel to the outer margin of the intrusion. The layered series is banked up against this surface (Fig. 3). Elsewhere the junction between the layered series and border group is indefinite.

The layered series shows conspicuously what has been called rhythmic layering; this is due to rhythmic variation in the proportions of the various minerals. The heavy and light constituents are gravity stratified. Throughout the layered series the minerals which are solid solutions undergo a continuous change, producing a variation in the series which cannot be detected without recourse to quantitative measurement; this has been called cryptic layering. Much of the layered series also shows a parallelism of the platy minerals and since this differs from the usual fluxion structures of igneous rocks it has been named igneous lamination. (These terms are defined more fully on pages 36—38.) The main part of the layered series, that which shows these various features in greatest perfection, is two thousand metres thick. Below and near the northern margin, layering is less perfect, and the rocks are distinguished as the transitional layered series since they are transitional to the border group. Above the main layered series is a discontinuous horizon of layered rocks reaching a maximum thickness of three hundred metres; these are devoid of igneous lamination and are distinguished as the Unlaminated Layered Series. (U. L. S.)

Towards the top of the main layered series in the Skaergaarden region, remarkable structures are encountered which have been named trough banding; these structures are described on page 45 et seq.

Section V. Form and Structure of the Intrusion.

The present form of the intrusion and the lie of the structures is partly the result of the flexuring which affected the intrusion subsequent to its consolidation. However, the original dips of the boundaries, layering, etc., may be estimated (see figure 14). The estimated original boundary dips indicate that the original Skaergaard intrusion had the form of an inverted cone, tilted northwards (Fig. 15); the term funnel intrusion has been used to describe this form. Evidence is given to show that the material which the intrusion has replaced was expelled upwards. Explosive action is postulated to account for the clean-cut, cone-shaped fracture which bounds the intrusion. The marginal border group varies in width in a regular way which indicates that it narrows when traced downwards from the level of the upper layered rocks. The layered rocks originally had the form of a pile of saucers which gradually increased in size upwards.

Section VI. Petrology of the Layered Series.

There is a steady trend in the composition of the various minerals forming the layered series. The plagioclase varies from An_{60} in the lowest exposed rocks, to An_{30} in the high rocks, the olivine from Fa_{36} to Fa_{97}, the clinopyroxene from a diopsidic type to a hedenbergite-clinoferrosilite solid solution. The orthopyroxene in the lowest layered rocks is approximately Fs_{40} and the composition changes gradually to Fs_{70} at a height of sixteen hundred metres in the layered series; above this it is not present. Olivines and clinopyroxenes have been separated from a number of the layered rocks; they have been analysed and subjected to detailed optical examination.

The transitional layered rocks and the lower part of the main layered series are olivine-gabbros containing a small amount of hypersthene. These are succeeded by the middle gabbros which are olivine-free except for narrow reaction rims of olivine between the pyroxene and iron ore. The upper part of the main layered series are rocks unusually rich in ferrous iron, and they have been called ferrogabbros (defined on page 98). The lower ferrogabbros contain hortonolite, iron-rich pigeonite and andesine. Iron-rich orthopyroxene is present in the lowest hortonolite-ferrogabbros but is absent in the upper, while quartz occurs as an interstitial mineral in the upper hortonolite-ferrogabbros and all higher rocks. The ferrohortonolite-ferrogabbros have more iron-rich olivine and clinopyroxene, and apatite is an important constituent. The fayalite-ferrogabbros represent this trend towards enrichment in ferrous iron carried to the extreme limit, the rocks of the purple band (p. 108) containing over thirty per cent. of total iron oxides. There is textural evidence which indicates that the hedenbergite-ferrosilite solid solution present in these rocks has inverted from a higher temperature mineral, presumably an iron-rich β-wollastonite. The unlaminated layered rocks even before erosion were of relatively small, total bulk. At the bottom of the series there are fayalite-ferrogabbros containing large fayalite crystals; upwards the fayalite-ferrogabbros pass into basic hedenbergite-granophyre (defined on page 112). The modal composition of average layered rocks is given in figure 18 while the composition of the chief minerals is presented in figure 25.

Rhythmic layering produces extreme rock types which consist of the same minerals as the adjacent, average, layered rocks although in different proportions. Textural evidence indicates that these rocks, and indeed the whole layered series, consisted at one time of an accumulation of discrete crystals surrounded by about twenty per cent. of magma filling the interstices between the grains. The precipitate of crystals has been called the primary precipitate and the material which subsequently crystallized from the magma which surrounded the primary crystals has been called the interprecipitate material (p. 122).

As the inner contact is approached the rocks of the layered series become increasingly rich in the heavier minerals and the layering changes to a rough banding. Again there is no change in the composition of the minerals but only in their proportions.

Evidence is given (pp. 125—27) that the layered series is due to fractional crystallization. The primary crystals are shown to represent a fraction successively accumulated as a precipitate at the bottom of the magma. The interprecipitate magma crystallized mainly as low temperature solid solutions of the same minerals as the primary precipitate, and they form narrow zones about the primary minerals. Despite slow cooling the primary crystals were not made over to homogeneous lower temperature solid solutions by reaction with the interprecipitate material. Phases may occur among the interprecipitate material of a particular rock which are not present as primary crystals but become so at higher horizons of the layered series.

A variation diagram for the analysed rocks of the layered series plotted against silica percentage is of no value as there is little variation in that constituent, but an approach to the ordinary SiO_2 variation diagram is obtained by plotting the oxides against height in the intrusion (Fig. 28). This diagram shows an approach to the variation in composition of the crystal fraction separating at successive moments. It does not show accurately the composition of the successive crystal fractions because of the small amount of interprecipitate material.

Section VII. Petrology of the Border Group.

The chilled marginal gabbro which is uniform round the whole intrusion, is an olivine-gabbro of a fairly common type. Evidence is given that this rock has the composition of the initial magma which filled the Skaergaard intrusion. Two analyses show that the composition of the chilled marginal gabbro resembles the Porphyritic Central magma of the British Tertiary igneous area. It differs from the latter in having a lower ratio of FeO to $FeO + Fe_2O_3$ (ie. the iron is in a more reduced state), and in having less abundant potash.

A small but integral part of the marginal border group along the western, eastern and northern margins consists of a rock having much-elongated, plagioclase crystals set at right angles to the cooling surface. The elongated felspars cannot have been orientated by magmatic flow and the evidence indicates that they were attached to already solidified material and grew inwards as the material necessary for their formation was carried to them from, the main body of the Skaergaard magma, by circulating currents.

The major unit of the marginal border group shows changes, in passing from the outside inwards, which are closely comparable with the upward changes in the layered series, although the regularity of the changes are effected by hybrid material formed from the acid inclusions. At any particular place the border rock, adjacent to the layered series, is comparable with rocks of the layered series lying just below the layered rock with which it is actually in contact. The gabbro-picrite (defined on page 162) which is found as blocks in the northern border group, is genetically connected with the other Skaergaard rocks and was apparently formed by mechanical concentration of olivine crystals. Along the south and south-east margins the marginal border group is adjacent to the upper border group. Here contamination with acid material complicates the sequence of types developed, but the essential change, on passing inwards, is from basic olivine-gabbro to more iron-rich, quartz-bearing gabbro.

The upper border group, like the high marginal border group, is partly the result of contamination with acid material. Where least contaminated it is a coarse quartz-gabbro, free from olivine. The pyroxene shows increasing richness in iron in descending in the series and, on the whole, there is increasing acidity of the plagioclase in the same direction. The rough parallelism between the changes of composition of the upper border group in descending in the series, and of the layered rocks in ascending, suggests that the upper border rocks were formed from material approximating in composition to the successive layered rocks although much contamination by acid gneiss has also taken place. Solidification of the upper border group was apparently from above downwards.

Section VIII. Inclusions.

Granophyre inclusions are found sporadically in the border group and in all the rocks of the layered series except the uppermost divisions. Consideration of the composition of these granophyres indicates that the inclusions were originally blocks of acid gneiss from the metamorphic complex. Complete re-fusion of the acid gneiss took place, but the reformed magma was highly viscous. Between the inclusion and the surrounding basic magma there is a zone of variable width consisting of extremely coarse quartz-gabbro which is hybrid in origin. A huge inclusion of spotted, sill gabbro forms the summit of Basistoppen, and another huge inclusion of acid gneiss occurs near the summit of Tinden. Small inclusions of basalt and sediment are occasionally found in the outer border group.

Section IX. Transgressive Granophyres.

Small transgressive granophyre veins in the upper and high marginal border group occur near inclusions of granophyre. On the evidence of their general mineral compositions and of their alkali content they are considered to be the acid magma formed by fusion of the acid gneiss inclusions which suffered injection into the surrounding heated rocks. Larger, transgressive, granophyre veins and a sill on Tinden have a more constant composition which is that of a soda-rich, acid granophyre. On Brödretoppen the upper border group contains a thick series of variable hedenbergite-granophyres which are believed to be later than the surrounding quartz-gabbros. These rocks are intermediate between the transgressive acid-granophyres and the basic hedenbergite-granophyres of the unlaminated layered series; they are regarded as late differentiates of the Skaergaard magma which formed after the unlaminated layered series. The acid granophyres are regarded as the result of further differentiation in the same direction.

Sections X and XI. The Successive Residual Magmas and the relationships to them of the various minerals and rock-types which have been developed.

The sequence of solidification of the different parts of the Skaergaard intrusion may be deduced from mineral compositions and the experimentally determined sequence of solidification of various solid solution series (Fig. 41). An unusual feature is that the layered series, which forms the bulk of the intrusion, can be proved to have formed from below upwards.

From the estimated volumes of the various rock types and from the deduced sequences of solidification, the composition of the residual magma at different stages during the differentiation of the Skaergaard intrusion has been calculated (Table XXXV). Layered rocks must exist below those now visible, and their average composition may be calculated on the assumption that the chilled marginal gabbro represents the composition of the initial Skaergaard magma; these hidden layered rocks (H.L.S.) are found to have an average composition corresponding to an olivine-eucrite. It is shown that all reasonable assumptions about the relative volumes of the different rock types will not significantly affect the general trend of the deduced composition of the successive residual magmas (pp. 222—224).

A composite variation diagram showing simultaneously the composition of the successive residual magmas and the layered series, is given in plate 27; this forms a basis for a discussion of the relationship between the composition of the magma and of the material separating from it. While ninety-five per cent. was solidifying the most important feature of the trend in composition of the magma is the increasing iron, and the decreasing magnesium content; these effects are due to the early separation of magnesium-rich olivine and pyroxene. The curves for P_2O_5 show the effect of a mineral suddenly changing from an interprecipitate mineral to a primary phase, and the curves for TiO_2 and Fe_2O_3 are similar. The trend in composition of the layered rocks and of the residual magmas is also shown by plotting normative components on a triangular diagram representing:— (a) the more refractory magnesium and lime components; (b) the medium refractory iron components; (c) the less refractory potash and soda felspars (Fig. 46). During the early and middle stages, that is, while the first ninety-five per cent. of the magma was solidifying, the change in composition of both the rocks and magma was mainly towards enrichment in the medium-refractory, iron compounds; the trend towards the least-refractory, potash and soda felspars only took place in the late stages, that is, while the last five per cent. of the magma was solidifying.

The change in composition of the minerals is also considered in relation to the trend in composition of the liquid; for the plagioclase, olivine and orthopyroxene series the effects are as expected from the experimental work on silicate melts. The absence of olivine in the middle gabbros is analogous to the sequence of phases separating during strong fractionation of certain mixtures in the system MgO—FeO—SiO_2. The trend in composition of the complex clinopyroxenes during crystal fractionation is from diopsidic types to a hedenbergite-clinoferrosilite solid solution. Both ortho- and clinopyroxenes were formed during the early part of the differentiation but with increasing richness in iron a

stage was reached when only clinopyroxene formed. The trend established for the Skaergaard pyroxenes is found among the pyroxenes
of basalts, and it suggests that olivine- and tholeiitic basalts are
genetically related.

Section XII. Mechanism of the Differentiation.

The fluxion structures of the border group, the igneous lamination
of the layered series, and the trough banding are regarded as evidence
of significant flow of the magma during the formation of the rocks
showing these features. The variation in size of the felspar crystals
relative to the margin, and the melanocratic rocks near the margin
suggest flow from the periphery towards the centre. The distribution
of the granophyre xenoliths also is most satisfactorily explained as the
result of powerful magmatic currents. These lines of evidence suggest
the theory of the convective circulation of the Skaergaard magma. It
is believed that downward currents were formed along the walls of the
intrusion due to the greater density of the cooled, and partly crystallized
magma which occurred there. Compensating currents are considered to
have swept across the floor of the intrusion; then to have risen at the
centre to the top of the intrusion where cooling and concomittant increase
in density again took place causing the circulation to continue. Consideration of the available data on the rise in melting point of minerals
with pressure, and on the probable thermal gradient established by
convection, show that, if a convective circulation occurred in the Skaergaard intrusion, the solid phases would separate at or near the bottom
of the circulating magma. The minerals denser than the magma, e.g. the
olivine and pyroxene, would sink through the current and be deposited
at the bottom of the liquid as the layered series. The available data
do not indicate whether the plagioclase would be more, or less dense
than the magma, but it is likely that, at any rate during the later
stages of the differentiation, the plagioclase would be less dense. The
deposition of plagioclase crystals with the other minerals of the layered
series may have been partly due to the plagioclase becoming entangled
in a numerically dense cloud of sinking, heavy minerals.

The igneous lamination is explained, as the result of orientation
of the platy minerals of the precipitate by magmatic currents; the same
hypothesis has been adopted by Grout for similar structure in the
Duluth Gabbros. The rhythmic layering must have been produced by
a rhythmic variation in the conditions of deposition, and it is suggested
that changes in velocity of the convection currents were responsible.
The trough banding is ascribed to the splitting of the convection currents
into narrow streams between which less powerful and more irregular
currents allowed the crystal precipitate to accumulate to a greater

height; thus the trough banding structures became banked up against the pile of crystals accumulated between the more powerful currents along the trough bandings.

In the sub-section on convection and the major variation it is emphasised that direct vertical sinking of crystals from the cooling upper part of the magma to the floor of the intrusion would produce the observed cryptic layering (or major variation), but such a process is not regarded as being so likely to have taken place as that involving convection. Various structural features, and the changes in the layered series relative to the margin, are not satisfactorily explained by the direct sinking hypothesis but are to be expected on the convection theory.

Since the lower marginal border group is closely comparable in composition with the layered series, it is considered to have formed in a similar way. The precipitate, however, accumulated on a steeper surface, and probably only where the velocity of the convection currents was reduced by friction with the already solidified material. The upper border group is considered to have been formed by solidification of the successive residual magmas, free from primary crystals. Its formation along the upper border of the intrusion is considered to be due to loss of heat from the stationary or almost stationary magma which would occur where convection currents had been slowed down by friction with the already solidified material. Abundance of granophyres and hybrid material in the upper border group is to be expected because the acid gneiss of the metamorphic complex especially when re-melted would have a less density than the Skaergaard magma. Crystal fractionation which gave rise to the layered series is regarded as the controlling factor in the differentiation of the Skaergaard intrusion.

The gabbro-picrite of the northern border group is interpreted as the result of direct sinking of olivine crystals before the convective circulation had been established. The unlaminated layered rocks at the top of the layered series show no evidence of magmatic currents and they were probably formed by direct sinking or floating of minerals in a thin residual layer of magma in which there was no significant convective circulation. In the formation of the main layered series, all the solid phases formed, were deposited at the bottom of the magma; this process has been named non-selective, gravitational fractionation. In the unlaminated layered series it seems that some minerals sank while others floated and this has been described as selective, gravitational fractionation (p. 282). The hedenbergite-granophyres and transgressive acid granophyres are considered to be the latest differentiates of the Skaergaard magma, and filter press action is believed to have contributed to their formation.

In a concluding sub-section the views of Becker, Pirrson and Grout
on convection are compared with those put forward for the Skaergaard
intrusion, and the structural features of other layered igneous complexes
(particularly the Ilimausak batholith described by Ussing) are briefly
compared with those of the Skaergaard intrusion.

Section XIII. The Trend of the Differentiation.

The original Skaergaard magma, as typified by the chilled marginal
gabbro, is shown to be very close to that which would be obtained by
mixing two parts of the Porphyritic Central magma of Mull and Ardnamurchan (SiO_2 percentage 48) with one part of the Non-porphyritic
Central magma (SiO_2 percentage 47). The only significant difference is
the lower state of oxidation of the iron and the lower percentage of
K_2O. However, among the Plateau basalts, types showing similar compositional features are occasionally found. It is considered that the
Skaergaard magma has certainly incorporated some acid magma from
the re-fused acid gneiss but the amount is believed to have been small
and not to have affected the trend of differentiation except in the late
stages.

The trend of fractional crystallization of the Skaergaard magma
during the early and middle stages supports Fenner's view that during
fractional crystallization of basalt there is absolute enrichment in iron.
It is pointed out that while ninety-five per cent. of the Skaergaard gabbro
magma solidified under conditions of strong fractionation, the silica
percentage of the solid fractions and of the successive residual magmas
did not rise beyond the limits of normal basic rocks, and that all the
products must be broadly classified as eucrites, gabbros and ferrogabbros.
Consideration of the analysed Greenland and Mull basalts indicates
that, in the magma reservoirs from which they were derived, differentiation has proceeded as in the Skaergaard intrusion. Among plutonic
complexes the Breven and Hallefors dykes described by Krokström
provide the only closely comparable types so far known.

The basic to acid granophyres produced during the late stage of
the differentiation, may be partly the result of the strong fractionation
of the original olivine-gabbro magma in the way suggested by Bowen,
but they must also be partly the result of contamination with acid
gneiss.

The original Skaergaard magma was very similar to normal basalt
yet it has produced iron-rich ferrogabbros by strong fractional crystallization, and this is not the result usually postulated. The low percentage of potash in the Skaergaard magma is believed to have had
no significant effect in producing the observed trend of differentiation,

and reasons are also given for believing that the unusually reduced state of the iron is not responsible for the trend to the ferrogabbros. Had the iron been in a more oxidized state, it would have only caused magnetite to form as a primary mineral at an earlier stage and, though the composition of the early solid fractions would differ from those of the Skaergaard magma, the later residual liquids would resemble the later Skaergaard liquids and be of ferrogabbro composition. The trend of the Skaergaard magma during differentiation is believed to be essentially the trend which would be followed by average basalt if subjected to the same degree of crystal fractionation.

The normal calc-alkaline series of igneous rocks is frequently considered to represent the result of crystal fractionation of basalt magma. From the evidence of the Skaergaard intrusion it appears that crystal fractionation of basalt magma leads to the ferrogabbros and not to intermediate rocks of the calc-alkaline series. This is considered strong evidence in support of the view that the calc-alkaline series of igneous rocks is, in the main, the result of the mixing of basic and acid material. The concept of convection currents in a mass of basic magma which is in contact with sialic crustal rocks, provides a possible mechanical method whereby hybridization between basic magma and granitic rocks might be effectively brought about. Mixing between basic magma and granitic rocks with the production of the calc-alkaline series is regarded as at least as important as hybridization of acid magma and basic rocks. The Non-porphyritic Central type of basalt is considered to have been produced by hybridization between sialic crustal rocks and iron-rich differentiates of normal olivine-gabbro. The amount of ferrogabbro material present in the known part of the Earth's crust is small; this provides a further indication that the bulk of post-Cambrian granites have not been formed by fractional crystallization of basalt magma.

TABLE XXXXI. AN

	Transitional	Layered Series									Unlaminated Layered S	
	I 4087	II 4077	III 3662	IV 3661	IVa	V 1907	VI 4145	VII 4142	VIII 1881	IX 4139	X 4137
SiO_2	45.48	46.37	48.15	45.65	46.90	44.81	44.61	44.13	48.27	45.19	52.13
Al_2O_3	16.41	16.82	18.02	15.08	16.55	13.96	11.70	17.88	8.58	9.37	15.87
Fe_2O_3	2.09	1.52	2.52	3.41	2.96	3.75	2.05	4.05	4.06	5.78	5.61
FeO	9.29	10.44	9.50	14.86	12.18	16.66	22.68	26.63	22.89	23.77	11.17
MgO	11.65	9.61	5.25	6.35	5.80	5.54	1.71	0.25	1.21	0.43	1.11
CaO	10.46	11.29	10.17	9.18	9.67	8.53	8.71	10.03	7.42	9.05	5.80
Na_2O	2.06	2.45	3.46	2.48	2.97	3.35	2.95	2.15	2.65	2.43	3.63
K_2O	0.27	0.20	0.14	0.28	0.21	0.33	0.35	0.47	0.34	0.49	1.38
H_2O+	0.77	0.29	0.20	0.22	0.21	0.34	0.22	0.30	1.13	0.57	0.86
H_2O-	0.26	0.09	0.02	0.08	0.05	0.19	0.20	0.19	0.37	0.31	0.25
TiO_2	0.94	0.79	2.64	2.59	2.61	2.55	2.43	2.48	2.20	1.67	1.14
MnO	0.06	0.09	0.12	0.15	0.14	0.17	0.21	0.48	0.26	0.32	0.30
P_2O_5	0.05	0.06	0.05	0.08	0.06	0.08	1.85	1.61	0.65	0.91	0.70
CO_2	..	..	0.03	..	..	..	0.04	..	..	..	..
S	..	..	0.14	..	..	..	0.31	..	..	..	..
ZrO_2	..	..	nil.	..	..	..	0.01	..	..	..	..
SrO	..	..	0.07	..	..	..	0.08	..	..	..	..
BaO	..	..	0.01	..	..	..	0.02	..	..	..	..
Cr_2O_3	..	..	nil.	..	..	..	trace	..	..	..	..
CuO	..	..	0.006	..	..	..	0.016	..	..	..	..
NiO	..	..	nil.	..	..	..	trace	..	..	..	..
	99.79	100.02	100.50	100.41	100.31	100.26	100.15	100.65	100.03	100.29	99.95
S.G.	3.05	3.04	3.01	3.19	..	3.18	3.22	3.39	3.18	3.30	2.97
$\dfrac{K_2O \times 100}{Na_2O + K_2O}$	12	8	4	10	7	9	11	22	9	17	28
$\dfrac{(FeO + Fe_2O_3) \times 100}{MgO + FeO + Fe_2O_3}$	49	55	70	74	72	79	94	99	96	99	94
$\dfrac{Fe_2O_3 \times 100}{FeO + Fe_2O_3}$	18	13	21	19	20	16	8	13	15	19	33

ALYSES OF ROCKS.

ted eries	Chilled Marginal Gabbro			Marginal Border Group					Upper Border Group			Transgressive Rocks		Gneiss
XI 4136	XII 1825	XIII 1724	XIIIa	XIV 1682	XV 1851	XVI 1837	XVII 4298	XVIII 2275	XIX 3052	XX 3050	XXI 4163	XXII 3047	XXIII 3058	XXIV 1867
55.30	48.01	47.83	47.92	41.27	46.15	47.01	45.51	50.41	48.02	49.16	57.00	58.81	75.03	68.17
18.52	19.11	18.62	18.87	8.71	21.34	19.12	14.36	18.30	15.40	15.83	13.02	12.02	13.17	16.13
2.18	1.20	1.16	1.18	2.69	0.79	0.71	5.04	1.96	4.35	4.13	2.15	5.77	1.56	0.58
7.47	8.44	8.87	8.65	10.52	8.21	9.12	12.65	8.13	9.30	8.17	9.25	9.38	0.58	2.09
0.21	7.72	7.92	7.82	27.09	9.97	8.98	5.38	4.25	4.06	4.36	2.64	0.72	0.15	1.82
8.20	10.33	10.59	10.46	6.59	10.24	10.88	9.91	11.20	10.05	10.81	5.75	5.03	0.69	2.07
6.01	2.34	2.54	2.44	0.69	1.72	2.33	2.43	2.74	2.62	2.90	4.11	3.91	4.24	4.40
0.78	0.17	0.20	0.19	0.13	0.12	0.28	0.35	0.46	0.22	0.35	0.99	2.39	3.85	3.22
0.41	0.55	0.27	0.41	0.87	0.27	0.24	0.12	0.32	0.94	0.62	0.97	0.21	0.28	0.40
0.06	0.05	0.16	0.10	0.07	0.10	0.09	0.10	0.19	0.07	0.06	0.41	0.19	0.13	0.16
0.94	1.51	1.29	1.40	1.54	0.55	1.47	3.96	1.75	3.88	3.59	1.78	1.26	0.31	0.63
0.09	0.12	0.09	0.11	0.16	0.15	0.08	0.20	0.16	0.25	0.21	0.12	0.21	0.01	0.05
0.42	0.07	0.06	0.07	0.02	0.05	0.06	0.32	0.25	0.92	0.39	2.11	0.71	0.02	0.31
..	0.11	0.01	0.06	..	0.13	..	..	..	0.19	..	..	..	..	..
..	0.29	0.25	0.27	..	0.12	..	..	..	0.14	..	..	..	..	..
..	nil.	nil.	nil.	..	nil.	..	..	..	0.01	..	..	..	..	..
..	0.21	0.19	0.20	..	0.07	..	..	..	0.12	..	..	..	..	..
..	0.02	0.02	0.02	..	0.03	..	..	..	0.04	..	..	..	..	..
..	trace	nil.	nil.	..	0.04	..	..	..	nil.	..	..	..	..	..
..	0.006	0.008	0.007	..	0.007	..	..	..	nil.	..	..	..	..	..
..	nil.	nil.	nil.	—	nil.	..	..	..	0.01	..	..	..	..	..
100.59	100.26	100.07	100.21	100.35	100.06	100.37	100.33	100.12	100.59	100.58	100.30	100.61	100.02	100.03
2.82	2.98	2.95	..	3.18	3.01	3.01	3.22	2.98	2.97	2.97	3.02	2.82	2.54	2.66
12	56	56	56	16	7	11	12	14	8	11	20	38	48	42
98	7	7	7	33	47	52	77	70	77	74	81	95	93	60
23	12	12	12	20	8	7	28	21	32	33	19	38	73	22

	Transitional	Layered			
	I 4087	I 4077	III 3662	IV 3661	IVa
Quartz..........	..	..	..	..	..
Orthoclase	1.67	1.11	0.83	1.67	1.25
Albite	17.29	17.82	29.32	20.96	25.15
Anorthite	34.75	34.82	32.94	29.19	31.06
Nepheline........	..	1.70	..	..	..
Corundum........	..	..	.. .	..	..
Diopside..... Wo	7.08	8.82	7.31	6.50	6.90
En	4.50	4.60	4.00	2.80	3.40
Fs	2.11	3.96	3.04	3.70	3.37
Hypersthene.. En	0.30	..	2.40	5.20	3.80
Fs	0.13	..	1.85	6.82	4.33
Olivine ...... Fo	16.94	13.58	4.69	5.60	5.15
Fa	9.18	10.00	4.59	7.55	6.07
Ilmenite	1.67	1.52	5.02	5.02	5.02
Magnetite........	3.02	2.09	3.71	4.87	4.29
Apatite..........	0.17	0.34	0.17	0.34	0.20
Calcite..........	..	..	0.05	..	0.25
Pyrites	..	..	0.24	..	0.12
Haematite	..	..	..	..	..
Water	1.03	0.38	0.22	0.30	0.26
Plag.	$Ab_{33}An_{67}$	$Ab_{34}An_{66}$	$Ab_{47}An_{53}$	$Ab_{40}An_{60}$	$Ab_{43}An_{57}$
Diop............	$Wo_{52}En_{33}Fs_{15}$	$Wo_{50}En_{27}Fs_{23}$	$Wo_{51}En_{28}Fs_{21}$	$Wo_{50}En_{22}Fs_{28}$	$Wo_{50}En_{25}Fs_{25}$
Hyp.	$En_{70}Fs_{30}$	..	$En_{57}Fs_{43}$	$En_{43}Fs_{57}$	$En_{50}Fs_{50}$
Oliv.	$Fo_{65}Fs_{35}$	$Fo_{58}Fa_{42}$	$Fo_{51}Fa_{49}$	$Fo_{43}Fa_{57}$	$Fo_{47}Fa_{53}$
Quartz..........	..	..	..	..	..
Orthoclase	..	..	..	..	..
Plagioclase.......	56	55	60	37	..
Clinopyroxene	29	21	33	} 51 {	..
Orthopyroxene ...	3	5	2		..
Olivine	11	17.5	..	..	..
Ore	0.7	1.5	5	12	..
Apatite..........	trace	trace	trace	trace	..
Biotite..........	..	..	..	..	..

1) Includes some quartz-felspar intergrowth.

2) Includes marginal felspar, probable perthite.

3) Includes some ore.

4) This figure is for material interstitial to the early well-shaped plagioclase and includes an inter border of felspar in graphic intergrowth with quartz and also clear quartz and some apatite and chlorite.

TABLE XXXXII.

Series				Unlaminated Layered Series		
V 1907	VI 4145	VII 4142	VIII 1881	IX 4139	X 4137	XI 4136
..	..	..	4.44	0.48	6.84	..
1.67	2.22	2.78	1.67	2.78	8.34	4.45
22.53	25.15	18.34	22.53	20.44	30.92	50.83
22.24	17.51	10.29	10.56	13.34	22.80	21.13
3.13	..	..	..	..	..	..
..	..	..	..	..	..	..
9.28	6.03	12.06	9.28	11.14	0.81	7.08
2.80	0.70	0.20	0.80	0.30	0.20	0.10
5.54	5.94	14.12	9.50	12.28	0.66	7.92
..	1.90	0.30	2.20	0.80	2.60	0.20
..	15.97	24.55	25.87	24.42	13.99	1.19
7.98	1.12	0.07	..	..	..	0.14
14.08	10.82	2.76	..	..	..	1.02
4.86	4.56	4.71	4.26	3.19	2.13	1.82
5.34	3.02	6.03	6.03	8.35	8.12	3.25
0.34	4.37	3.70	1.68	2.02	1.68	1.01
..	0.10	..	..	..	..	..
..	0.54	..	..	..	..	..
..	..	..	..	..	..	..
0.53	0.42	0.49	1.50	0.88	1.11	0.47
$Ab_{50}An_{50}$	$Ab_{59}An_{41}$	$Ab_{65}An_{35}$	$Ab_{68}An_{32}$	$Ab_{60}An_{40}$	$Ab_{58}An_{42}$	$Ab_{71}An_{29}$
$Wo_{53}En_{16}Fs_{31}$	$Wo_{48}En_5Fs_{47}$	$Wo_{45}En_1Fs_{54}$	$Wo_{47}En_4Fs_{49}$	$Wo_{47}En_1Fs_{52}$	$Wo_{50}En_{12}Fs_{38}$	$Wo_{47}En_1Fs_{52}$
..	$En_{11}Fs_{89}$	En_1Fs_{99}	En_8Fs_{92}	En_3Fs_{97}	$En_{16}Fs_{84}$	$En_{14}Fs_{86}$
$Fo_{36}Fa_{64}$	$Fo_{10}Fa_{90}$	Fo_2Fa_{98}	..	..	..	$Fo_{12}Fa_{88}$

V 1907	VI 4145	VII 4142	VIII 1881	IX 4139	X 4137	XI 4136
..	1	6[1]	11[2]	9[3]	39[4]	26[4]
..	..	..	..	..	..	..
56	45	24	29	26	28	56
20	28	37	34	33	22	15
..	..	..	..	..	..	..
16	17	21	17	18	..	..
8	6.5	9	8	12	10	2
trace	2.5	3	1	2	1	1
..	..	..	..	..	..	..

[5]) Perhaps a little low as in doubtful cases assumed to be clinopyroxene.
[6]) Includes some uralite.
[7]) Includes some serpentine.
[8]) Includes little chlorite.

[9]) Excluding ore from the decompositio
[10]) Includes little talc.
[11]) Includes micropegmatite.
[12]) Micropegmatite with chlorite, apatite

I. NORMS AND MODES.

	Chilled Marginal Gabbro			Marginal Border Group			
	XII 1825	XIII 1724	XIIIa	XIV 1682	XV 1851	XVI 1837	XVII 4298
	..	..	..	..	..	..	..
	1.11	1.11	1.11	0.56	0.56	1.67	2.22
	19.91	21.48	20.69	3.14	14.15	19.91	20.44
	40.87	38.64	39.75	20.29	49.76	40.59	27.24
	..	..	..	1.42	..	..	..
	..	..	..	..	0.20	..	..
	4.18	5.80	4.99	6.61	..	5.34	8.47
	2.50	3.30	2.90	4.90	..	3.20	4.20
	1.45	2.24	1.85	1.06	..	1.85	4.09
	9.60	5.40	7.50	..	9.40	1.60	8.90
	5.28	3.56	4.42	..	5.28	0.92	8.58
	5.04	7.70	6.37	43.96	10.92	12.46	0.28
	3.88	5.10	4.59	10.61	5.92	8.36	0.41
	2.89	2.43	2.66	2.89	1.06	2.89	7.45
	1.86	1.86	1.86	3.94	1.16	0.93	7.19
	0.34	0.17	0.20	..	0.17	0.34	0.67
	0.20	..	0.10	..	0.20	..	..
	0.54	0.48	0.51	..	0.20	..	..
	..	..	..	..	..	..	..
	0.60	0.43	0.51	0.94	0.37	0.33	0.22
	$Ab_{32}An_{68}$	$Ab_{36}An_{64}$	$Ab_{34}An_{66}$	$Ab_{13}An_{87}$	$Ab_{22}An_{78}$	$Ab_{33}An_{67}$	$Ab_{43}An_{57}$
	$Wo_{53}En_{30}Fs_{17}$	$Wo_{50}En_{30}Fs_{20}$	$Wo_{51}En_{30}Fs_{19}$	$Wo_{52}En_{40}Fs_{8}$		$Wo_{51}En_{31}Fs_{18}$	$Wo_{51}En_{25}Fs_{24}$
	$En_{64}Fs_{36}$	$En_{60}Fs_{40}$	$En_{62}Fs_{38}$		$En_{68}Fs_{32}$	$En_{64}Fs_{36}$	$En_{51}Fs_{49}$
	$Fo_{57}Fa_{43}$	$Fo_{60}Fa_{40}$	$Fo_{59}Fa_{41}$	$Fo_{80}Fa_{20}$	$Fo_{54}Fa_{46}$	$Fo_{60}Fa_{40}$	$Fo_{41}Fa_{59}$

Modes (in volume percentage)

	XII 1825	XIII 1724	XIIIa	XIV 1682	XV 1851	XVI 1837	XVII 4298
	..	..	..	..	..	..	..
	..	..	..	..	..	..	..
	65.5	57.2	..	15.8	54.2	64	56
	21.2	25.5[6]	..	13.8	22.7	8[8]	35[8]
	1.7[5]	..	..	4.7	..	2	..
	11.1	16.3[7]	..	65.0	22.8	25[10]	..
	0.5	1.0	..	0.7[9]	0.3	1	9
	..	..	..	..	..	..	trace
	..	..	..	..	..	..	..

tion of olivine.

ite and quartz.

13) About one tenth replaced by chlorite.
14) Mainly micropegmatite also includes dusty borders to plagioclase.
15) Includes micropegmatite and dusty borders to plagioclase.
16) Includes some iron ore and chlorite.

	Upper Border Group				Transgressive Rocks		Gneiss
XVIII 2275	XIX 3052	XX 3050	XXI 4163	XXII 3047	XXIII 3058	XXIV 1867
1.74	6.72	4.32	13.14	14.58	33.96	22.74
2.78	1.11	2.22	6.12	14.46	22.80	18.90
23.58	22.01	24.63	34.58	33.01	35.63	37.20
35.86	29.75	28.36	13.90	8.06	3.61	8.90
..	..	..	..	..	..	..
..	..	..	..	..	0.92	2.14
7.97	5.80	9.51	0.93	5.22	..	..
4.40	3.20	5.80	0.30	0.80	..	..
3.17	2.38	3.17	0.66	4.88	..	..
6.20	6.90	5.10	6.30	1.00	0.40	4.50
7.26	4.88	2.90	11.88	5.94	..	2.38
..	..	..	..	..	..	..
3.34	7.45	6.84	3.34	2.43	0.61	1.22
3.02	6.26	6.03	3.02	8.35	0.93	0.93
0.34	2.02	1.01	5.04	1.68	..	1.01
..	0.50	..	..	..	..	..
..	0.24	..	..	..	..	..
..	..	..	..	..	0.96	..
0.51	1.01	0.68	1.38	0.40	0.41	0.56
$Ab_{40}An_{60}$	$Ab_{42}An_{58}$	$Ab_{46}An_{54}$	$Ab_{71}An_{29}$	$Ab_{80}An_{20}$	$Ab_{91}An_{9}$	$Ab_{81}An_{19}$
$Wo_{51}En_{28}Fs_{21}$	$Wo_{51}En_{28}Fs_{21}$	$Wo_{51}En_{31}Fs_{18}$	$Wo_{49}En_{16}Fs_{25}$	$Wo_{48}En_{7}Fs_{45}$	En_{100}	..
$En_{46}Fs_{54}$	$En_{60}Fs_{40}$	$En_{64}Fs_{36}$	$En_{35}Fs_{65}$	$En_{14}Fs_{86}$	..	$En_{66}Fs_{34}$
..	..	..	..	..	..	..

XVIII	XIX	XX	XXI	XXII	XXIII	XXIV
6[11])	21.9[12])	3[12])	21[14])	36[15])		28
..	..	..	..	..		51
53	55.5	57	42	40		13
27[8])	20.6[13])	37	28[13])	13		..
..	..	..	..	..		..
4	..	..	..	..	nd	..
10	2.0	3	9	11[10])		..
trace	..	..	..	..		..
..	..	..	..	..		8[16])

LIST OF REFERENCES

1893. BAYLEY, W. S. "The Eruptive and Sedimentary Rocks on Pigeon Point, Minnesota and their Contact Phenomena." Bull. U. S. Geol. Surv. No. 109, pp. 1—121.

1894. HARKER, A. "Carrock Fell: a Study in the Variation of Igneous Rock Masses—Part I. The Gabbro." Quart. Journ. Geol. Soc. Vol. 50, pp. 311—37.

1895. HARKER, A. "Carrock Fell: a Study in the Variation of Igneous Rock Masses —Part II. The Carrock Fell Granophyre. Part III. The Grainsgill Greisen." Quart. Journ. Geol. Soc. Vol. 51, pp. 125—47.

1897. BECKER, G. F. "Some Queries on Rock Differentiation." Amer. Journ. Sci. 4th Series. Vol. III, pp. 21—40.

1897. BECKER, G. F. "Fractional Crystallization of Rocks." Amer. Journ. Sci. 4th Series. Vol. IV, pp. 257—61.

1897. NEWTON, E. T. and J. J. H. TEALL. "Notes on a Collection of Rocks and Fossils from Franz Josef Land, made by the Jackson-Harmsworth Expedition during 1894—1896." Quart. Journ. Geol. Soc. Vol. LIII, pp. 477—518.

1897. TEALL, J. J. H. "Differentiation in Igneous Magmas as a Result of Progressive Crystallization." Geol. Mag. Vol. IV, pp. 553—55.

1898—1900. AMDRUP, G. "Carlsbergfondets Ekspedition til Østgrønland udført i Aarene 1898—1900 under Ledelse af G. Amdrup." Meddel. om Grønland. Vols. XXVII, XXVIII, XXIX and XXX.

1898. HENDERSON, J. A. L. "Norites, Gabbros and Pyroxenites and other South African Rocks." pp. 1—56. London.

1904. HARKER, A. "The Tertiary Igneous Rocks of Skye." Mem. Geol. Surv. Scot., pp. 1—481.

1904. RAVN, J. P. J. "The Tertiary Fauna at Kap Dalton in East Greenland." Meddel. om Grønland. Vol. XXIX, pp. 95—140.

1904. WEIDMAN, S. "Widespread occurrence of Fayalite in certain Igneous Rocks of Central Wisconsin." Journ. Geol. Vol. XII, pp. 551—61.

1905. PIRSSON, L. V. "Petrology and Geology of the Igneous Rocks of the Highwood Mountains." U. S. Geol. Surv. Bull. 237, pp. 1—208.

1908. HARKER, A. "The Natural History of Igneous Rocks." London, pp. 1—384.

1908. HARKER, A. "Geology of the Small Isles of Inverness-shire." Mem. Geol. Surv. Scot., pp. 1—210.

1909. CLOUGH, C. T., H. B. MAUFE and E. B. BAILEY. "The Cauldron-Subsidence of Glen Coe and the Associated Igneous Phenomena." Quart. Journ. Geol. Soc., vol. LXV, pp. 611—76.

1909—1910. SHAND., S. J. "On Borolanite and its Associates in Assynt." Trans. Edin. Geol. Soc. Vol. IX, Part III, pp. 202—15, and Part V, pp. 376—419.

1912. USSING, N. V. "Geology of the Country around Julianehaab." Meddel. om Grønland. Vol. XXXVIII, pp. 1—376.

1914. DALY, R. A. "Igneous Rocks and their Origin." New York and London, pp. 1—563.

1914. DAY, A. L., R. B. SOSMAN and J. C. HOSTETTER. "Determination of Mineral and Rock Densities at High Temperatures." Amer. Journ. Sci., 4th Series Vol. 37, pp. 1—39.

1915. BOWEN, N. L. "The Later Stages of the Evolution of the Igneous Rocks." Journ. Geol. Vol. XXIII, Supplement, pp. 1—91.

1915. BOWEN, N. L. "Crystallization—Differentiation in Silicate Liquids." Amer. Journ. Sci. 4th Series. Vol. XXXIX, pp. 175—91.

1916. TYRRELL, G. W. "The Picrite-Teschenite Sill of Lugar (Ayrshire)." Quart. Journ. Geol. Soc. Vol. LXXII, pp. 84—131.

1917. BOWEN, N. L. "The Problem of the Anorthosites." Journ. Geol. Vol. 25, pp. 209—43.

1917. WASHINGTON, H. S. "Chemical Analyses of Igneous Rocks." U. S. Geol. Surv. Prof. Paper, No. 99, pp. 1—1201.

1918. GOODCHILD, W. H. "The Evolution of Ore Deposits from Igneous Magmas." Mining Mag. Vol. XVIII, pp. 20—29, 75—82 etc.

1918. GROUT, F. F. "Internal Structures of Igneous Rocks; their significance and origin; with special reference to the Duluth Gabbro." Journ. Geol. Vol. 26, pp. 439—58.

1918. GROUT, F. F. "Two-phase convection in igneous magmas." Journ. Geol. Vol. 26, pp. 481—99.

1918. GROUT, F. F. "A Type of Igneous Differentiation." Journ. Geol. Vol. 26, pp. 626—58.

1918. HOLMES, A. "The basaltic rocks of the Arctic Regions." Min. Mag. Vol. 18, pp. 180—223.

1919. BOWEN, N. L. "Crystallisation Differentiation in Igneous Magmas." Journ. Geol. Vol. 27, pp. 393—430.

1921. BOWEN, N. L. "Diffusion in silicate melts." Journal of Geology. Vol. XXIX, pp. 295—317.

1921. FOSLIE, S. "Field Observations in Northern Norway bearing on Magmatic Differentiation." Journ. Geol. Vol. 29, pp. 701—19.

1921. "Report of the Committee on British Petrographic Nomenclature, Min. Mag. Vol. 19, pp. 137—47.

1921—1923. VOGT, J. H. L. "The Physical Chemistry of the Crystallization and Magmatic Differentiation of Igneous Rocks." Journ. Geol. Vols. 29, pp. 318—50, 426—43, 515—39 and 627—49; Vol. 30, pp. 611—30 and 659—72; Vol. 31, pp. 233—52 and 407—19.

1922. DIXEY, F. "The Norite of Sierra Leone." Quart. Journ. Geol. Soc. Vol. LXXVIII, pp. 299—347.

1922. TUTTON, A. E. H. "Crystallography and Practical Crystal Measurement." London. Vol. 2, pp. 959—65.

1922. WASHINGTON, H. S. "Deccan Traps and Other Plateau Basalts." Bull. Geol. Soc. Amer. Vol. 33, pp. 765—803.

1924. ADAMS, L. H. "Temperatures at moderate depths within the Earth." Journ. Washington Acad. Sci. Vol. 14, pp. 459—72.

1924. BAILEY, E. B., H. H. THOMAS etc. "The Tertiary and Post-Tertiary Geology of Mull, etc." Mem. Geol. Surv., Scotland, pp. 1—445.

1924. HAWKES, L. "Olivine-dacite in the Tertiary Volcanic Series of Eastern Iceland." Quart. Journ. Geol. Soc. Vol. LXXX, pp. 549—67.

1924. VOGT, J. H. L. "The Physical Chemistry of the Magmatic Differentiation of Igneous Rocks." Videns. Skrifter, Oslo. Kl. 1. No. 15, pp. 1—132.

1924. WAGNER, P. "On Magmatic Nickel Deposits of the Bushveld Complex, etc." Geol. Surv. S. Africa, Mem. no. 21, pp. 14—181.

1925. ASKLUND, B. "Petrological Studies in the Neighbourhood of Stavsjo at Kolmården." Årsbok Sveriges Geol. Undersök. Vol. 17, No. 6, pp. 1—122.

1925. FERMOR, L. L. "On the Basaltic Lavas penetrated by the deep boring for coal at Bhusawal, Bombay Presidency." Rec. Geol. Survey India. Vol. 58, pp. 93—240.

1925. GROUT, F. F. "The Vermilion Batholith of Minnesota." Journ. Geol. Vol. XXXIII, pp. 467—87.

1926. BRIDGMAN, P. W. "The effect of pressure on the viscosity of forty three pure liquids." Proc. Amer. Acad. Arts and Sciences. Vol. 61, p. 57—99.

1926. FENNER, C. N. "The Katmai Magmatic Province." Journ. Geol. Vol. 34, pp. 675—772.

1927. BALK, R. "Die primare Struktur des Nordmassivs von Peekskill." Neues Jarhbuch, Beilage-Band. Abt. B. Vol. 57, pp. 249—303.

1927. GREIG, J. W. "Immiscibility in Silicate Melts." Part I and Part II. Amer. Journ. Sci., 5th Series. Vol. 13, pp. 1—44 and 133—54.

1927. LIGHTFOOT, B. "Traverses along the Great Dyke of Rhodesia." S. Rhodesia Geol. Surv. No. 21, pp. 1—8.

1928. BOWEN, N. L. "The Evolution of the Igneous Rocks." Princeton, pp. 1—332.

1928. SIMPSON, J. B. "Notes on the Geology of the Faeröe Islands." Geol. Mag. Vol. LXV, pp. 510—17.

1928. DALY, R. A. "Bushveld Igneous Complex of the Transvaal." Bull. Geol. Soc. Amer. Vol. XXXIX, pp. 703—68.

1928. TYRRELL, G. W. "Geology of Arran." Mem. Geol. Surv. Scotland, pp. 1—292.

1929. DU TOIT, A. L. "The Volcanic Belt of the Lebombo—a Region of Tension." Trans. Roy. Soc. S. Africa. Vol. XVIII, pp. 189—217.

1929. FENNER, C. N. "The Crystallisation of Basalts." Amer. Journ. Sci. Ser. 5. Vol. 18, pp. 225—53.

1929. JEFFREYS, H. "The Earth." 2nd Ed. pp. 1—278. Cambridge.

1929. TOMKEIEFF, S. I. "A Contribution to the Petrology of the Whin Sill." Min. Mag. Vol. 22, pp. 100—120.

1929. WAGNER, P. A. "Platinum Deposits and Mines of South Africa etc." Edinburgh, pp. 1—326.

1930. BOWEN, N. L. and J. F. SCHAIRER and H. W. V. WILLEMS. "The Ternary System Na_2SiO_3—Fe_2O_3—SiO_2." Amer. Journ. Sci. Vol. XX, 5th Series, pp. 405—55.

1930. DALY, R. A. and T. F. W. BARTH. "Dolerites associated with the Karroo System, S. Africa." Geol. Mag. Vol. 67, pp. 97—110.

1930. HOLMES, A. "Petrographic Methods and Calculations." 2nd Ed. London.

1930. JUNNER, N. R. "The Norite of Sierra Leone." Compte Rendu, Cong. Geol. Internat. South Africa. Vol. 2, pp. 417—33.

1930. RICHEY, J. E., H. H. THOMAS, etc. "The Geology of Ardnamurchan, North-West Mull and Coll." Mem. Geol. Surv. Scotland, pp. 1—393.

1930. WALKER, F. "The Geology of the Shiant Isles (Hebrides)." Quart. Journ. Geol. Soc. Vol. 86, pp. 355—98.

1930. WALKER, F. "A tholeiitic phase of the quartz-dolerite magma of Central Scotland." Min. Mag. Vol. 22, pp. 368—76.

22*

1931. DAY, A. L. "Annual Report of the Director of the Geophysical Laboratory."
Year Book No. 30 for 1930—31 of Carnegie Institution of Washington,
pp. 75—100.

1931. FENNER, C. N. "The residual liquids of crystallising magmas." Min. Mag.
Vo!. 22, pp. 539—60.

1931. GUPPY, E. M., H. H. THOMAS, etc. "Chemical Analyses of Igneous Rocks etc."
Mem. Geol. Surv. Gt. Brit., pp. 1—166.

1931A. HOLMES, A. "The Problem of the Association of Acid and Basic Rocks in
· Central Complexes." Geol. Mag. Vol. LXVIII, pp. 241—55.

1931B. HOLMES, A. "Radioactivity and Earth Movement." Trans. Geol. Soc. Glasgow. Vol. XVIII, pp. 559—606.

1931. KENNEDY, W. Q. "The Parent Magma of the British Tertiary Province."
Summary of Progress of the Geol. Surv. of Great Britain for 1930. Part II,
pp. 61—73.

1931. OSBORNE, F. F., and E. J. ROBERTS. "Differentiation in the Shonkin Sag
Laccolith, Montana." Amer. Journ. Sci. 5th Ser. Vol. 22, pp. 331—53.

1931. WOLFF, F. v. "Der Vulkanismus." II Band. Spezieller Teil. Stuttgart, pp. 829—
1111.

1932. BACKLUND, H. G. "On the Mode of Intrusion of Deep-Seated Alkaline Bodies."
Bull. Geol. Inst. Upsala. Vol. XXIV, pp. 1—24.

1932. BACKLUND, H. G. and D. MALMQUIST. "Zur Geologie und Petrographie der
Nordostgrönländischen Basaltformation." Pt. 1. Die Basische Reihe.
Meddel. om Grønland. Bd. 87, no. 5, pp. 1—61.

1932. HALL, A. L. "The Bushveld Igneous Complex of the Central Transvaal."
Mem. Geol. Surv. S. Africa. No. 28, pp. 1—554.

1932. KROKSTRÖM, T. "The Breven Dolerite Dyke—a Petrogenetic Study." Bull.
Geol. Instit. Upsala. Vol. 23, pp. 243—330.

1932. KROKSTRÖM, T. "On the Ophitic Texture and the Order of Crystallization in
Basaltic Magmas." Bull. Geol. Instit. Upsala. Vol. 24, pp. 197—216.

1932. RICHEY, J. E. "Tertiary Ring Structures in Britain." Trans. Geol. Soc. Glasgow. Vol. XIX, Part I, pp. 42—140.

1932. TSUBOI, S. "On the Course of Crystallisation of Pyroxenes from Rock Magmas."
Jap. Journ. Geol. & Geogr. Vol. 10, pp. 67—82.

1932. WATKINS, H. G. "The British Arctic Air-Route Expedition, 1930—31." Geogr.
Journ. Vol. LXXIX, pp. 353—67, pp. 466—501.

1933. BOWEN, N. L., J. F. SCHAIRER and E. POSNJAK. "The System CaO—FeO—
SiO_2." Amer. Journ. Sci. 5th Ser. Vol. 26, pp. 193—284.

1933. BRAMMALL, A. "Syntexis and Differentiation." Geol. Mag. Vol. 70, pp. 97—
107.

1933. DALY, R. A. "Igneous Rocks and the Depths of the Earth." New York and
London, pp. 1—598.

1933. KENNEDY, W. Q. "Trends of Differentiation in Basaltic Magmas." Amer.
Journ. Sci. Vol. 25, pp. 239—56.

1933. MIKKELSEN, E. "The Blosseville Coast of East Greenland." Geogr. Journ.
Vol. LXXXI, pp. 385—402.

1933. MIKKELSEN, E. "The Scoresby Sound Committee's 2nd East Greenland
Expedition in 1932 to King Christian IX's Land—Report on the Expedition." Meddel. om Grønland. Vol. CIV, pp. 3—71.

1933. NOCKOLDS, S. R. "Some Theoretical Aspects of Contamination in Acid Magmas."
Journ. Geol. Vol. 41, pp. 561—89.

1933. RAVN, J. P. J. "New Investigations of the Tertiary at Kap Dalton, East
Greenland." Meddel. om Grønland. Vol. CV, pp. 1—15.

1933. SPENDER, M. "Map-making during the Expedition." Meddel. om Grønland. Vol. CIV, No. 2, pp. 1—20.

1933. WINCHELL, A. N. "Elements of Optical Mineralogy." New York, pt. 2, pp. 1—442.

1934. ANDERSON, B. W. and C. J. PAYNE. "Liquids of High Refractive Index." Nature. London. Vol. 133, pp. 66—67.

1934. ANDERSON, B. W. and C. J. PAYNE. "The Refractometer and other Refractive Index Methods." Gemnologist. London. Vol. 3, pp. 216—27.

1934. NOCKOLDS, S. R. "The Production of Normal Rock Types by Contamination and their bearing on Petrogenesis." Geol. Mag. Vol. 71, pp. 31—39.

1934. PHEMISTER, T. C. "The Role of Water in Basaltic Magma." Tschermak's Min. Pet. Mitt. Vol. 45, pp. 19—132.

1934. TOMITA, T. "Variations in optical properties, according to chemical composition, in the pyroxenes of the clinoenstatite-clinohypersthene-diopside-hedenbergite system." Journ. Shanghai Sci. Inst. Sect. 2. Vol. 1, pp. 41—58.

1934. WAGER, L. R. "Geological Investigations in East Greenland, Part I, General Geology from Angmagssalik to Kap Dalton." Meddel om Grønland. Vol. CV, No. 2, pp. 1—46.

1935. BACKLUND, H. G. and D. MALMQUIST. "Zur Geologie und Petrographie der Nordostgrönlandischen Basaltformation. Pt. 2. Die Sauren Ergussgesteine von Kap Franklin." Meddel. om Grønland. Bd. 95, No. 3, pp. 1—84.

1935. BOWEN, N. L. "Ferrosilite as a Natural Mineral." Amer. Journ. Sci. Ser. 5. Vol. 30, pp. 481—94.

1935. BOWEN, N. L. and J. F. SCHAIRER. "The System, $MgO—FeO—SiO_2$." Amer. Journ. Sci. Ser. 5. Vol. 29, pp. 151—217.

1935. DEER, W. A. "The Cairnsmore of Carsphairn Igneous Complex." Quart. Journ. Geol. Soc. Vol. 91, pp. 47—76.

1935: ELDER, S. "Comparison of Three Scottish Magmas." Geol. Mag. Vol. LXXII, pp. 80—85.

1935. HENRY, N. F. M. "Some data on the iron-rich hypersthenes." Min. Mag. Vol. 24, pp. 221—26.

1935. INGERSON, E. "Layered Peridotite Laccoliths of the Trout River Area, Newfoundland." Amer. Journ. Sci. 5th Series. Vol. 29, pp. 422—40.

1935. KOCH, L. "Geologie von Grönland," pp. 1—159. Berlin.

1935. KUNO, H. "Petrology of Alaid Volcano, North Kurile." Jap. Journ. Geol. & Geogr. Vol. 12, pp. 153—62.

1935. LOMBAARD, B. V. "On the Differentiation and Relationships of the Rocks of the Bushveld Complex." Trans. Geol. Soc. S. Africa. Vol. XXXVII, pp. 5—52.

1935. TRÖGER, W. E. "Spezielle Petrographie der Eruptivgesteine. Ein Nomenklatur-Kompendium." Berlin, pp. 1—360.

1935. WAGER, L. R. "Geological Investigations in East Greenland, Pt. II. Geology of Kap Dalton." Meddel. om Grønland. Vol. CV, No. 3, pp. 1—32.

1935. WALKER, F. "The late Palaeozoic quartz-dolerites and tholeiites of Scotland." Min. Mag. Vol. 24, pp. 131—59.

1936. ALLING, H. L. "Interpretative Petrology of the Igneous Rocks." New York and London, pp. 1—353.

1936. ANDERSON, E. M. "The Dynamics of the Formation of Cone-sheets, Ringdykes, and Caldron-subsidences." Proc. Roy. Soc. Edinburgh. Vol. LVI, pt. II, pp. 128—57.

1936. BARTH, T. F. W. "The Crystallization Process of Basalts." Amer. Journ. Sci. Series 5. Vol. XXXI, pp. 321—51.

1936. Buddington, A. F. "Gravity Stratification as a criterion in the interpretation of the structure of certain intrusives of the North-Western Adironacks." Rep. of XVI Internat. Geol. Congr. Washington. 1933.

1936. Cooper, J. R. "Geology of the Southern Half of the Bay of Islands Igneous Complex." Bull. 4, Dept. of Natural Resources (Geol. sect.) of Newfoundland, pp. 1—62.

1936. Courtauld, A. "A Journey in Rasmussen Land". Geogr. Journ. Vol. LXXXVIII pp. 193—208.

1936. Krokström, T. "The Hällefors Dolerite Dyke and Some Problems of Basaltic Rocks." Bull. Geol. Instit. Upsala. Vol. 26, pp. 115—263.

1936. Peoples, J. W. "Gravity Stratification as a Criterion in the Interpretation of the Structure of the Stillwater Complex, Montana." Report XVI International Geol. Cong., Washington, 1933, pp. 353—60.

1936. Walker, F. and C. F. Davidson. "A Contribution to the Geology of the Faeroes." Trans. Roy. Soc. of Edinburgh. Vol. LVIII, Part III, pp. 869—97.

1937. Balk, R. "Structural Behaviour of Igneous Rocks." Geol. Soc. Amer. Mem. 5, pp. 1—177.

1937. Barksdale, J. D. "The Shonkin Sag Laccolith." Amer. Journ. Sci. Ser. 5. Vol. 33, pp. 321—59.

1937. Buddington, A. F. and H. H. Hess. "Layered Peridotitic Laccoliths in the Trout River Area, Newfoundland." Amer. Journ. Sci. Vol. 33, pp. 380—88.

1937. Ingerson, E. "Layered Peridotitic Laccoliths in the Trout River Area, Newfoundland—A Reply." Amer. Journ. Sci. Vol. 33, pp. 389—392.

1937. Krokström, T. "On the Association of granite and dolerite in igneous bodies." Bull. Geol. Instit. Upsala. Vol. XXVI, pp. 265—77.

1937. Kuno, H. "Fractional Crystallisation of Basaltic Magmas." Jap. Journ. Geol. & Geogr. Vol. 14, pp. 189—208.

1937. Nockolds, S. R. "On the Fields of Association of Certain Common Rock Forming Minerals." Tschermak's Min. Pet. Mitt. Vol. 49, pp. 101—16.

1937. Phemister, T. C. "A Review of the Problems of the Sudbury Irruptive." Journ. Geol. Vol. LXV, No. 1, pp. 1—47.

1937. Richey, J. E. "Variation in the Amount of Apatite in British Tertiary Igneous Rocks." Summary of Progress of Geol. Surv. of Great Britain for 1935 —Part II, pp. 46—52.

1937. Wager, L. R. "The Kangerdlugssuaq Region of East Greenland." Geogr. Journ. Vol. XC, No. 5, pp. 393—425.

1937. Wegmann, E. "Sur la genèse des roches alcalines de Julianehaab (Groenland) Note. Comptes rendus des séances de l'Academie des Sciences, t. 204, p. 1125—26.

1938. Bowen, N. L. and J. F. Schairer. "Crystallization Equilibrium in Nepheline-Albite-Silica Mixtures with Fayalite." Journ. Geol. Vol. XLVI, No. 3, Part II, pp. 397—411.

1938. Deer, W. A. and L. R. Wager. "Two new pyroxenes included in the system clinoenstatite, clinoferrosilite, diopside and hedenbergite." Min. Mag. Vol. 25, pp. 15—22.

1938. Edwards, A. B. "The Tertiary Volcanic Rocks of Central Victoria." Quart. Journ. Geol. Soc. Vol. 94, pp. 243—320.

1938. Fenner, C. N. "Olivine Fourchites from Raymond Fosdick Mountains, Antarctica." Bull. Geol. Soc. Amer. Vol. 49, pp. 367—400.

1938. Hess, H. H. and A. H. Phillips. "Orthopyroxenes of the Bushveld Type." Amer. Min. Vol. 23, pp. 450—56.

1938. HUTTON, C. O. "The Stilpnomelane Group of Minerals." Min. Mag. Vol. 25, pp. 172—206.

1938. KENNEDY, W. Q. and E. M. ANDERSON. "Crustal Layers and the Origin of Magmas." Bulletin Volcanalogique de l'Union Geodesique et Géophysique Internationale Serie II. Vol. III, pp. 24—82.

1938. KOOMANS, C. and P. H. KUENEN. "On the Differentiation of the Glen More Ring-dyke, Mull." Geol. Mag. Vol. LXXV, pp. 145—60.

1938. WAGER, L. R. & W. A. DEER. "A Dyke Swarm and Crustal Flexure in East Greenland." Geol. Mag. Vol. 75, pp. 39—46.

1938. WEGMANN, C. E. "Geological Investigations in Southern Greenland. Pt. 1. On the Structural Divisions of Southern Greenland." Meddel. om Grønland. Vol. 113, No. 2, pp. 1—148.

1939. DEER, W. A. and L. R. WAGER. "Olivines from the Skaergaard Intrusion, Kangerdlugssuaq, East Greenland." Amer. Min. Vol. 24, pp. 18—25.

Note. — While this paper was in the press there appeared the following "Primary banding in norite and gabbro" by H. H. Hess [Trans. American Geophysical Union, Ninetheenth Meeting, 1938]. In this paper it is suggested that the rythmically varying factor which produces the successive gravity stratified layers is repeated alteration of the state of the magma from quiescence to turbulence. In this hypothesis as in ours currents in the magma play an important part; while we have postulated rythmic variations in the velocity of the magmatic currents as a cause of this special banding, Hess has suggested a variation from a state whithout magmatic currents to one to one in which there are turbulent currents of moderate strength.

CONTENTS

EXPLANATION OF PLATES

Plate 1.

Southern part of the Skaergaard Intrusion from the west. Air photograph by H. G. Watkins, August, 1930. Gn, metamorphic complex; B, basalts and tuffs; BG, marginal border group; U, upper border group; HOG, hypersthene-olivine-gabbros; MG, middle gabbros; FG, ferrogabbros; PB, purple band; ULS, unlaminated layered series; R, Basistoppen raft of sill gabbro; Gr, granophyre and intermediate rocks of Tinden and Brödretoppen; MD, macrodyke.

Plate 2.

Northern part of the Skaergaard Intrusion from the west. Air photograph by II. G. Watkins, August, 1930. Gn, metamorphic complex; B, basalts and tuffs; BG, marginal border group; U, upper border group; HOG, hypersthene-olivine-gabbros; MG, middle gabbros; FG, ferrogabbros; PB, purple band; ULS, unlaminated layered series; R, Basistoppen raft of sill gabbros; Gr, granophyre and intermediate rocks of Tinden and Brödretoppen; MD, macrodyke.

Plate 3.

Fig. 1. Perpendicular felspar rock 20 metres from outer margin, Mellemö. Elongated felspar are arranged roughly perpendicular to the contact at this point.

Fig. 2. Wavy layers of coarse pyroxene and felspar rock in the olivine gabbro of the outer border group, topmost exposure on Kraemers Island, and about 20 metres from the margin. The layers of coarse pyroxene and felspar extend roughly perpendicular to the margin and have probably an origin similar to the perpendicular felspars of Pl. 3, fig. 1. Width of rock face shown about 1 metre.

Plate 4.

Fig. 1. Banding in the main zone of the border group, centre of Mellemo. On the more level surfaces the banding is torturous while it is more even on steeper faces.

Fig. 2. Another example of banding in the main zone of the border group showing how in places it is interrupted by coarse patches and in others the coarse material is streaked out, thus contributing to the banded appearance. Southeast corner of Ivnarmiut.

Plate 5.

Fig. 1. Inner contact, dipping inwards at 70°, between the border group (right) and the layered series (middle and left). The dip of the layering can be seen and is roughly parallel to the boards of the boat. The dip of the fluxion structures in the border group is parallel to the inner contact. The layered rocks are rich in melanocratic constituents and the layering is rather uneven. South-east corner of Ivnarmiut.

Fig. 2. Small penecontemporaneous faults in the layered series near, and parallel to, the inner contact, Mellemö.

Plate 6.

Rhythmic layering in the south-west face of Gabbrofjaeld (Peak 1277 m). The triple group is conspicuous, and below is a finer-scale layering. Height of the face about 300 metres.

Plate 7.

Fig. 1. Gabbrofjaeld from the Base. The triple group is conspicuous below the summit of Peak 1277 m. and can just be detected about half way up Pukugaq-ryggen. Layering is also seen in the foreground rocks. August, 1936.

Fig. 2. East face of Gabbrofjaeld (Peak 1277 m) overlooking Gabbro Mountain Glacier and Pukugaqryggen. The triple group can be traced by eye for $2^{1}/_{2}$ kilometres. June, 1936.

Plate 8.

Fig. 1. Gravity differentiated layers separated by layers of average rock, 300 metres west of the Main House. Towards the top centre incipient trough banding can be seen.

Fig. 2. Differentiated layers separated by layers of average rock, 400 yards west-north-west of Main House.

Plate 9.

Fig. 1. Wispy banding of the outermost transitional layered rocks in a face parallel to the strike of the banding, Uttentals Plateau. One set of bands sometimes cuts off a lower set, crudely simulating the false bedding of sediments.

Fig. 2. Closer view of the wispy banding shown in Fig. 1.

Plate 10.

Fig. 1. Irregular layering approaching the ordinary banding of basic rocks, west-north-west Ridge of Gabbrofjaeld (Peak 1300 metres) just above Uttentals Plateau.

Fig. 2. West face of Basistoppen from Ivnarmiut. The prominent shoulder at about 500 metres is due to the resistant purple band and the ferrogabbros immediately below. Above lie the less resistant unlaminated layered series, while the top 150 metres is composed of the raft of spotted sill gabbro. East-west dykes give gulleys and in two cases show up as light streaks.

Plate 11.

Fig. 1. Trough banding F north of the Main House. The structure is symmetrical about an axial plane dipping at 70—80°.

Fig. 2. A closer view showing the gravity stratified character of the thicker bands.

Plate 12.

Fig. 1. Longitudinal section of trough banding E. Strongly gravity differentiated bands are usually separated by bands of approximately average rock, as shown in Pl. 11, fig. 2.

Fig. 2. Trough banding F overlying even layering. The thick gravity stratified band with a conspicuous light component near the base of the trough banding is apparently interrupted but this is due to a projecting ledge of a resistant melanocratic part.

Plate 13.

Fig. 1. Trough banding E, showing great height compared with width and the massive rock on both sides. Some bands extend further into the massive rock than others.

Fig. 2. Flank of a zone of trough banding N, on the neck of the Skaergaard Halvö. The banding is on a finer scale than in the examples further from the margin. Again the banding extends variable distances into the adjacent massive rock.

Plate 14.

Fig. 1. Granophyre inclusion in the olivine-gabbro of the marginal border group, 30 metres from the contact, Mellemö. Large plagioclase and pyroxene crystals are developed marginally. The small elongated white patch 20 centimetres above the hammer is lichen.

Fig. 2. Much digested acid inclusion with acid material preserved only along upper margin of inclusion, about 40 metres from the contact, Mellemö.

Plate 15.

Fig. 1. Hypersthene-olivine-gabbro, 4077, at 500 metres in the layered series. Collected where Uttentals Plateau joins the west-north-west ridge from Peak 1,300 m. of Gabbrofjaeld at a height of 350 metres above the sea and 1,400 metres from the nearest point of the Sund. Plagioclase, olivine and clinopyroxene are the main minerals. A little orthopyroxene fringes the centre olivine and a little iron ore is also present (p. 89). Magnification × 16.

Fig. 2. Middle gabbro, 3661 at 1,200 metres in the layered series. Collected at the west foot of Pukugaqryggen, 75 metres above the sea. Plagioclase, clino-and orthopyroxenes in about equal amounts and iron ore are the chief minerals. The orthopyroxene can be distinguished in some cases by the parallel striations due to sheet inclusions of clinopyroxene. A little olivine is present in a reaction rim between the iron ore and pyroxenes. Some feathery reaction rim is developed at certain places between the iron ore and felspar (p. 96). Magnification × 16.

350   L. R. Wager & W. A. Deer.   IV

Plate 16.

Fig. 1. Ferrogabbro, 4145, from 2,150 metres in the layered series. Collected 250 metres above the Base Houses on the west ridge of Basistoppen. Plagioclase, iron-rich clinopyroxene, ferrohortonolite (with abundant ore inclusions) and iron ore are the chief minerals. Apatite is abundant and interstitial micropegmatite occurs but does not show in the photograph (p. 104). Magnification × 16.

Fig. 2. Fayalite-ferrogabbro, 1881, from the purple band at 2,400 metres in the layered series. Collected near summit of second nunatak east of Basistoppen. The light minerals consist of andesine with a cloudy, more acid border and interstitial quartz. The dark minerals are fayalite, brown and green, iron-rich clinopyroxene, iron ore and dark brown chlorite (p. 109). Magnification × 16.

Plate 17.

Fig. 1. Fayalite-ferrogabbro, 1974, from the purple band at the top of the rock mass to the north-east of the seaward end of Basis Glacier. The chief minerals are plagioclase, iron-rich clinopyroxene (cloudy with inclusions), fayalite (almost opaque with inclusions) and iron ore. Apatite is fairly abundant and a little micropegmatite is present (p. 110).

Fig. 2. The same under crossed nicols showing that the pyroxene areas consist of a mass of small, variously orientated crystals.

Plate 18.

Fig. 1. Fayalite-ferrogabbro, 4139, from the lower part of the unlaminated layered series, west face of Basistoppen. Large fayalite crystals, iron-rich clinopyroxene and iron ore make up the dark minerals. Andesine, a considerable amount of micropegmatite visible in the photograph, and apatite are the light minerals (p. 113). Magnification × 16.

Fig. 2. Basic hedenbergite-granophyre, 4137, from the middle part of the unlaminated layered series, west face of Basistoppen. The dark minerals are pyroxene close to hedenbergite in composition and iron ore; the light are andesine (clear) surrounded by a fine-grained micropegmatite (p. 113). Magnification × 16.

Plate 19.

Fig. 1. Hedenbergite-andesinite, 4136, from the top of the unlaminated layered series, immediately below the Basistoppen Raft, west face of Basistoppen. Mainly crystals of andesine some of which are up to one centimetre long, surrounded by fine-scale micropegmatite with conspicuous acicular crystals of quartz. Hedenbergite, iron ore, a little apatite and dark chloritic decomposition products are also present (p. 116). Magnification × 16.

Fig. 2. Gabbro-picrite, 1682, from the northern border group, 100 metres above the Sund. The large olivine crystals are slightly decomposed along the cracks which are thus conspicuous. Surrounding them is plagioclase, clinopyroxene in small rectangular crystals, a little iron ore and considerable masses of poikilitic orthopyroxene (p. 160). Magnification × 16.

Plate 20.

Fig. 1. Chilled marginal olivine gabbro, 1825, 3 metres from the contact at the head of Udløberen. The olivine is slightly decomposed marginally and along cracks. The pyroxene is both ortho and clino but both have the same habit and cannot easily be distinguished. One or two grains of iron ore can be seen (p. 137). Magnification × 16.

Fig. 2. Perpendicular felspar rock, 1851, 20 metres from the margin, Mellemö Part of one of the much elongated felspars crosses the field. The remainder of the field is essentially the same as the chilled marginal olivine-gabbro though finer-grained (p. 144). Magnification × 16.

Plate 21.

Fig. 1. Olivine-gabbro, 4289, from the outer part of the marginal border group, Ivnarmiut. Plagioclase, olivine and part of an extensive ophitic clino-pyroxene crystal are shown (p. 151). Magnification × 16.

Fig. 2. Olivine-free gabbro, 4298, from the inner part of the marginal border group, Ivnarmiut. The minerals present are plagioclase, ortho- and clino-pyroxene and iron ore (p. 154). Magnification × 16

Plate 22.

Fig. 1. Quartz-gabbro, 3050, from the upper border group, 100 metres below the summit of Brödretoppen. The main minerals are plagioclase, ophitic clino-pyroxene and iron ore with interstitial micropegmatite, chlorite and apatite (p. 172). Magnification × 16.

Fig. 2. Foliated quartz-gabbro, 4158, from the upper border group, 100 metres below the saddle between Osttoppen and Brödretoppen. The minerals are the same as in the previous rock (p. 176). Magnification × 16.

Plate 23.

Fig. 1. Quartz-rich gabbro, 4163, 200 metres below the saddle between Osttoppen and Brödretoppen. Plagioclase (in some places cloudy with decomposition products), euhedral iron-rich pyroxene and iron ore are surrounded by much fine-scale granophyric quartz and felspar (p. 177). Magnification × 16.

Fig. 2. Hedenbergite-granophyre, 3047, west face of Brødretoppen. Fayalite, greenish iron-rich pyroxene, a little iron ore and dark chlorite are surrounded by granophyric material (p. 210). Magnification × 16.

Plate 24.

Fig. 1. Transgressive granophyre vien, 4064, shores of Uttentals Sund, north of the Base Houses. The rock consists of some clear plagioclase surrounded by cloudy felspar, quartz, decomposed biotite, a little ore and accessory minerals not to be clearly distinguished (p. 206). Magnification × 16.

Fig. 2. Granophyre inclusion, 1841, 40 metres from the contact, Mellemö. Porphyritic plagioclase cloudy with decomposition products is surrounded by quartz, cloudy felspar and stilpnomelane. A little ilmenite is present in the porphyritic plagioclase and much stilpnomelane projects into the drusy cavity (p. 187). Magnification × 16.

Plate 25.

Fig. 1. Heights of the more important described specimens of the Layered Series.

Fig. 2. Localities of the more important described specimens from Skaergaard Intrusion (numbers referring to the marginal border group are written outside the intrusion).

Plate 26.

Vertical Sections through the Skaergaard Intrusion.

Plate 27.

Variation diagram (plotted against percentage solidified) for the rocks of the layered series (black) and the estimated contemporaneous residual magmas (red).

Færdig fra Trykkeriet den 21. August 1939.

Tracing for Plate 1.

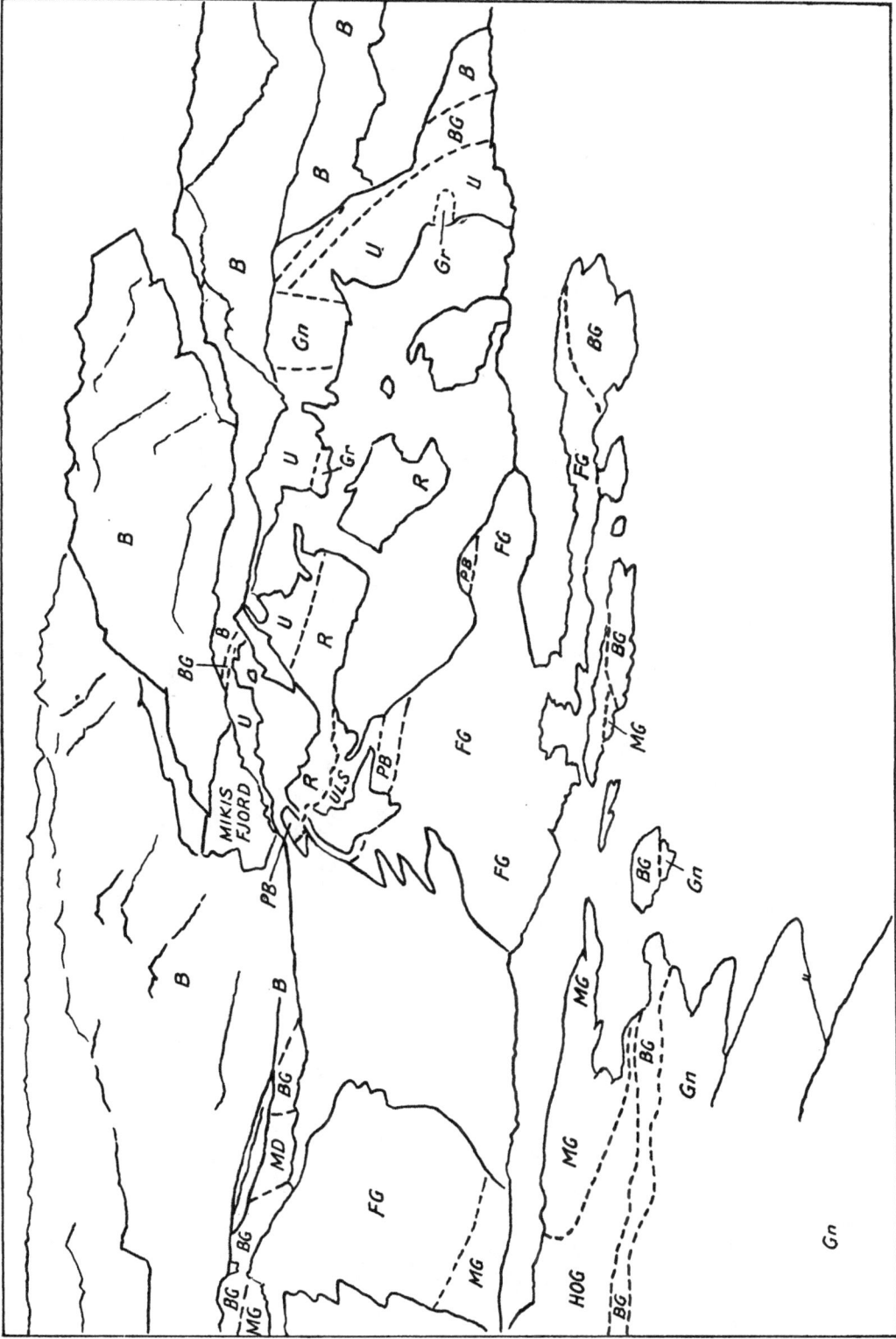

Tracing for Plate 2.

Fig. 1.

Fig. 2.

Fig. 1.

Fig. 2.

Fig. 1.

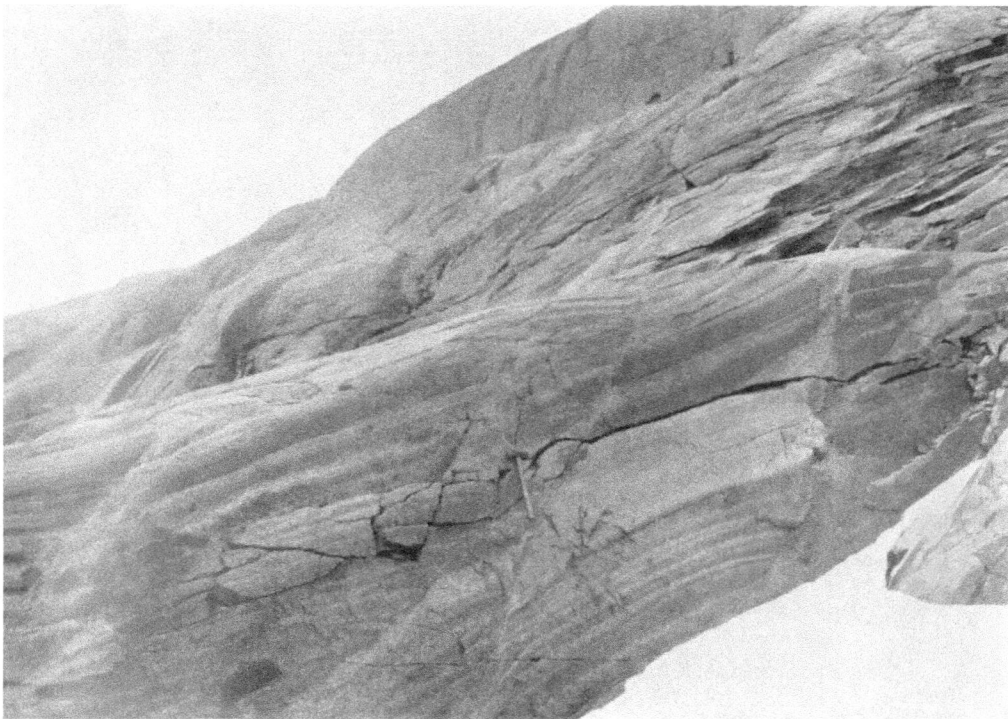

Fig. 2.

Fig. 1.

Fig. 2.

Fig. 1.

Fig. 2.

Fig. 1.

Fig. 2.

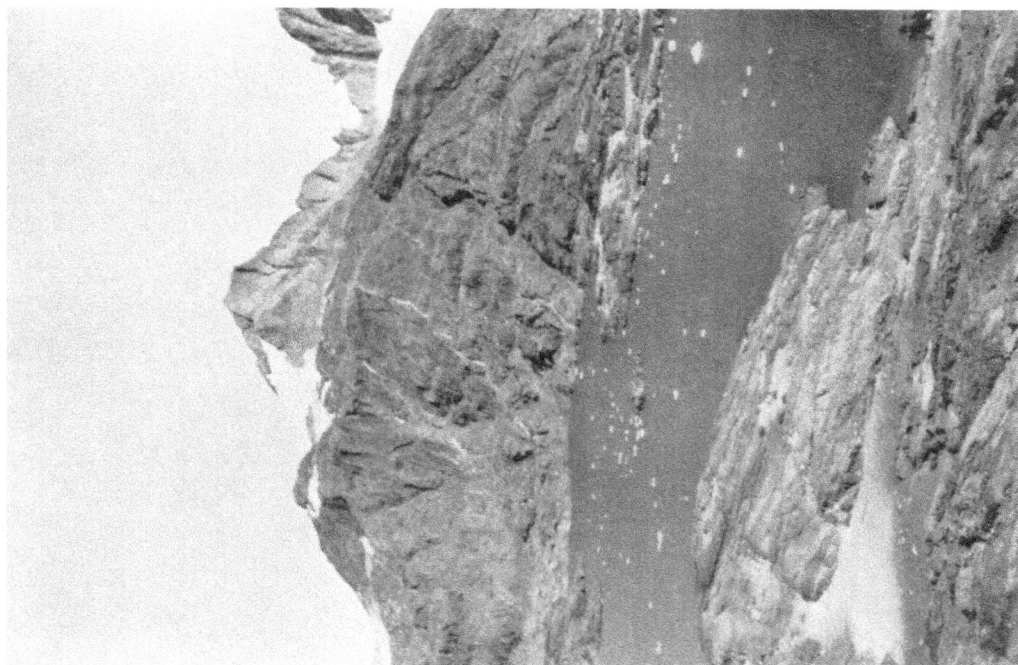

Fig. 2.

Fig. 1.

Fig. 1.

Fig. 2.

Fig. 1.

Fig. 2.

Fig. 1.

Fig. 2.

Fig. 2.

Fig. 1.

Fig. 2.

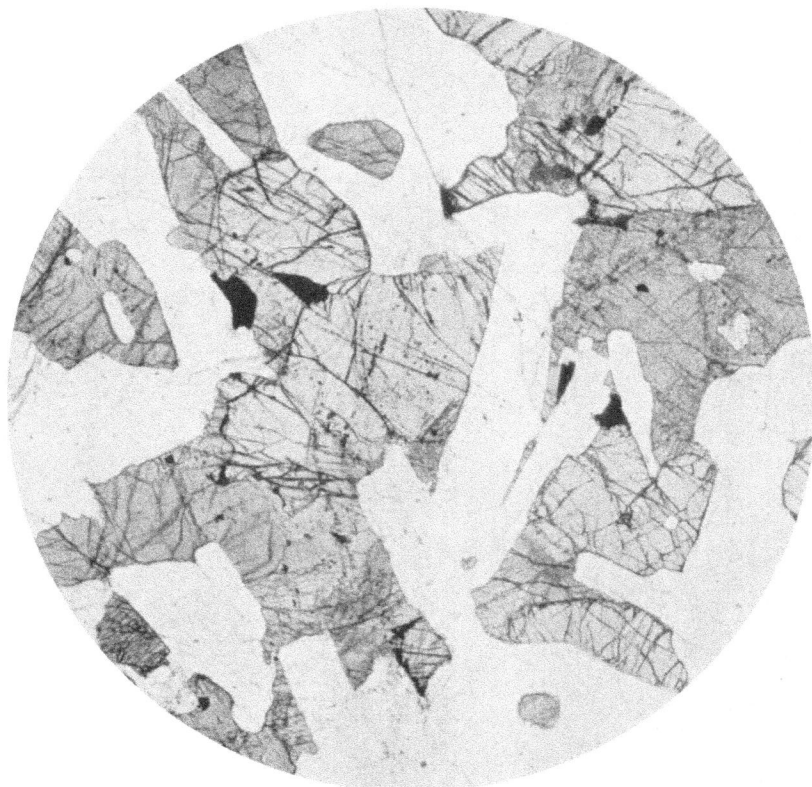

Fig. 1.

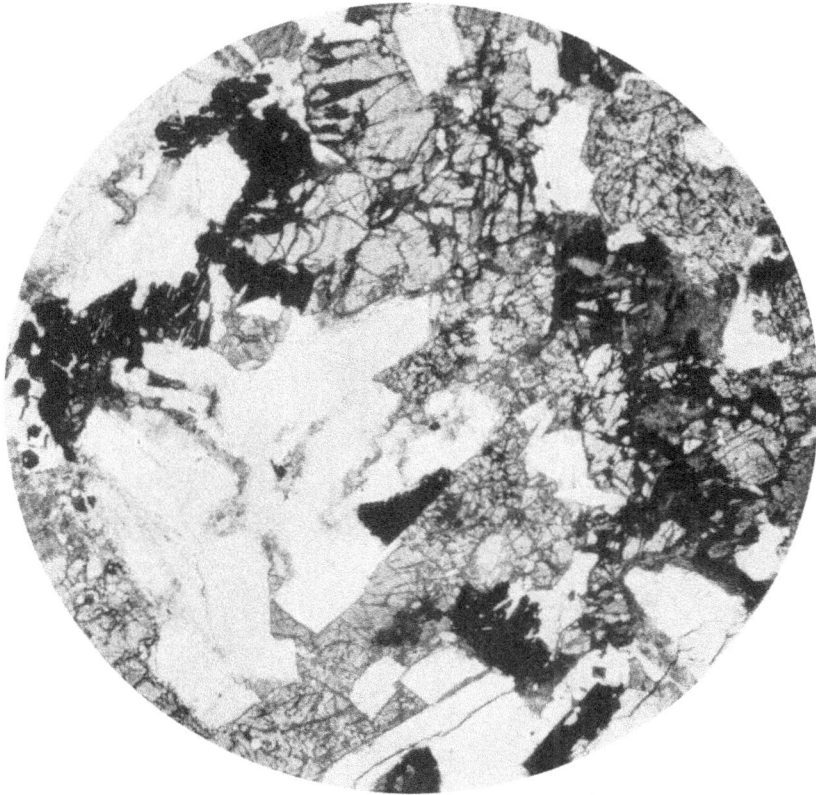

Fig. 2.

Fig. 1.

Fig. 2.

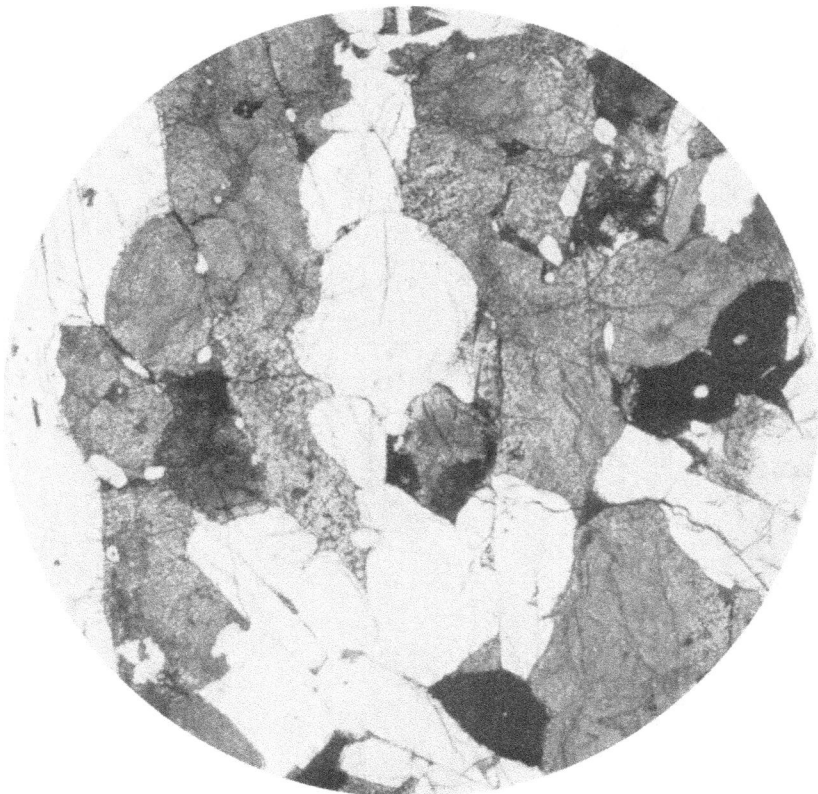

Fig. 1.

Fig. 2.

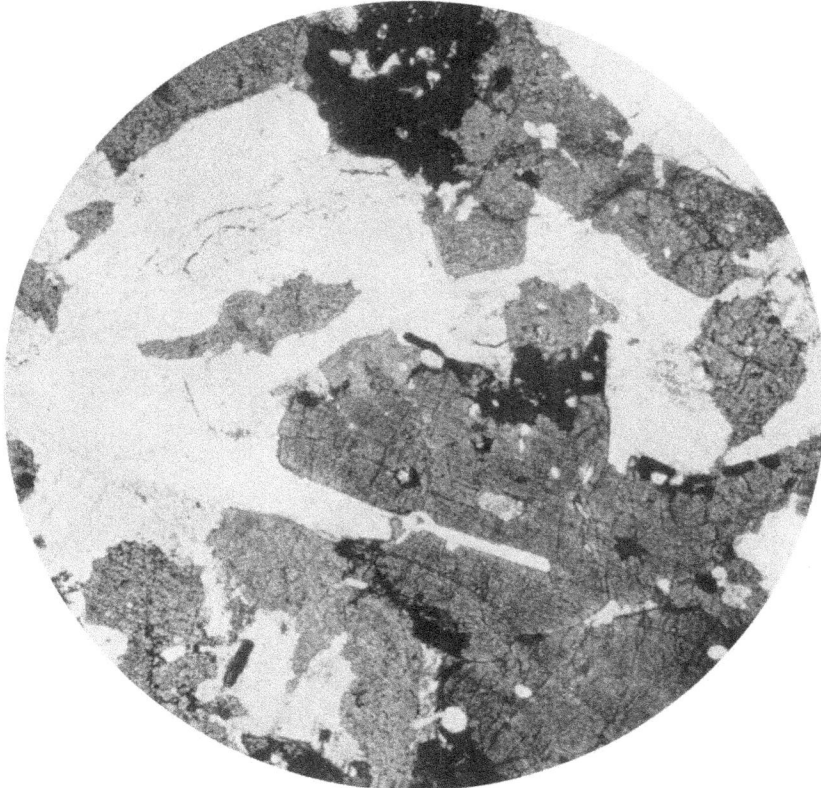

Fig. 1.

Fig. 2.

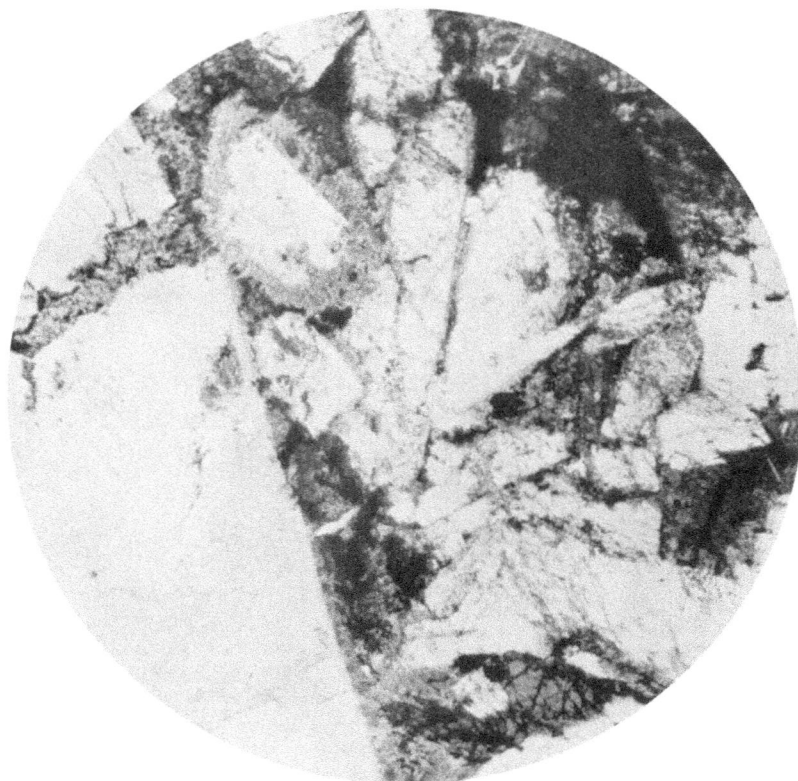

Fig. 1.

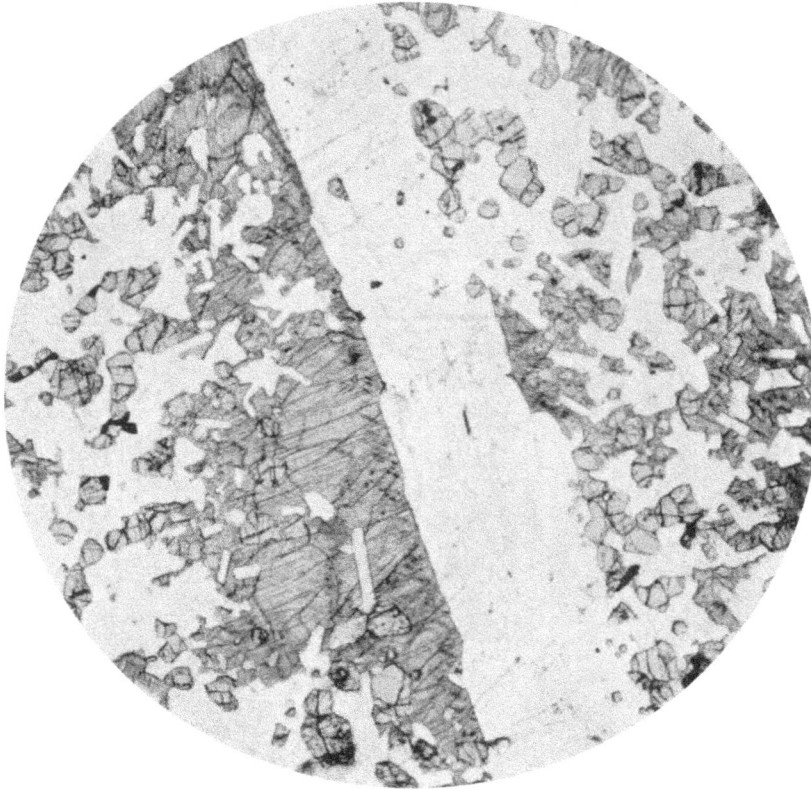

Fig. 2.

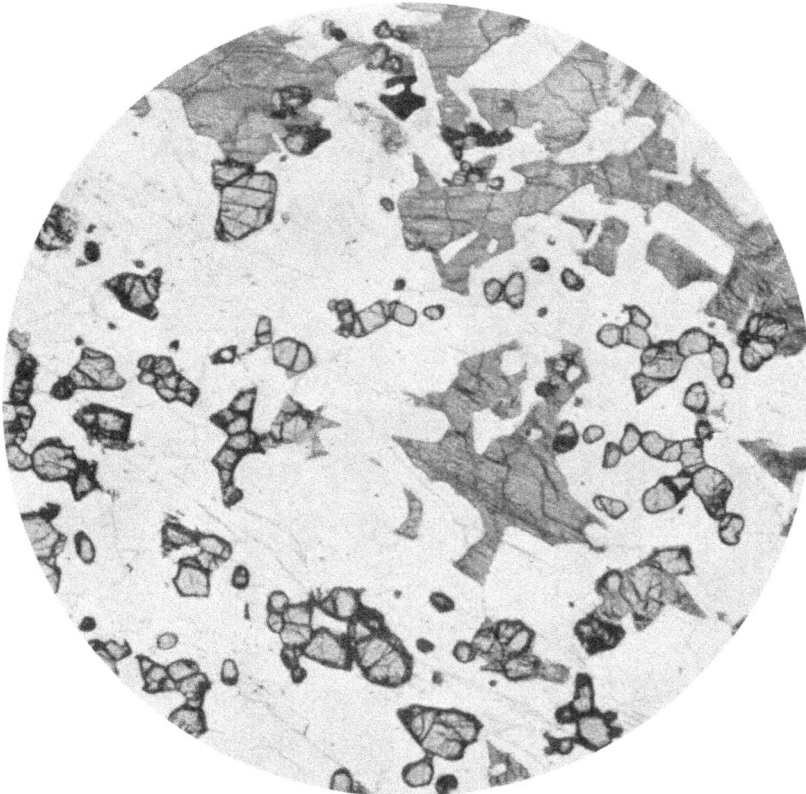

Fig. 1.

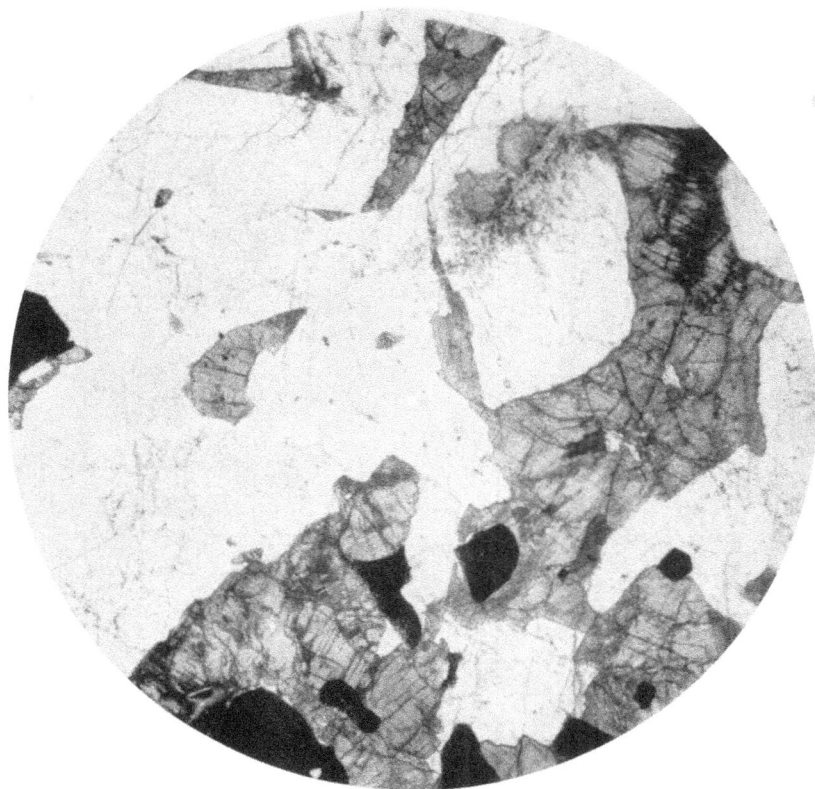

Fig. 2.

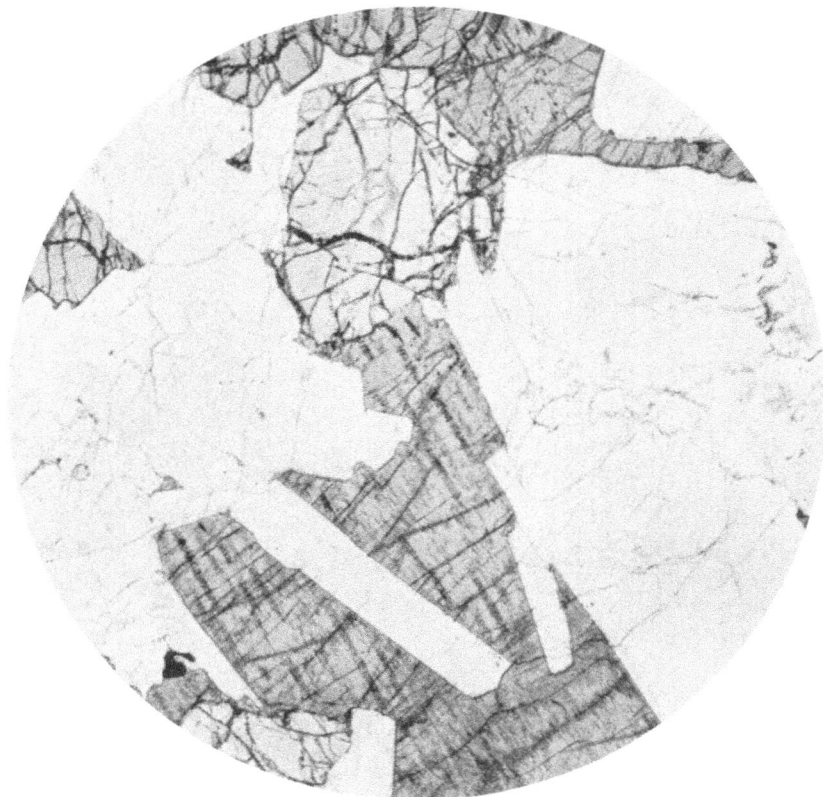

Fig. 1.

Fig. 2.

Fig. 1.

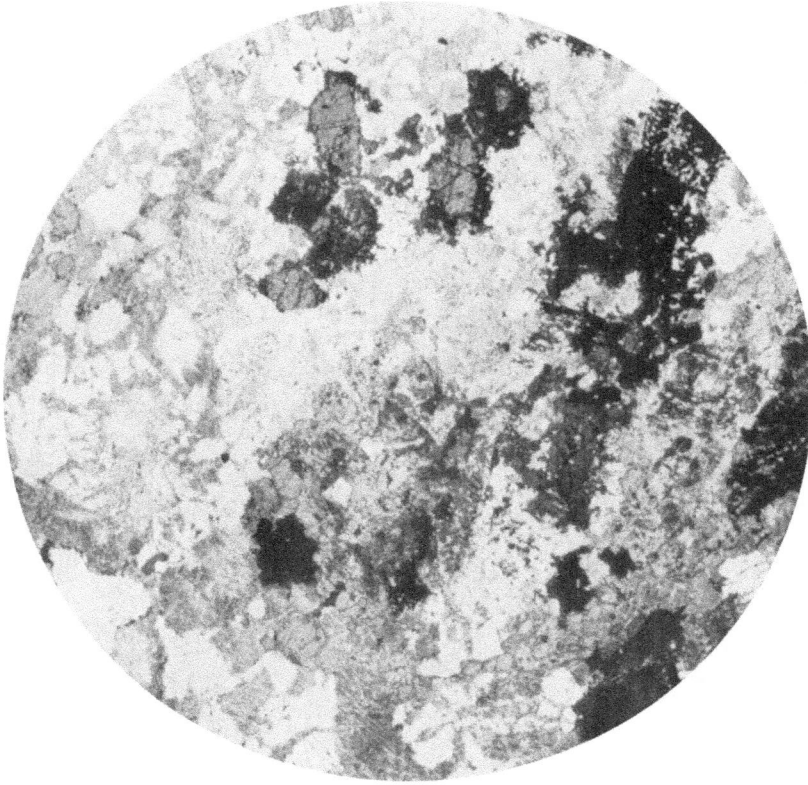

Fig. 2.

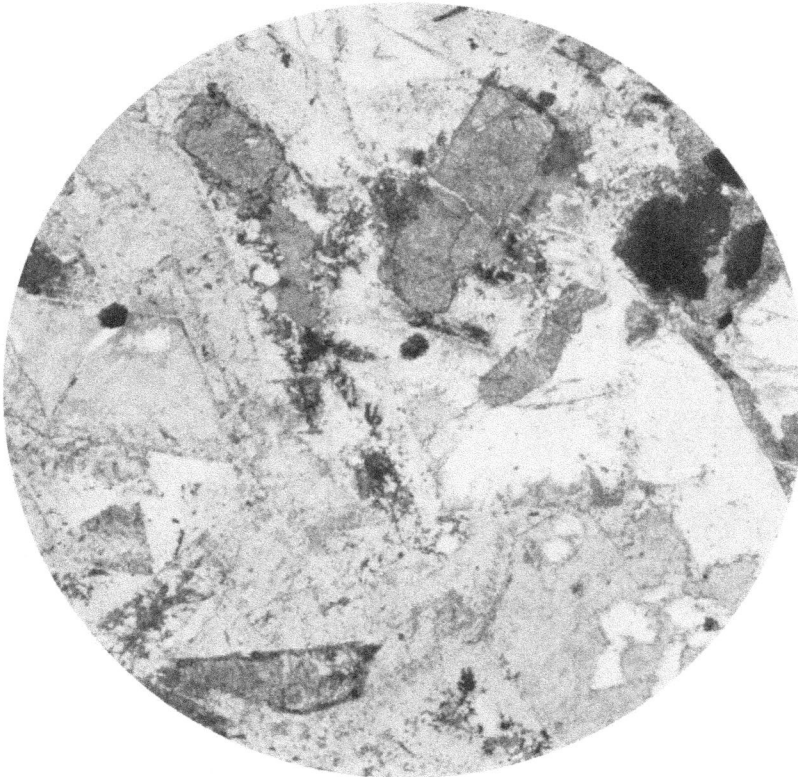

Fig. 1.

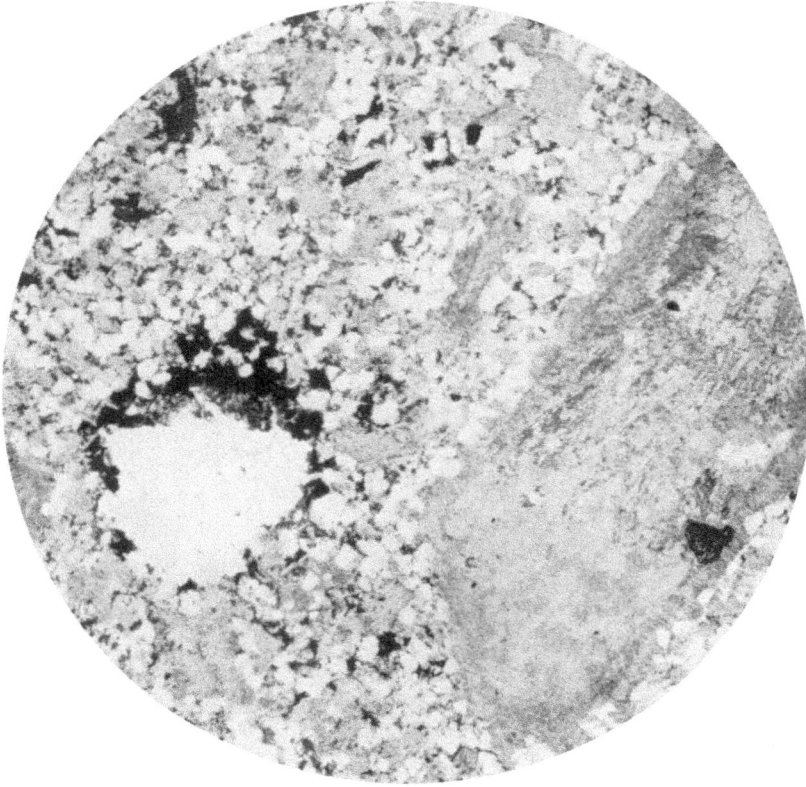

Fig. 2.

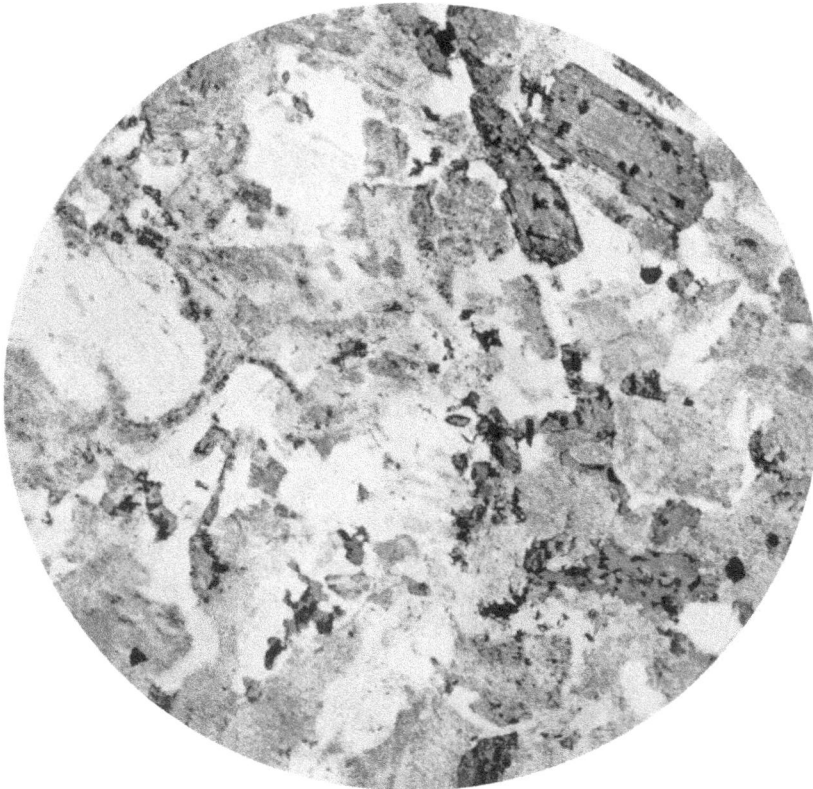

Fig. 1.

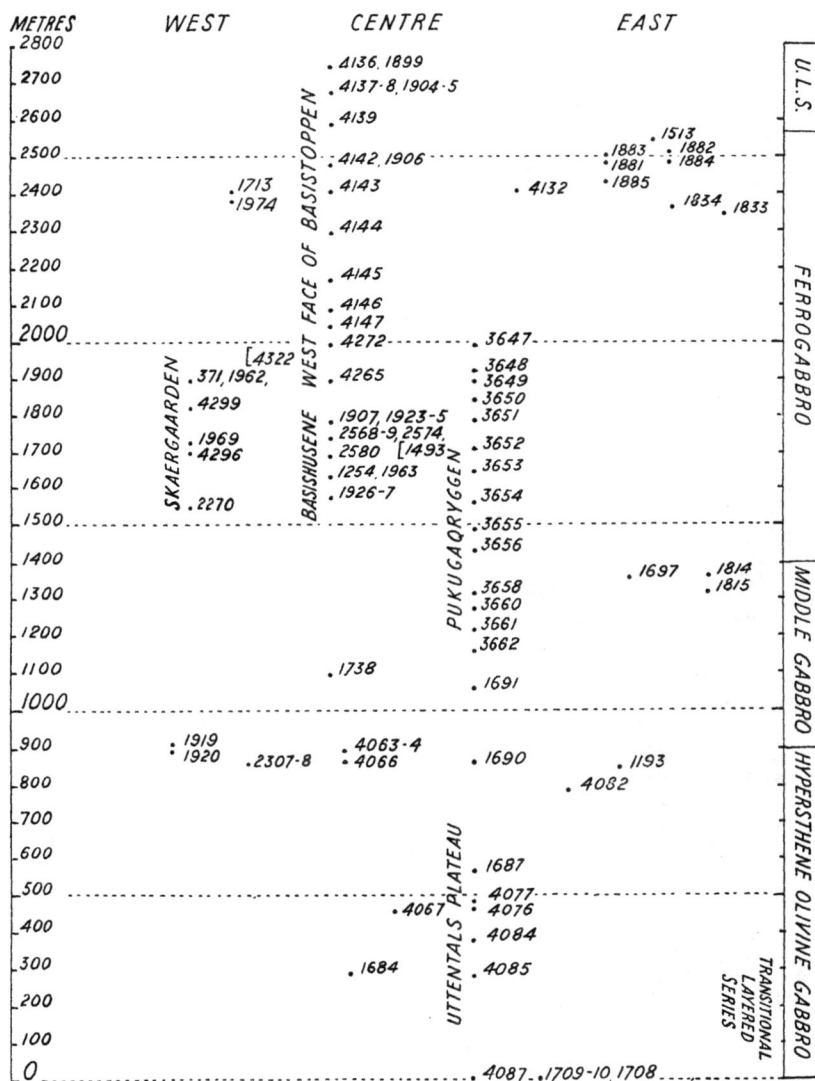

Fig. 1.

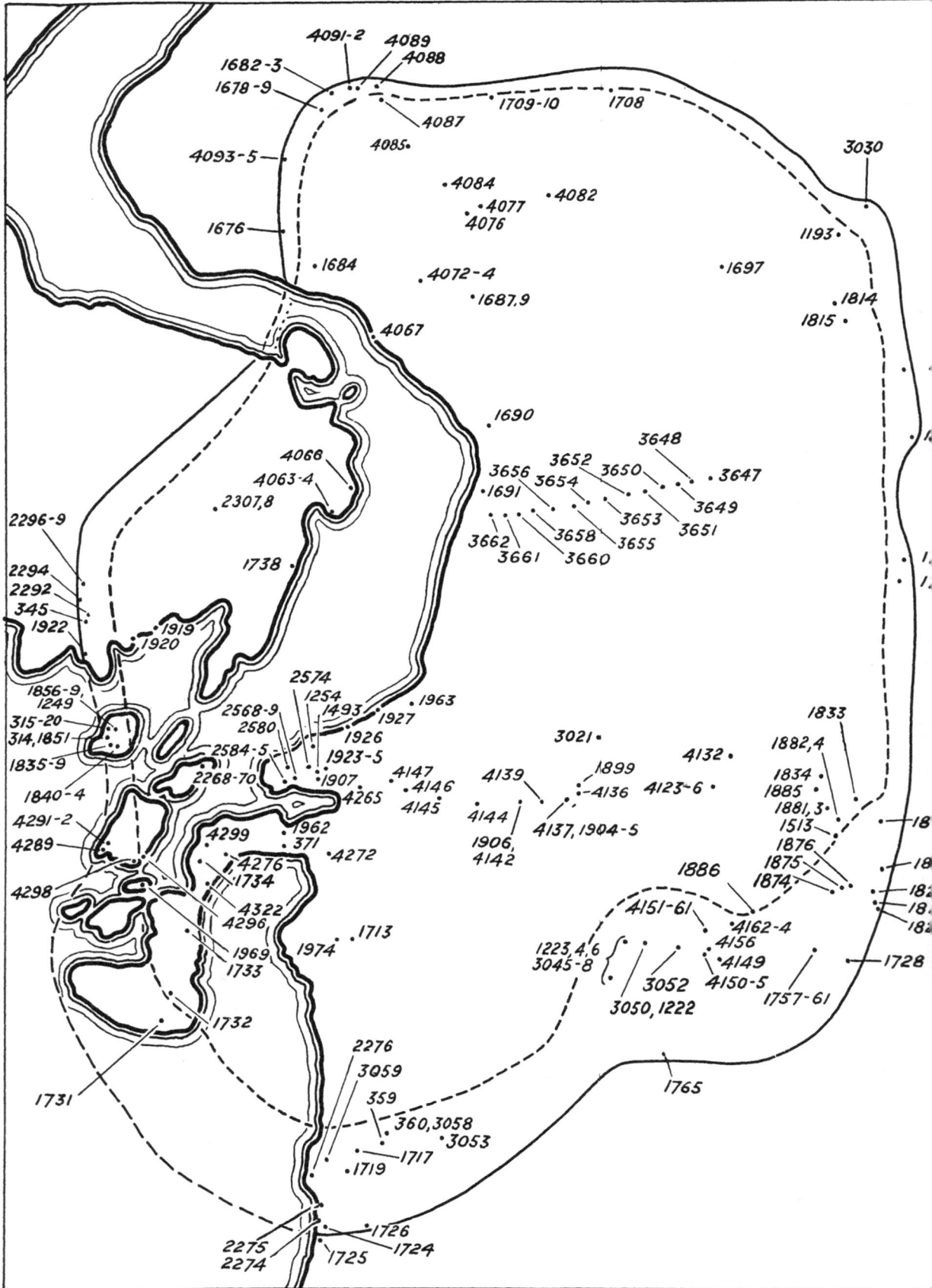

4091-2 4089
1682-3 4088
1678-9
4087 1709-10 1708
4085.
4093-5
4084 .4082 3030
.4077
4076 1193.
1676
1684 .4072-4 .1697 .1814
.1687,9 1815.
4067
1690
4066 3648
4063-4 3652 3650 .3647
2307,8 3656 3654 3649
2296-9 .1691 3653 3651
3662 3658 3655
2294 3661 3660
2292
345 1738
1922 1919
1920
2574
1856-9, 1254 .1963
1249 2568-9 1493 1927
315-20 2580 3021. 1833
314,1851 1926 1882,4
1835-9 2584-5 1923-5 4132 1834
2268-70 1907 4139 1899 4123-6 1885
1840-4 4147 -4136 1881,3,
4265 .4146 4144 4137,1904-5 1513
4291-2 4145 1906, 1876
4289 1962 4142 1875
371 .4272 1886 1874
4298 4299 4151-61 4162-4
4276 4156
1734 4149
4322 1223,4,6 4150-5 1728
4296 3045-8 3052 1757-61
1969 .1713 3050,1222
1733 1974
1732
2276
3059
1731 359
360,3058
3053
1717
1719
1726
2275 1724
2274 1725

Fig. 2.

H

rs

Gabbrofjaeld

1277 Metres

Forbindelsgletscher

TRANSITIONAL

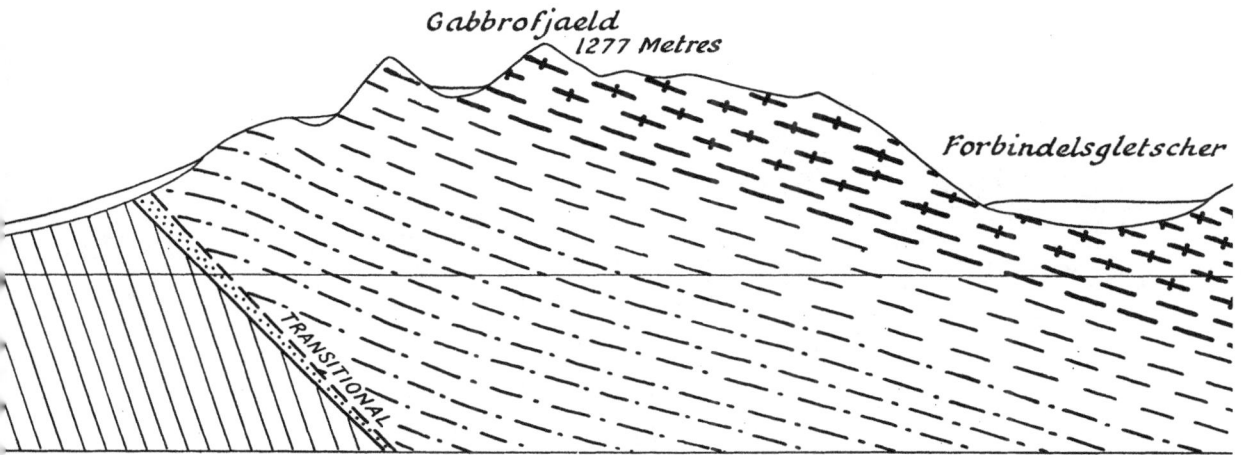

North-South Section through Skaergaard I

of W.

Brodretoppen

Hamn

narmiut Skaergaard Halvö

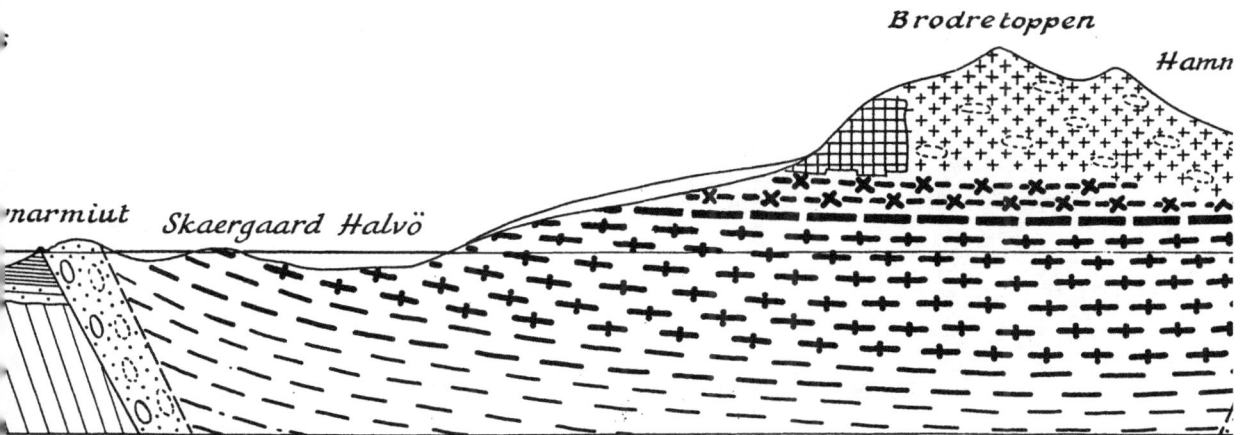

Approximately East-West Section through Skaergaard Intrusion

SOUTH

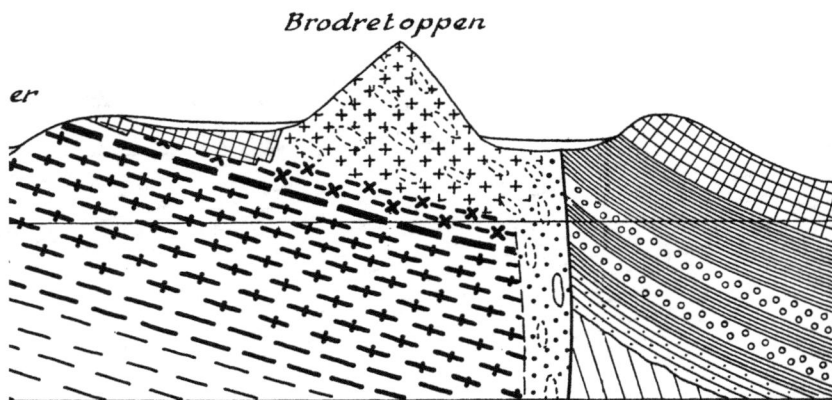

COUNTRY ROCKS

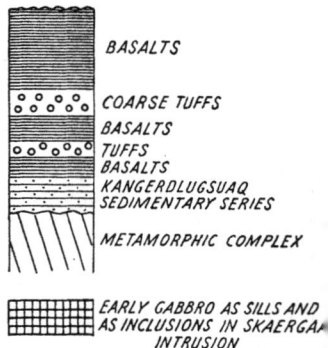

BASALTS

COARSE TUFFS

BASALTS

TUFFS

BASALTS

KANGERDLUGSUAQ
SEDIMENTARY SERIES

METAMORPHIC COMPLEX

EARLY GABBRO AS SILLS AND
AS INCLUSIONS IN SKAERGAA
INTRUSION

Brodretoppen

er

Intrusion

10° S. of E.

mmers Pas

SKAERGAARD INTRUSION

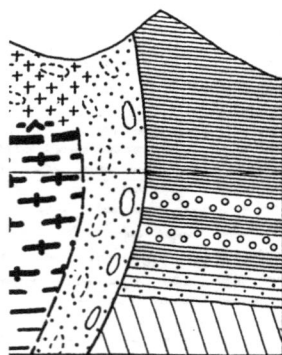

QUARTZ GABBRO WITH GRANOPHYRE
AND HYBRID INCLUSIONS

MAINLY OLIVINE GABBRO WITH GRANOPHYRE
AND HYBRID INCLUSIONS

OLIVINE GABBRO AND GABBRO-PICRITE OF
NORTHERN BORDER

BORDER GROUP

UPPER MARGINAL

THE HORIZONTAL AND VERTICAL SCALES ARE
THE SAME

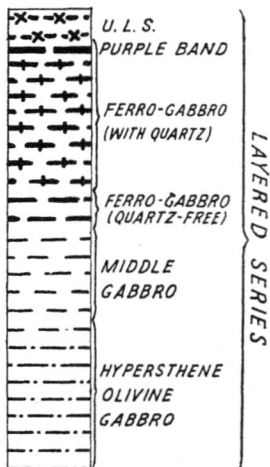

1   0   1   2

—Kilometres—

U. L. S.

PURPLE BAND

FERRO-GABBRO
(WITH QUARTZ)

FERRO-GABBRO
(QUARTZ-FREE)

MIDDLE
GABBRO

HYPERSTHENE
OLIVINE
GABBRO

LAYERED SERIES

'17

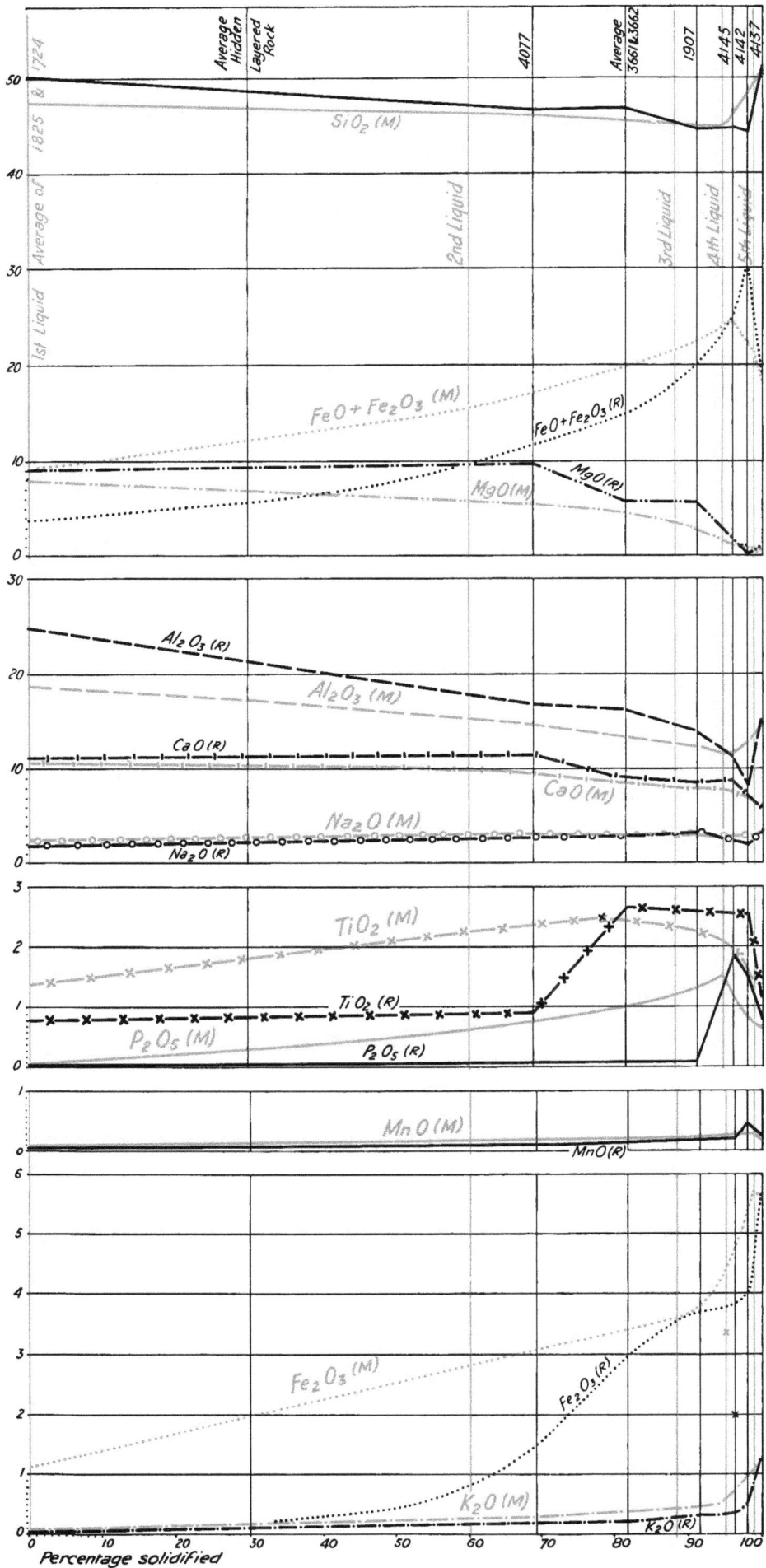

Percentage solidified

VANDFALDSDALEN

GABBROFJÆLD

PUKUGAQRYGGEN

FORBINDELSESGLETSCHER

UTTENTALS PLATEAU

Uttentals Sund

Puns Sund

Uttentals Sund

KRAEMERS Ö

Strömstedet

Mellemö

Porphyroblastic schists in Gneiss

Large Inclusion of Sediment

Granophyre Veins Abundant

Steep Contact

Basishusene

Basistoppen

Nunatak I

Nunatak II

Basisnæsset

Jvnarmiut

Thin Sediments along Gneiss

Többeren

Skærgaard Intrusion (Tertiary)

1 cm = 720 m

Layered Series
- Unlaminated — Purple Band, Triple Group — Quartz
- Ferro-Gabbro — Olivine
- Middle Gabbro — Hypersthene
- Hypers-Thene Olivine Gabbro — Olivine

Border Group
- Upper — Quartz Gabbro without Olivine
- Marginal — Gabbro usually Olivine Bearing
- Olivine Gabbro and Gabbro Picrite
- Granophyre of Tinden

Observed Boundaries
Inferred Boundaries
Merging Boundaries

1 cm = 400 m

Tertiary
- B — Basalts
- T — Main Tuffs
- B — Basalts
- T B B — Agglomerates Basalts Tuffs Basalts
- S — Kangerdlugssuak Sediments (Upper Cretaceous)
- Gn — Metamorphic Complex (Pre-Cambrian)

Tertiary Intrusions (Pre-Skærgaard)
- Gabbro and Peridotite of Sills
- Gabbro of Macro-Dykes

Dips of Basalts, Layering and Fluxion Structures
Dip and Strike of Dyke Swarm
Contact Dips

Kilometers
0 1 2 3 4

Mikis Fjord
Eskimo Ruiner
Mikis Hus
Hammers Bjærg
910
Hammers Pas
Kilen
Uttidal
Basispasset
Östtoppen
Brödretoppen
Sydtoppen
1120
Basisgletscher
Brödregletscher
Skærgaardsbugt
Skærgaardshalvö
Tinden
Tindegletscher
Pilespidsen
Hængefjældet
763
Hammers Gletscher
Contact vertical
Thun Sill
Gn

GEOLOGICAL INVESTIGATIONS IN EAST GREENLAND, PART I-XI

All titles are out of print. Plans are made for digitization of these titles, making them available as e-books and as print copies (facsimile editions). Part III is the first title made available this way.

More information at www.mtp.dk/MoG

PART I
General geology from Angmagsalik to Kap Dalton, by L.R. Wager, 1934, 46 pp., Monographs on Greenland | Meddelelser om Grønland, vol. 105, no. 2

PART II
Geology of Kap Dalton, by L.R. Wager, 1935, 2 pp., Monographs on Greenland | Meddelelser om Grønland, vol. 105, no. 3

PART III
The petrology of the Skaergaard Intrusion, Kangerdlugssuaq, East Greenland, by L.R. Wager and W.A. Deer, 1939, 352 pp., Monographs on Greenland | Meddelelser om Grønland, vol. 105, no. 4, ISBN 978 87 635 1385 2 (available, 2010)

PART IV
The stratigraphy and tectonics of Knud Rasmussens ‹Land and the Kangerdlugssuaq Region, by L.R. Wager, 1947, 64 pp., Monographs on Greenland | Meddelelser om Grønland, vol. 134, no. 5

PART V
The petrography of the Prinsen af Wales Bjerge Lavas, by Y.M. Anwar, 1955, 31 pp., Monographs on Greenland | Meddelelser om Grønland, vol. 135, no. 1

PART VI
A differentiated basic Sill enclosed in the Skærgaard Intrusion, East Greenland and related Sills injecting the lavas, by C.J. Hughes, Monographs on Greenland | Meddelelser om Grønland, vol. 137, no. 2

PART VII
The basistoppen sheet. A differentiated basic Intrusion into the upper part of the Skaergaard complex, East Greenland, by J.A.V. Douglas, 1964, 69 pp., Monographs on Greenland | Meddelelser om Grønland, vol. 164, no. 5

PART VIII
The petrology of the Kangerdlugssuaq Alkaline intrusion, east Greenland, by D.R.C. Kempe, W.A. Deer and L.R. Wager, 1970, 49 pp., Monographs on Greenland | Meddelelser om Grønland, vol. 190, no. 2

PART IX
The mineralogy of the Kangerdlugssuaq Alkaline intrusion, east Greenland, by D.R.C. Kempe and W.A. Deer, 1970, 95 pp., Monographs on Greenland | Meddelelser om Grønland, vol. 190, no. 3

PART X
The Gabbro cumulates of the Kap Edvard Holm lower layered series, by D. Abbott and W.A. Deer, 1972, 42 pp., Monographs on Greenland | Meddelelser om Grønland, vol. 190, no. 3

PART XI
The Minor peripheral intrusions, Kangerdlugssuaq, East Greenland, by W.A. Deer and D.R.C. Kempe, 1976, 25 pp., Monographs on Greenland | Meddelelser om Grønland, vol. 197, no. 4

MONOGRAPHS ON GREENLAND | MEDDELELSER OM GRØNLAND

ABOUT THE SERIES

Monographs on Greenland | Meddelelser om Grønland (ISSN 0025 6676) has published scientific results from all fields of research on Greenland since 1878. The series numbers more than 345 volumes comprising more than 1250 titles.

In 1979 Monographs on Greenland | Meddelelser om Grønland was developed into a tripartite series consisting of Bioscience (ISSN 0106-1054), Man & Society (ISSN 0106-1062), and Geoscience (ISSN 0106-1046).

Monographs on Greenland | Meddelelser om Grønland was renumbered in 1979 ending with volume no. 206 and continued with volume no. 1 for each subseries. As of 2008 the original Monographs on Greenland | Meddelelser om Grønland numbering is continued in addition to the subseries numbering.

Further information about the series, including addresses of the scientific editors of the subseries, can be found at www.mtp.dk/MoG.

MANUSCRIPTS SHOULD BE SENT TO

Museum Tusculanum Press
University of Copenhagen
126 Njalsgade, DK-2300 Copenhagen S
DENMARK
info@mtp.dk | www.mtp.dk
Tel. +45 353 29109 | Fax +45 353 29113
VAT no.: 8876 8418

ORDERS

Books can be purchased online at www.mtp.dk, via order@mtp.dk, through any of our distributors in the US, UK, and France or via online retailers and major booksellers. Museum Tusculanum Press bank details: Amagerbanken, DK-2300 Copenhagen S, BIC: AM BK DK KK, IBAN: DK10 5202 0001 5151 08.

DISTRIBUTORS

USA & Canada: ISBS International Specialized Book Services, 920 NE 58th Ave. Suite 300 - Portland, OR 97213, Phone: +1 800 944 6190 (toll-free), Fax: +1 503 280 8832, orders@isbs.com

United Kingdom: Gazelle Book Services Ltd., White Cross Mills, High Town, GB-Lancaster LA1 4XS, United Kingdom, Phone: +44 1524 68765, Fax: +44 1524 63232, sales@gazellebooks.co.uk

France: Editions Picard, 82, rue Bonaparte, F-75006 Paris, France, Phone: +33 (0) 1 4326 9778, Fax: +33 1 43 26 42 64, livres@librairie-picard.fr

www.ingramcontent.com/pod-product-compliance
Lightning Source LLC
Chambersburg PA
CBHW081800200326
41597CB00023B/4094